T0186632

INTELLIGENT SYSTEMS

Technology and Applications

VOLUME VI

Control and Electric Power Systems

Edited by
Cornelius T. Leondes

INTELLIGENT SYSTEMS

Technology and Applications

VOLUME VI

Control and Electric Power Systems

CRC PRESS

Boca Raton   London   New York   Washington, D.C.

Library of Congress Cataloging-in-Publication Data

Intelligent systems : technology and applications / edited by Cornelius T. Leondes.
      p. cm.
    Includes bibliographical references and index.
    Contents: v. 1. Implementation techniques -- v. 2. Fuzzy systems, neural networks, and
expert systems -- v. 3. Signa, image, and speech processing -- v. 4. Database and
learning systems -- v. 5. Manufacturing, industrial, and management systems -- v. 6.
Control and electric power systems.
     ISBN 0-8493-1121-7 (set : alk. paper)
     1. Intelligent control systems. I. Leondes, Cornelius T.

TJ217.5 I5448 2002
629.8--dc21
                                               2002017473

Visit the CRC Press Web site at www.crcpress.com

© 2003 by CRC Press LLC

No claim to original U.S. Government works
International Standard Book Number 0-8493-1121-7
Library of Congress Card Number 2002017473
Printed in the United States of America 1 2 3 4 5 6 7 8 9 0
Printed on acid-free paper

Foreword

Intelligent Systems: Technology and Applications is a significant contribution to the artificial intelligence (AI) field. Edited by Professor Cornelius Leondes, a leading contributor to intelligent systems, this set of six well-integrated volumes on the subject of intelligent systems techniques and applications provides a valuable reference for researchers and practitioners. This landmark work features contributions from more than 60 of the world's foremost AI authorities in industry, government, and academia.

Perhaps the most valuable feature of this work is the breadth of material covered. Volume I looks at the steps in implementing intelligent systems. Here the reader learns from some of the leading individuals in the field how to develop an intelligent system. Volume II covers the most important technologies in the field, including fuzzy systems, neural networks, and expert systems. In this volume the reader sees the steps taken to effectively develop each type of system, and also sees how these technologies have been successfully applied to practical real-world problems, such as intelligent signal processing, robotic control, and the operation of telecommunications systems. The final four volumes provide insight into developing and deploying intelligent systems in a wide range of application areas. For instance, Volume III discusses applications of signal, image, and speech processing; Volume IV looks at intelligent database management and learning systems; Volume V covers manufacturing, industrial, and business applications; and Volume VI considers applications in control and power systems. Collectively this material provides a tremendous resource for developing an intelligent system across a wide range of technologies and application areas.

Let us consider this work in the context of the history of artificial intelligence. AI has come a long way in a relatively short time. The early days were spent in somewhat of a probing fashion, where researchers looked for ways to develop a machine that captured human intelligence. After considerable struggle, they fortunately met with success. Armed with an understanding of how to design an intelligent system, they went on to develop useful applications to solve real-world problems. At this point AI took on a very meaningful role in the area of information technology.

Along the way there were a few individuals who saw the importance of publishing the accomplishments of AI providing guidance to advance the field. Among this small group I believe that Dr. Leondes has made the largest contribution to this effort. He has edited numerous books on intelligent systems that provide a wealth of information to individuals in the field. I believe his latest work discussed here is his most valuable contribution to date and should be in the possession of all individuals involved in the field of intelligent systems.

Jack Durkin

Preface

For most of our history the wealth of a nation was limited by the size and stamina of the work force. Today, national wealth is measured in intellectual capital. Nations possessing skillful people in such diverse areas as science, medicine, business, and engineering, produce innovations that drive the nations to a higher quality of life. To better utilize these valuable resources, intelligent systems technology has evolved at a rapid and significantly expanding rate to accomplish this purpose. Intelligent systems technology can be utilized by nations to improve their medical care, advance their engineering technology, and increase their manufacturing productivity, as well as play a significant role in a very wide variety of other areas of activity of substantive significance.

Intelligent systems technology almost defines itself as the replication to some effective degree of human intelligence by the utilization of computers, sensor systems, effective algorithms, software technology, and other technologies in the performance of useful or significant tasks. Widely publicized earlier examples include the defeat of Garry Kasparov, arguably the greatest chess champion in history, by IBM's intelligent system known as "Big Blue." Separately, the greatest stock market crash in history, which took place on Monday, October 19, 1987, occurred because of a poorly designed intelligent system known as computerized program trading. As was reported, the Wall Street stockbrokers watched in a state of shock as the computerized program trading system took complete control of the events of the day. Alternatively, a significant example where no intelligent system was in place and which could have, indeed no doubt would have, prevented a disaster is the Chernobyl disaster which occurred at 1:15 A.M. on April 26, 1987. In this case the system operators were no doubt in a rather tired state and an effectively designed class of intelligent system known as "backward chaining" EXPERT System would, in all likelihood, have averted this disaster.

The techniques which are utilized to implement Intelligent Systems Technology include, among others:

Knowledge-Based Systems Techniques
EXPERT Systems Techniques
Fuzzy Theory Systems
Neural Network Systems
Case-Based Reasoning Methods
Induction Methods
Frame-Based Techniques
Cognition System Techniques

These techniques and others may be utilized individually or in combination with others.

The breadth of the major application areas of intelligent systems technology is remarkable and very impressive. These include:

Agriculture	Law
Business	Manufacturing
Chemistry	Mathematics
Communications	Medicine
Computer Systems	Meteorology
Education	Military
Electronics	Mining
Engineering	Power Systems
Environment	Science
Geology	Space Technology
Image Processing	Transportation
Information Management	

It is difficult now to find an area that has not been touched by Intelligent Systems Technology. Indeed, a perusal of the tables of contents of these six volumes, *Intelligent Systems: Technology and Applications*, reveals that there are substantively significant treatments of applications in many of these areas.

Needless to say, the great breadth and expanding significance of this field on the international scene requires a multi-volume set for an adequately substantive treatment of the subject of intelligent systems technology. This set of volumes consists of six distinctly titled and well-integrated volumes. It is appropriate to mention that each of the six volumes can be utilized individually. In any event, the six volume titles are:

1. Implementation Techniques
2. Fuzzy Systems, Neural Networks, and Expert Systems
3. Signal, Image, and Speech Processing
4. Database and Learning Systems
5. Manufacturing, Industrial, and Management Systems
6. Control and Electric Power Systems

The contributors to these volumes clearly reveal the effectiveness and great significance of the techniques available and, with further development, the essential role that they will play in the future. I hope that practitioners, research workers, students, computer scientists, and others on the international scene will find this set of volumes to be a unique and significant reference source for years to come.

Cornelius T. Leondes
Editor

About the Editor

Cornelius T. Leondes, B.S., M.S., Ph.D., Emeritus Professor, School of Engineering and Applied Science, University of California, Los Angeles, has served as a member or consultant on numerous national technical and scientific advisory boards. Dr. Leondes served as a consultant for numerous Fortune 500 companies and international corporations. He has published over 200 technical journal articles and has edited and/or co-authored more than 120 books. Dr. Leondes is a Guggenheim Fellow, Fulbright Research Scholar, and IEEE Fellow, as well as a recipient of the IEEE Baker Prize award and the Barry Carlton Award of the IEEE.

Contributors

S. Barro
University of Santiago de
    Compostela
Santiago, Spain

Guanrong Chen
City University of Hong Kong
Hong Kong

J. Correa
University of Santiago de
    Compostela
Santiago, Spain

Elmer P. Dadios
De La Salle University
Manilla, Philippines

Raul Garduno-Ramirez
Electrical Research Institute
Temixco, Mexico

N.D. Hatziargyriou
National Technical University
    of Athens
Athens, Greece

R. Iglesias
University of Santiago de
    Compostela
Santiago, Spain

Sung-Kwan Joo
University of Washington
Seattle, Washington

E.S. Karapidakis
National Technical University
    of Athens
Athens, Greece

Kwang Y. Lee
Pennsylvania State University
University Park, Pennsylvania

Han-Xiong Li
City University of Hong Kong
Hong Kong

Chen-Ching Liu
University of Washington
Seattle, Washington

O.P. Malik
University of Calgary
Calgary, Canada

D.L. Moreno
University of Santiago de
    Compostela
Santiago, Spain

Pan-Wei Ng
Rational Software
Singapore

Michel Pasquier
Nanyang Technological University
Singapore

Chai Quek
Nanyang Technological University
Singapore

C.V. Regueiro
University of A Coruña
Coruña, Spain

M. Rodriguez
University of Santiago de
    Compostela
Santiago, Spain

Marino Sforna
GRTN — Italian Independent
    System Operator
Milan, Italy

Dejan Sobajic
Electric Power Research Institute
Palo Alto, California

Zita A. Vale
Polytechnic Institute of Porto (IPP)
Porto, Portugal

Contents

Volume VI: Control and Electric Power Systems

1 Power System Stabilizers Based on Adaptive Control
and Artificial Intelligence Techniques
O.P. Malik VI-1

2 Machine Learning Applied to Power System Dynamic Security Assessment
N.D. Hatziargyriou and E.S. Karapidakis VI-35

3 Power Systems Control Centers
Zita A. Vale VI-63

4 Intelligent Fault Diagnosis in Power Systems
Marino Sforna, Chen-Ching Liu, Dejan Sobajic, and Sung-Kwan Joo VI-113

5 Short Term and Online Electric Load Forecasting
Marino Sforna VI-135

6 Intelligent Multiagent Control for Power Plants
Raul Garduno-Ramirez and Kwang Y. Lee VI-181

7 Applying Architectural Patterns to the Design of Supervisory Control Systems
Pan-Wei Ng, Chai Quek, and Michel Pasquier VI-211

8 Feature Based Integrated Design of Fuzzy Control Systems
Han-Xiong Li and Guanrong Chen VI-253

9 Intelligent Controllers for Flexible Pole Cart Balancing Problem
Elmer P. Dadios VI-285

10 A Control Architecture for Mobile Robotics Based on Specialists
C.V. Regueiro, M. Rodriguez, J. Correa, D.L. Moreno, R. Iglesias, and S. Barro VI-337

Index VI-361

Contents

Volume I: Implementation Techniques

1   The Quest for the Intelligent Machine
    Jack Durkin

2   Soft Computing Framework for Intelligent Human-Machine System Design,
    Simulation, and Optimization
    Xuan F. Zha and Samuel Y.E. Lim

3   Visualization and Modeling for Intelligent Systems
    Andrew W. Crapo, Laurie B. Waisel, William A. Wallace, and Thomas R. Willemain

4   Online Adaptation of Intelligent Decision-Making Systems
    Hisao Ishibuchi, Ryoji Sakamoto, and Tomoharu Nakashima

5   Knowledge-Intensive Collaborative Design Modeling and Decision
    Support Using Distributed Web-Based KBS Tools
    Xuan F. Zha and He Du

6   A Multi-Agent Framework for Collaborative Reasoning
    Chunyan Miao, Angela Goh, and Yuan Miao

7   Architecting Multi-Agent Systems: A Middleware Perspective
    Zhonghua Yang, Robert Gay, and Chunyan Miao

8   Applied Intelligent Techniques in Assisted Surgery Planning
    Chua Chee-Kai and Cheah Chi-Mun

9   Intelligent Software Systems for CNC Machining
    George-C. Vosniakos

10 Intelligent Systems Techniques and Applications in Product Forecasting
    James Jiang, Gary Klein, and Roger A. Pick

11 Neural Network Systems and Their Application in Software Sensor
    Systems for Chemical and Biotechnological Processes
    Svetla Vassileva and Xue Z. Wang

12 From Simple Graphs to Powerful Knowledge Representation: The Conceptual
    Graph Formalism
    Guy W. Mineau

13 Autonomous Mental Development by Robots: Vision, Audition, and Behaviors
    Juyang Weng, Wey-Shiuan Hwang, and Yilu Zhang

Index

Contents

Volume II: Fuzzy Systems, Neural Networks, and Expert Systems

1 Neural Network Techniques and Their Engineering Applications
  Ming Zhou

2 Recurrent Neural Fuzzy Network Techniques and Applications
  Chia-Feng Juang

3 Neural Network Systems Techniques and Applications in Signal Processing
  Ying Tan

4 Neural Network and Fuzzy Control Techniques for Robotic Systems
  Kazuo Kiguchi, Keigo Watanabe, and Toshio Fukuda

5 Processing of Linguistic Symbols in Multi-Level Neural Networks
  Jong Joon Park and Abraham Kandel

6 Applied Fuzzy Intelligence System: Fuzzy Batching of Bulk Material
  Zimic Nikolaj, Mraz Miha, Virant Jernej, and Ficzko Jelena

7 Designing and Refining Linguistic Fuzzy Models to Improve Their Accuracy
  Rafael Alcalá, Jorge Casillas, Oscar Cordón, and Francisco Herrera

8 GA-FRB: A Novel GA-Optimized Fuzzy Rule System
  W.J. Chen and Chai Quek

9 Fuzzy Cellular Automata and Fuzzy Sequential Circuits
  Mraz Miha, Zimic Nikolaj, Virant Jernej, and Ficzko Jelena

10 PACL-FNNS: A Novel Class of Falcon-Like Fuzzy Neural Networks
   Based on Positive and Negative Exemplars
   W.L. Tung and Chai Quek

11 Maximum Margin Fuzzy Classifiers with Ellipsoidal Regions
   Shigeo Abe

12 Neural Networks: Techniques and Applications in Telecommunications Systems
   Si Wu and K.Y. Michael Wong

13 A Fuzzy Temporal Rule-Based Approach for the Design
   of Behaviors in Mobile Robotics
   *Manuel Mucientes, Roberto Iglesias, Carlos V. Regueiro,
   Alberto Bugarín, and Senen Barro*

Index

Contents

Volume III: Signal, Image, and Speech Processing

1   Artificial Intelligence Systems Techniques and Applications in Speech Processing
    Douglas R. Campbell, Colin Fyfe, and Mark Girolami

2   Fuzzy Theory, Methods, and Applications in Nonlinear Signal Processing
    Kenneth E. Barner

3   Intelligent System Modeling of Bioacoustic Signals Using
    Advanced Signal Processing Techniques
    Leontios J. Hadjileontiadis, Yannis A. Tolias, and Stavros M. Panas

4   Automatic Recognition of Multichannel EEG Signals Using a Committee
    of Artificial Neural Networks
    Bjorn O. Peters, Gert Pfurtscheller, and Henrik Flyvbjerg

5   Morphological Color Image Processing: Theory and Applications
    I. Andreadis, Maria I. Vardavoulia, G. Louverdis, and Ph. Tsalides

6   The Future of Artificial Neural Networks and Speech Recognition
    Qianhui Liang and John G. Harris

7   Advanced Neural-Based Systems for 3D Scene Understanding
    Gian Luca Foresti

8   Shape Representation and Automatic Model Inference in Character
    Recognition Systems
    Hirobumi Nishida

9   Novel Noise Modeling Using AI Techniques
    Abdul Wahab and Chai Quek

Index

Contents

Volume IV: Database and Learning Systems

1   Model Selection in Knowledge Discovery and Data Mining
    Ho Tu Bao and Nguyen Trong Dung

2   Mining Chaotic Patterns in Relational Databases and Rule Extraction by Neural Networks
    Chao Deng

3   Topological Relations in Spatial Databases
    Eliseo Clementini

4   Intelligent Image Retrieval for Supporting Concept-Based Queries: A
    Knowledge-Based Approach
    Jae Dong Yang

5   The Techniques and Experiences of Integrating Deductive Database Management
    Systems and Intelligent Agents for Electronic Commerce Applications
    Chih-Hung Wu, Shing-Hwang Doong, and Chih-Chin Lai

6   A Method for Fuzzy Query Processing in Relational Database Systems
    Shyi-Ming Chen and Yun-Shyang Lin

7   A Method for Generating Fuzzy Rules from Relational Database Systems
    for Estimating Null Values
    Shyi-Ming Chen and Ming-Shiou Yeh

8   Adaptation of Cluster Sizes in Objective Function Based Fuzzy Clustering
    Annette Keller and Frank Klawonn

9   Learning Rules about the Development of Variables over Time
    Frank Höppner and Frank Klawonn

10 Integrating Adaptive and Intelligent Techniques into a Web-Based Environment
    for Active Learning
    Hongchi Shi, Othoniel Rodriguez, Yi Shang, and Su-Shing Chen

11 The System Perspective of an Intelligent Tutoring System Based
    on Inquiry Teaching Approach
    L.H. Wong and Chai Quek

12 An Intelligent Educational Metadata Repository
    Nick Bassiliades, Fotis Kokkoras, Ioannis Vlahavas, and Dimitrios Sampson

Index

Contents

Volume V: Manufacturing, Industrial, and Management Systems

1   Integrated Intelligent Systems in the Design of Products and Processes
Xuan F. Zha and He Du

2   Secure Identification and Communication in Distributed Manufacturing Systems
István Mezgár and Zoltán Kincses

3   Concurrent Engineering (CE) Product Design
San Myint and M.T. Tabucanon

4   Multi-Product Batch Processing Systems
Samuel Chung, Jae Hak Jung, and In-Beum Lee

5   Forecasting Applications
Christine W. Chan and Hanh H. Nguyen

6   An Agent Approach for Intelligent Business Forecasting
Zhiqi Shen, Robert Gay, Xiang Li, and Zhonghua Yang

7   Chemical and Process Systems
Barry Lennox and Gary Montague

8   Nonlinear Chemical Processes
Bruce Postlethwaite

9   Reinforcement Learning Applications to Process Control and Optimization
Ernesto C. Martinez and Osvaldo A. Somaglia

10   Analyzing the Make-or-Buy Decision in the Automobile Industry Using
the Fuzzy Analytic Hierarchical Process
Andre de Korvin, Robert Kleyle, and Mohamed E. Bayou

11   The Digital Factory Becomes Reality
Engelbert Westkämper, Jörg Pirron, Rüdiger Weller, and Jörg Niemann

12   Automated Inspection of Printed Circuit Boards
Hyung Suck Cho

13   Chemical Process Control Systems
M.A. Hussain, C.W. Ng, N. Aziz, and I.M. Mujtaba

Index

1

Power System Stabilizers Based On Adaptive Control and Artificial Intelligence Techniques

1.1 Introduction VI-1
1.2 Fixed Parameter Controllers VI-2
     Conventional PSS • Linear Quadratic and H-Infinity Optimal Controllers • Variable Structure Control • Rule-Based and Fuzzy Logic Controllers • Neural-Network Based Controllers
1.3 Adaptive Controllers VI-10
     Analytical-Approach Based Adaptive PSS • Adaptive PSS with ANN Identifier and Pole-Shift Control • Adaptive PSS with ANN Identifier and ANN Controller • Adaptive-Network Based Fuzzy Logic Controller
1.4 Concluding Remarks VI-30
     Acknowledgments VI-30
     References VI-31

O.P. Malik
University of Calgary

Abstract. Conventional power system stabilizers containing a phase lag/lead network for phase compensation have played a very significant role in enhancing the stability of power systems. Various new approaches based on modern control and artificial intelligence techniques to improve the performance of the power system stabilizer have been proposed during the past 30 years. Although it is feasible to develop a satisfactory stabilizer using any one of these techniques, each has its unique strengths and drawbacks. Descriptions of a number of new algorithms are given in this chapter. To bring out the focal points, descriptions of the various algorithms are given in general terms only. The interested reader can pursue the details of the algorithms through the extensive list of references provided.

Key words: Power system stabilizers, self-tuning control, adaptive control, neural networks, fuzzy logic

1.1 Introduction

Improvement of power system stability by controlling the field excitation of a synchronous generator has been an important topic of investigation since the 1940s. Introduction of the high-gain, continuously acting automatic voltage regulator (AVR) helped improve the dynamic limits of

0-8493-1121-7/03/$0.00+$1.50

power networks. However, AVRs could also introduce negative damping, particularly in large, weakly coupled systems, and consequently make the system unstable. To overcome this problem, supplementary stabilizing signals were introduced in the excitation system.

The supplementary stabilizing signal enhances system damping by producing a torque in phase with the speed of the synchronous generator. In the conventional arrangement, the stabilizing signal is usually derived by processing any one of a number of possible signals, e.g., speed, acceleration, power, or frequency, through a suitable phase lag/lead circuit, called the power system stabilizer (PSS), to obtain the desired phase relationship.[1] The output of the PSS, i.e., the stabilizing signal, is introduced into the excitation system at the input to the AVR/exciter along with the voltage error. The effectiveness of damping produced by excitation control has been demonstrated by simulation, field tests, and in practice.[2–5]

The evolution and development of the PSS from the fixed parameter analog type to that using adaptive control and artificial-intelligence (AI) based algorithms are outlined in this article. Mathematical details are kept to a minimum and only a limited number of illustrative results are given. Further details can be obtained from the references.

1.2   Fixed Parameter Controllers

A common feature of the fixed parameter controllers is that their design is done off-line. Using the state and/or output feedback, optimal structure and gains of the controller that minimize a certain performance index or meet design specifications are determined. Various approaches proposed to design fixed parameter PSSs are extensively reported in the relevant literature.

1.2.1   Conventional PSS

A conventional PSS (CPSS) is based on the use of a linear transfer function designed by applying the linear control theory[5] to the system model linearized at a preassigned operating point.

An IEEE type PSS1A CPSS has the transfer function

$$U_{pss}(s) = K_s \frac{1}{1+sT_6} \frac{sT_5}{1+sT_5} \frac{1+sT_1}{1+sT_2} \frac{1+sT_3}{1+sT_4} \bullet \dot{\delta} \; or \; \Delta P_e(s) \qquad \text{(VI.1.1)}$$

It contains a network to compensate for the phase difference from the excitation controller input to the damping torque output, i.e., the gain and phase characteristics of the excitation system, the generator, and the power system, which collectively determine the open-loop transfer function. Algorithms are available to calculate the parameters of the CPSS.

By appropriately tuning the phase and gain characteristics of the compensation network during the simulation studies at the design stage and further during commissioning, it is possible to set the desired damping ratio. Various tuning techniques have been introduced to effectively tune the CPSS parameters.[6] CPSSs are in wide use, and they have played an important role in improving the dynamic stability of power systems.

The CPSS is designed for a particular operating point for which the linearized model of the generator is obtained. Power systems are non-linear and operate over a wide range. For example, the gain of the plant increases with generator load. Also, the phase lag of the plant increases as the AC system becomes stronger. Due to non-linear characteristics, wide operating conditions, and unpredictability of perturbations in a power system, the CPSS, a linear controller, generally cannot maintain the same quality of performance under all conditions of operation.

The parameter settings of a CPSS are a compromise that provides acceptable, though not optimal, performance over the full range of operating conditions. An illustrative set of results with a CPSS

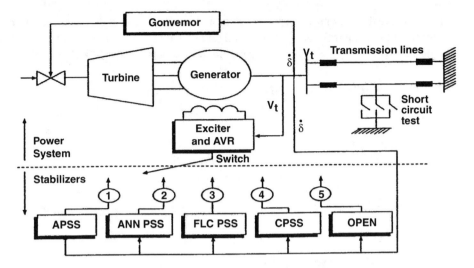

FIGURE VI.1.1   System model used in the simulation studies.

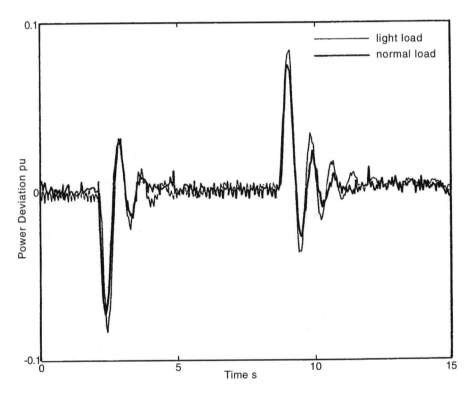

FIGURE VI.1.2   System response to a 0.1 pu torque disturbance with a CPSS for $P = 0.88$ pu and $P = 0.5$ pu, 0.85 pf lag.

providing a supplementary excitation control signal to a synchronous generator connected to a constant voltage bus through a double circuit transmission line (Figure VI.1.1) is shown in Figure VI.1.2. These experimental results, taken on a physical model in the laboratory, show the response to a 0.1 pu torque disturbance at two operating conditions, i.e., the operating condition of 0.88 pu load at which the parameters of the CPSS were optimized, and a light load condition of 0.5 pu.

To further improve the power system performance and stability, various other approaches based on linear quadratic optimal control,[7] H-infinity,[8,9] variable structure,[10] rule-based,[11] fuzzy logic, neural network, and adaptive control have been proposed in the literature to design a PSS.

1.2.2   Linear Quadratic and H-Infinity Optimal Controllers

The linear quadratic optimal[7] and H-infinity[8,9] based PSSs also fall in the fixed parameter controller category. The design of these controllers is done off-line on a linearized model of the power system. Using the state and/or output feedback, gains that minimize a certain performance index are determined. The H-infinity control design differs from the linear optimal control design in that it provides for uncertainties over a prespecified range in the system parameters and disturbances. However, because of the fixed feedback gains, variations in the system structure and/or characteristics cannot be tracked. It is, therefore, not possible to provide optimal performance over the entire operating range.

1.2.3   Variable Structure Control

To solve the parameter tracking problem, design of a CPSS based on the variable structure control theory has been proposed.[10] Although it is an elegant design technique, its design procedures share some commonality with that of the linear optimal control. Because of the absence of any formal procedures, the weights in the performance index of the linear optimal control and the weights of the switching vector for the variable structure algorithm have to be determined by trial and error.

1.2.4   Rule-Based and Fuzzy Logic Controllers

Unlike the conventional control techniques, which require complicated mathematical models derived from a deep understanding of a system, exact equations, and precise numeric values, fuzzy logic control techniques are rule-based systems. In these systems, a set of fuzzy rules represents a control decision mechanism to adjust the effects of certain causes coming from the controlled system.[12,13]

The basic feature of fuzzy logic control is that a process can be controlled without the knowledge of its underlying dynamics. The operator can simply express the control strategy, learned through experience, by a set of rules. These rules describe the behavior of the controller using linguistic terms. The controller then infers the proper control action from this rule base which plays the role of the human operator.

The theme of fuzzy logic is to relate the numeric variables to linguistic variables, where dealing with the linguistic variables is closer to the human spirit. Each linguistic variable represents a fuzzy subset. Each fuzzy subset has a membership function that defines how far this measurement belongs to this linguistic variable. The basic configuration of a fuzzy logic controller (FLC), as shown in Figure VI.1.3, comprises four principal components: a fuzzification module, a knowledge base, an inference mechanism, and a defuzzification module. The simplicity of the FLC concept makes it easy to implement and, hence, much faster to develop.

Some of the major features of the FLC are:

- This method does not require the exact mathematical model of the system.
- It offers ways to implement simple but robust solutions that cover a wide range of system parameters and can cope with major disturbances.
- The simplicity of the concept makes it easy to implement and requires writing less software code.
- Because the control strategy mimics the human way of thinking, the experience of a human operator can be implemented through an automatic control method.

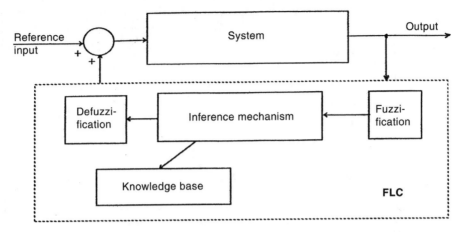

FIGURE VI.1.3 Basic structure of a fuzzy logic controller.

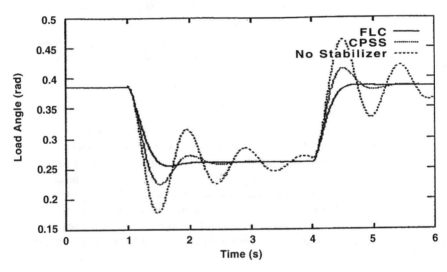

FIGURE VI.1.4 Response to a 0.1 pu step decrease in torque and return to initial condition (power 0.3 pu, power factor 0.9 lead).

Satisfactory results have been obtained with PSSs designed based on FLC.[14,15] Two illustrative results, one for a simulation study and one for an experimental study, are shown in Figures VI.1.4 and VI.1.5, respectively. For the experimental study, the FLC PSS was implemented on an 8-bit microcontroller system.

Although FLC introduces a good tool to deal with complicated non-linear and ill-defined systems, it suffers from the drawback of parameter tuning for the controller. Proper decision rules cannot easily be derived by expertise for too complex systems, making fine tuning or achieving the optimal FLC a difficult task. Some significant operating conditions, i.e., disturbances or parameter changes, may be outside the expert's experience. Design and tuning of an FLC for a multi-input multi-output system is extremely tedious. Often, the approach adopted is to define membership functions and decision rules subjectively by studying an operating system or an existing controller.

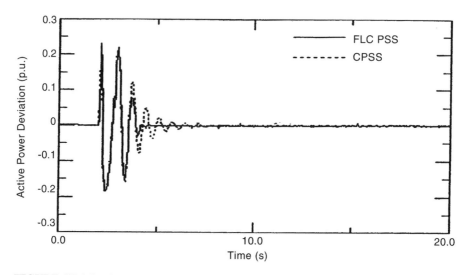

FIGURE VI.1.5   System response for three phase short circuit (power 0.81 pu, 0.87 pf).

Genetic algorithm, a global optimization method, can be used to help in the optimization and tuning of an FLC. However, it also has its limitations because it can fall into a local optimal point if the parameters are not selected properly.

1.2.5   Neural-Network Based Controllers

Artificial neural networks (ANNs) attempt to achieve good performance by interconnection of simple computational elements. They offer many advantages by virtue of their characteristics, which are the capability to synthesize complex and transparent mappings, speed due to the parallel mechanisms, robustness and fault tolerance, and adaptive adjustability to the new environment. The success of ANNs in controlling unknown systems under significant uncertainties makes them very attractive.

Among the many properties of a neural network, the property that is of primary significance is the ability of the network to learn from training data and to improve its performance through learning. Basic classes of learning paradigms are supervised learning, reinforced learning, and unsupervised learning.

To model the input/output behavior of a dynamic system, the ANN is trained using input/output data, and the weights of the neural network are adjusted using a training algorithm. Because the typical application involves non-linear systems, the ANN is trained for classes of inputs and initial conditions. The underlying assumption is that the non-linear static map generated by the ANN can adequately represent the system behavior in the ranges of interest for the particular application. For this to be possible, the ANN must be provided with information about the history of the system—typically, delayed inputs and outputs.

How much history is needed depends on the desired accuracy. There is a trade-off between accuracy and computational complexity of training, since the number of inputs used affects the number of weights in the ANN and, subsequently, the training time. One sometimes starts with as many delayed input signals as the order of the system and then modifies the ANN accordingly. It is also believed that using a two hidden-layer ANN instead of a one hidden-layer ANN has certain advantages.[16] The number of neurons in the hidden layer(s) is typically chosen based on empirical criteria, and one may iterate over a number of networks to determine the ANN that has a reasonable number of neurons and accomplishes the desired degree of approximation.

There are different control schemes to train a neural network to control a plant that is too complex or about which very little is known. In a typical control problem, the desired plant output may be

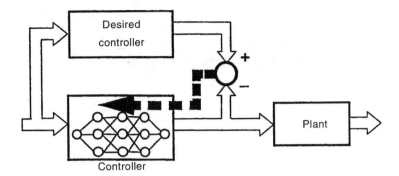

FIGURE VI.1.6   Copying an existing controller with a network. (From Hariri, A. and Malik, O.P., *Excitation Control in Power Systems*, Encyclopedia of Electrical and Electronic Engineering, Vol. 7, 1995. Used with permission of John Wiley & Sons, New York.)

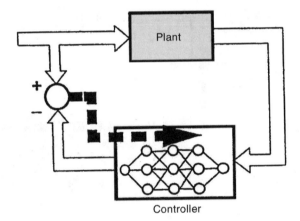

FIGURE VI.1.7   Inverse plant modeling using a network. (From Hariri, A. and Malik, O.P., *Excitation Control in Power Systems*, Encyclopedia of Electrical and Electronic Engineering, Vol. 7, 1995. Used with permission of John Wiley & Sons, New York.)

known but not the desired controller output, i.e., the control signal. Three basic ways in which the training information required for supervised learning can be obtained are:

- Copying an existing controller.[17] This approach, illustrated in Figure VI.1.6, is very useful where the desired controller may be a device that is impractical to use or may use very complicated algorithms to calculate the control signal.
- Identification of system inverse.[18–20] This approach, depicted in Figure VI.1.7, requires that an inverse of the plant is feasible. During its operation as a controller, the ANN acts as the inverse of the plant, producing from the desired response y_{pd} a signal x_p that drives the output of the plant to $y_p \approx y_{pd}$. In order for the control effects to be fed back to the ANN controller, the output of the plant, $y_p(t)$ and its delays are used instead of $y_{pd}(t)$ and its delays, as shown in Figure VI.1.8.
- Differentiating a model. The application of this idea requires that a plant model be available in a form that can be differentiated. The plant model is in the form of a layered network, as shown in Figure VI.1.9. This approach is discussed in further detail later on.

Power system stabilizers have been developed using each of these approaches. A few results with the inverse input/output mapped controller used as an ANN PSS are given here.

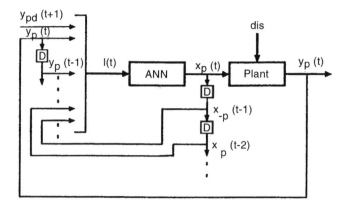

FIGURE VI.1.8   ANN controller architecture in operation mode.

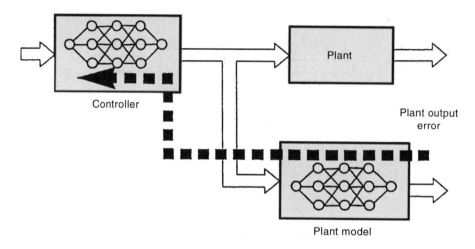

FIGURE VI.1.9   Back propagating through a forward model of the plant. (From Malik, O.P. and Hariri, A., *Trans. Inst. Meas. Control*, 24(2), 87–107, 2002. With permission.)

The input vector to the ANN PSS consisted of the speed deviation, $\Delta\omega(t)$, or the generator power output deviation, $\Delta P(t)$, as the plant output, its expected response, $\Delta\omega(t + 1)$ or $\Delta P(t + 1)$, n delayed values of $\Delta\omega$ or ΔP, and m delayed values of the supplementary control signal generated by the PSS, $u(t - 1), u(t - 2), \ldots$. The output of the ANN PSS is the supplementary control signal, $u(t)$. The architecture of the ANN PSS operation is shown in Figure VI.1.10.

In order to get the generator input/output pairs, which were used as the ANN training data, an adaptive PSS was used to control the generator in a range that covered most of the typical working conditions. Single machine constant voltage bus system responses with the ANN PSS, CPSS, and without PSS for a 0.05 pu step increase in input torque and removal after 4 s with the generator operating at 0.3 pu power, 0.85 pf lag, are shown in Figure VI.1.11. In this simulation study, $\Delta\omega$ was used as the supplementary signal.

Results of two experimental studies on the single machine system, using ΔP_e as the supplementary signal, are shown in Figures VI.1.12 and VI.1.13. With the generator operating at 0.793 pu power, 0.93 pf lag, a 4.5% step increase in the reference voltage was applied at 5s and removed at 20s. Since the CPSS was tuned at this operating condition, both PSSs gave approximately the same response, as seen in Figure VI.1.12. The operating condition of the generating unit was then changed to 1.36 pu

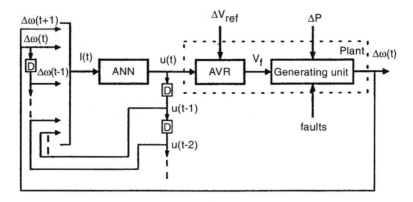

FIGURE VI.1.10   Inverse input/output mapped ANN PSS operation.

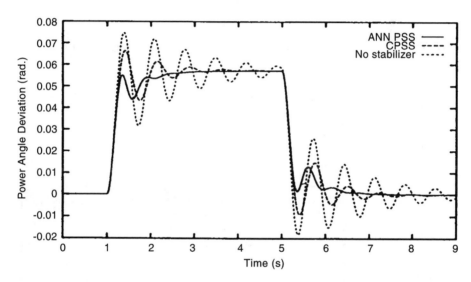

FIGURE VI.1.11   ANN PSS response to 0.05 pu step increase in torque with the generator operating at 0.2 pu power, 0.85 pf lag.

power, 0.94 pf lag. Results for a 4.3% step decrease in reference voltage applied at 5 s and removed at 15 s, as depicted in Figure VI.1.13, show that the ANN PSS can still give as good a response at this heavily loaded condition. However, the CPSS could not give the same performance as in the operating condition for which it had been tuned.

Most applications of the ANN based controllers as PSSs reported in the literature are the supervised learning algorithms that require a desired controller as reference for training purposes. Selection of the number of neurons and the number of layers in multi-layer networks is not a trivial task. It is, to a large extent, a process of trial and error. The ANN is trained off-line. Once trained, the implementation is done using the off-line trained weights which are subsequently held constant.

Some drawbacks to the use of conventional ANNs are that:

- It is difficult for an outside observer to understand or modify the network decision-making process.
- They may require a long training time to get the desired performance.

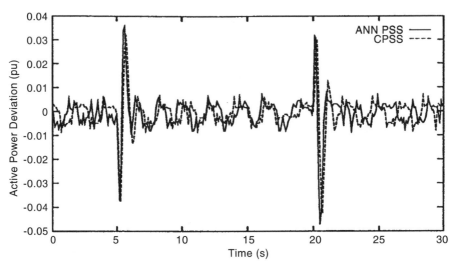

FIGURE VI.1.12   Comparison between ANN PSS and CPSS under voltage reference disturbances.

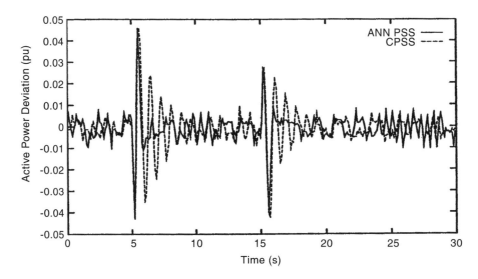

FIGURE VI.1.13   Comparison between ANN PSS and CPSS in new operating condition under voltage reference disturbances.

1.3   Adaptive Controllers

Adaptive control can be described as the changing of controller parameters based on the changes in system operating conditions. Whenever an adaptive controller detects changes in system operating conditions, that controller responds by determining a new set of control parameters.

The adaptive control theory provides a possible way to solve many of the problems associated with the CPSS. Two distinct approaches—direct adaptive control and indirect adaptive control—can be used to control a plant adaptively. In direct control approach, the parameters of the controller are directly adjusted to reduce some norm of the output error. In indirect control, the parameters of the plant are estimated as the elements of a vector at any instant k, and the parameter vector of the controller is adapted based on the estimated plant parameter vector.

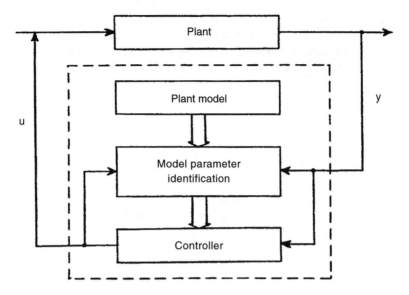

FIGURE VI.1.14   Block diagram of a self-tuning controller.

A general configuration of the indirect adaptive control as a self-tuning controller is shown in Figure VI.1.14. At each sampling instant, the input and output of the generating unit are sampled, and a plant model is obtained by some online identification algorithm to represent the dynamic behavior of the generating unit at that instant in time. It is expected that the model obtained at each sampling instant can track the system operating conditions.

The required control signal for the generating unit is computed based on the identified model. Various control techniques can be used to compute the control. All control algorithms assume that the identified model is the true mathematical description of the controlled system.

1.3.1   Analytical-Approach Based Adaptive PSS

In the analytical approach to the design of an adaptive PSS, sampled data design techniques are used to compute control in the following way:

- Select a sampling frequency, f, about ten times the normal frequency of oscillation to be damped.
- At each sampling interval $T (= 1/f)$, update the system model parameters. A number of identification algorithms have been developed using discrete domain mathematics. The least squares or extended least squares technique, in recursive form, is usually used to identify the system, i.e., the discrete transfer function of the controlled plant.
- Use the updated estimates of the parameters to compute the control based on the control strategy chosen. Various control strategies, among them optimal, minimum variance, pole-zero assignment, pole assignment, and pole shift, have been proposed.

An extensive amount of work has been done to develop and implement an adaptive PSS based on the above strategy. Such a PSS can adjust its parameters online according to the environment in which it works and can provide good damping over a wide range of operating conditions of the power system.

To keep the sampling period small enough for online control, there must be a compromise between the order of the identified model and the computation time for parameter identification and optimization. Therefore, the identified model is generally a low-order discrete model. Since the power system is a high-order, non-linear, continuous system, care has to be taken to ensure that the low-order discrete identified model can properly describe the dynamic behavior of the power system. Thus, there must be a compromise between the order of the discrete model and the computation time for parameter identification and optimization. With the present high-speed microprocessors, this does not impose a major constraint.

1.3.1.1   System Model

The generating unit is identified by a discrete ARMA model of the form

$$A(z^{-1})y(t) = B(z^{-1})u(t) + e(t) \qquad \text{(VI.1.2)}$$

where $A(z^{-1})$ and $B(z^{-1})$, polynomials in the delay operator of z^{-1}, are of the form

$$A(z^{-1}) = 1 + a_1 z^{-1} + \cdots + a_i z^{-i} + \cdots + a_{n_a} z^{-n_a} \qquad \text{(VI.1.3)}$$

$$B(z^{-1}) = b_1 z^{-1} + \cdots + b_i z^{-i} + \cdots + b_{n_b} z^{-n_b} \qquad \text{(VI.1.4)}$$

$$n_a \geq n_b$$

the variables $y(t)$ and $u(t)$ are the system output and system input, respectively, and $e(t)$ is assumed to be a sequence of independent random variables with zero mean.

1.3.1.2   System Parameter Estimation

The control is computed based on the identified model parameters, a_i and b_i. Thus, to compute the control appropriate to the varying operating conditions, the system parameters have to be estimated online. The correctness of the identification determines the preciseness of the identified model which tries to reflect the true system. For a time-varying system, the tracking ability of the identification method is very desirable.

There are several methods that can be used to obtain an estimate for the model parameters.[21] A commonly used technique of achieving a continuous tracking of the system behavior is the recursive least squares (RLS) parameter identification method.

Coefficients of the polynomials $A(z^{-1})$ and $B(z^{-1})$ in Equation VI.1.3 can be calculated recursively online by the RLS technique, as follows.

Rewrite Equation VI.1.3 in the following form suitable for identification:

$$y(t) = \hat{\theta}^T(t)\Psi(t) + e(t) \qquad \text{(VI.1.5)}$$

where

$$\hat{\theta}(t) = [a_1 a_2 \ldots a_{n_a} b_1 b_2 \ldots b_{n_b}]^T$$

is the parameter vector and

$$\Psi(t) = [-y(t-1) \ldots - y(t - n_a)u(t-1) \ldots u(t - n_b)]^T$$

is the measurement variable vector.

The system parameter vector $\hat{\theta}(t)$ can be calculated by the following recursive set of equations:[21]

$$\hat{\theta}(t+1) = \hat{\theta}(t) + K(t)[y(t) - \hat{\theta}^T(t)\Psi(t)] \qquad \text{(VI.1.6)}$$

$$K(t) = \frac{P(t)\Psi(t)}{\rho(t) + \Psi^T(t)P(t)\Psi(t)} \qquad \text{(VI.1.7)}$$

$$P(t+1) = \frac{1}{\rho(t)}[P(t) - K^T(t)P(t)\Psi(t)] \qquad \text{(VI.1.8)}$$

where $\rho(t)$ is the forgetting factor, $P(t)$ is the covariance matrix, and $K(t)$ is the modifying gain vector.

The forgetting factor $\rho(t)$ is used to enhance the ability of the identifier to track the operating conditions of the actual system as it discounts the importance of the older data. It can be chosen as a constant or a variable. A variable forgetting factor is employed to improve the tracking ability, especially under large disturbances. The online calculation of the forgetting factor $\rho(t)$ is given in the following form:

$$\rho(t) = \rho_0 \rho(t-1) + (1 - \rho_0) \quad (0 < \rho_0 < 1) \tag{VI.1.9}$$

It is clear from Equation VI.1.9 that $\rho(t)$ is an exponential function of time t. The coefficient ρ_0 determines its increasing rate and $\rho(0)$ determines its initial value.

When the power system is under transient conditions, such as configuration change or operating condition change, the identifier is required to track such changes quickly. This can be achieved with the variable forgetting factor. Define the identification error as:

$$\xi(t) = y(t) - \hat{\theta}^T \Psi(t) \tag{VI.1.10}$$

The absolute value of $\xi(t)$ indicates the accuracy of the identified model and simultaneously indicates the instant when the environment condition changes. It can be used as a criterion to determine the value of the forgetting factor $\rho(t)$:

$$\rho(t) = \min\left\{\rho_0 \, \rho(t-1) + (1 - \rho_0), 1 - \frac{\xi^2(t)}{\Sigma_0}\right\} \tag{VI.1.11}$$

where Σ_0 is a constant and can be determined using the method proposed by Cheng et al.[22] It should be pointed out that in order to get smooth parameter identification, a minimum forgetting factor ρ_{min} is specified. If one wishes to keep the identifier more sensitive to the system changes under steady-state operation, a maximum forgetting factor $\rho_{max}(< 1.0)$ can be specified, i.e.,

$$\rho_{min} < \rho(t) < \rho_{max} \tag{VI.1.12}$$

1.3.1.3 Adaptive Optimal Control Algorithm

In real-time control, it is acceptable to express a generator by a third-order model.[22] Thus, Equation VI.1.3 can be rewritten as:

$$\frac{y(z)}{u(z)} = \frac{z^{-1}(b_1 + b_2 z^{-1} + b_3 z^{-2})}{1 + a_1 z^{-1} + a_2 z^{-2} + a_3 z^{-3}} \tag{VI.1.13}$$

where a_i and b_i are time-varying parameters.

It is easy to identify the transfer function of a controlled system using the recursive least squares technique.[21]

From Equation VI.1.13, one obtains

$$zy(z) + a_1 y(z) + a_2 z^{-1} y(z) + a_3 z^{-2} y(z) = b_1 u(z) + b_2 z^{-1} u(z) + b_3 z^{-2} u(z) \tag{VI.1.14}$$

Let the state variables be

$$\left.\begin{aligned} x_1 &= z^{-2} u(z) \\ x_2 &= z^{-1} u(z) \\ x_3 &= z^{-2} y(z) \\ x_4 &= z^{-1} y(z) \\ x_5 &= y(z) \end{aligned}\right\} \tag{VI.1.15}$$

Substituting Equations VI.1.15 and VI.1.13 gives in time domain

$$X(k+1) = AX(k) + Bu(k) \tag{VI.1.16}$$

where

$$X = (x_1 \ x_2 \ x_3 \ x_4 \ x_5)^T$$

$$A = \begin{vmatrix} 0 & 1 & 0 & 0 & 0 \\ 0 & 0 & 0 & 0 & 0 \\ 0 & 0 & 0 & 1 & 0 \\ 0 & 0 & 0 & 0 & 1 \\ b_3 & b_2 & -a_3 & -a_2 & -a_1 \end{vmatrix}$$

$$B = (0 \ 1 \ 0 \ 0 \ b_1)^T$$

Based on linear optimal control theory,[23] for the performance index

$$J = \sum_{k=1}^{\infty} (X_k^T Q X_k + u_{k-1}^T R u_{k-1}) \tag{VI.1.17}$$

the optimal control law is

$$u = -KX \tag{VI.1.18}$$

and

$$K = (R + B^T P B)^{-1} B^T P A \tag{VI.1.19}$$

where P is the solution of the discrete Riccati equation,

$$P = Q + A^T P A - A^T P B (R + B^T P B)^{-1} B^T P A \tag{VI.1.20}$$

As matrices A and B have many zero elements, the solution of matrix P is easy.[24]

In time domain, the optimal control law, Equation VI.1.18, becomes

$$u(k) = -[k_1 u(k-2) + k_2 u(k-1) + k_3 y(k-2) + k_4 y(k-1) + k_5 y(k)] \tag{VI.1.21}$$

where k_1, k_2, k_3, k_4, and k_5 are feedback gains, the elements of matrix K.

It can be seen from the above derivation that no observer is required for the calculation of control output; the output and control are directly used as feedback. Thus, the controller can respond quickly to changes in the output.

1.3.1.4 Adaptive Optimal Power System Stabilizer

In this case, the function of the adaptive controller as a PSS is to provide a stabilizing signal to supplement the AVR output, as shown in Figure VI.1.1. After extensive simulation studies and experimental studies in the laboratory on a physical model, the adaptive optimal power system stabilizer was installed and tested on one 400 MW thermal unit in a two 400 MW unit Keephills generating station with the cooperation of TransAlta Utilities. In the interest of safety, a hardware limiter provided by TransAlta, just before the output of the controller was fed into the AVR summing junction, was used for these tests.

The input signal to the stabilizer was the deviation in the machine electrical power output. As the power transducer output is 1 mA for the rated output of 400 MW, the power signal was converted into a voltage signal and filtered. A gain stage at the input to the controller allowed the range of the signal variation to be adjusted during the tests.

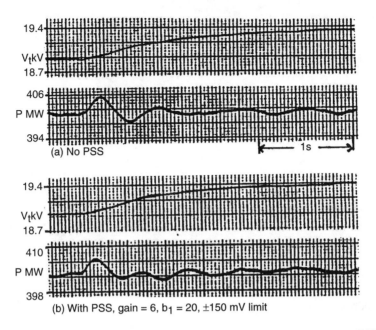

FIGURE VI.1.15 Generator response to a 3% step increase in voltage reference; (a) no PSS; (b) with PSS, gain = 6, $b_1 = 20$, +150 mv limit. (From Malik, O.P. and Mao, C., *Int. J. Electrical Power Energy Sys.*, 15(3), 169, 1993. With permission.)

Because of the limited time left at the end of another series of tests, only a few tests for 3% step decrease in voltage reference and a few tests with 3% step increase in voltage reference for various limits on the controller output could be performed. During these tests, the generating unit was operating at an output of 405 MW. For all tests, V_t, V_f, I_f, P, Q, and control signals before and after the hardware limiter were recorded on a chart recorder. For each test, 250 samples of the controller input and calculated control signal could be stored in the ROM. The data was then transferred to the PC for storage on a disk.

The TransAlta system is inherently a very stable system. Although conventional stabilizers were installed initially on the Keephills units, they are normally operated without stabilizers. Additionally, these stabilizers were not available for use at the time of the tests. Therefore, no comparison could be made between the performance of the adaptive optimal stabilizer and the conventional stabilizer.

1. **No PSS control** — at the start, only a noise signal was injected into the AVR. The controller output was monitored on the operator console for step changes in voltage reference to verify the controller response. The controller output was found to be stable, and, in their experience, the station operating personnel seemed to be satisfied with its behavior.
2. **3% step changes in voltage with control** — in the next series of tests, both the noise and the optimal stabilizer output were fed into the AVR. Studies were performed with the hardware limiter changed from ±50 mV to ±100 mV to ±150 mV. For this machine, a control signal of 210 mV corresponds to a 1% change in terminal voltage reference.

Terminal voltage response, power variation, and the control before and after the hardware limiter for a few tests with 3% step increase in voltage reference are shown in Figure VI.1.15. The results are also tabulated in Table VI.1.1.

A 3% step change in voltage reference is a rather small disturbance and did not generate much control action. Despite this, the adaptive optimal PSS did reduce the peak-to-peak variation in power fluctuation and the time to reach the final value of the terminal voltage.

TABLE VI.1.1   Three Percent Step Increase in Voltage Reference

	No PSS Control	Control Limit		
		$\pm 100\,\mathrm{mV}$ Gain $= 6, b_1 = 1$	$\pm 100\,\mathrm{mV}$ Gain $= 6, b_1 = 20$	$\pm 150\,\mathrm{mV}$ Gain $= 6, b_1 = 20$
First peak-to-peak variation in power (MW)	7.36	6.93	6.07	6.07
Time to reach final value(s)	2.64	2.34	2.04	2.44

1.3.1.5 Self-Optimizing Pole Shifting Control Algorithm

An extensive amount of work has been done to develop and implement a pole-shift based adaptive PSS as reported in the literature.[22,25–32]

Once the system model is identified as in Equation VI.1.3, the control signal is calculated based on this model and the self-optimizing pole shifting control algorithm.[28] The basic idea is summarized below.

If a linear feedback control law

$$\frac{u(t)}{y(t)} = -\frac{G(z^{-1})}{F(z^{-1})} \tag{VI.1.22}$$

is chosen to generate the control signal $u(t)$ so that the closed-loop characteristic polynomial of the system in Equation VI.1.3 takes the desired form $T(z^{-1})$, i.e.,

$$A(z^{-1})F(z^{-1}) + B(z^{-1})G(z^{-1}) = T(z^{-1}) \tag{VI.1.23}$$

where

$$F(z^{-1}) = 1 + f_1 z^{-1} + \cdots + f_i z^{-1} + \cdots + f_{n_f} z^{-n_f}$$

$$G(z^{-1}) = g_0 + g_1 z^{-1} + \cdots + g_i z^{-1} + \cdots + g_{n_g} z^{-n_g}$$

and

$$n_f = n_b - 1, \ n_g = n_a - 1$$

It can be seen that once $T(z^{-1})$ is specified, $F(z^{-1})$ and $G(z^{-1})$ can be determined by Equation VI.1.23 and, thus, the control signal $u(t)$ from Equation VI.1.22.

For the proposed algorithm, the most important point is that the desired closed-loop polynomial $T(z^{-1})$ is not chosen manually. It takes the form of $A(z^{-1})$ with poles shifted radially towards the center of the unit circle in the z-plane by a factor α, called the pole-shifting factor.[22] Thus,

$$T(z^{-1}) = A(\alpha z^{-1}) \tag{VI.1.24}$$

Obviously, α is the only parameter to be determined, and its value reflects the stability of the closed-loop system.

If the pole-shifting factor α is fixed, the pole-shifting control algorithm degenerates into a special case of the pole assignment control algorithm. It is evident that the rule determining the pole-shifting factor is very important. For optimum performance, it is desirable to modify α online according to the operating conditions of the controlled system.

To consider the time domain performance of the controlled system, it is often desirable that the system output $y(t + 1)$ follow a prespecified reference $y_r(t + 1)$ as close as possible. A performance index J can be formed to measure the difference between the predicted system output $\hat{y}(t + 1)$ and its reference.

$$J = E[\hat{y}(t + 1) - y_r(t + 1)]^2 \qquad \text{(VI.1.25)}$$

By prediction theory, $\hat{y}(t + 1)$ is determined by its system parameter polynomial $A(z^{-1})$, $B(z^{-1})$, control signal sequence $\{u(t)\} = \{u(t), u(t - 1), \ldots\}$ and its past history $\{y(t)\} = \{y(t), y(t - 1), \ldots\}$. Considering that $u(t)$ is a function of the pole-shifting factor α, the performance index J becomes:[28]

$$J = W[A(z^{-1}), B(z^{-1}), C(z^{-1}), \{u(t)\}, \{y(t)\}, \alpha, y_r(t + 1)] \qquad \text{(VI.1.26)}$$

Equation VI.1.26 is of great value since it represents the self-optimizing property of the proposed algorithm. The pole-shifting factor α is the only unknown variable in Equation VI.1.26 and, thus, can be determined by minimizing J. The control signal $u(t)$ can then be determined. Investigations have shown that the relation between J and α is quite explicit and straightforward.[29] This makes it possible to implement this algorithm without special hardware requirements and considerations.

Since α reflects the stability of the closed-loop system, it is desirable to assign the pole-shifting factor α to a specified value under steady state, such as α_0. To accommodate this ability, Equation VI.1.26 can be modified as

$$J = W[(z^{-1}), B(z^{-1}), C(z^{-1}), \{u(t)\}, \{y(t)\}, \alpha] + \mu(\alpha - \alpha_0)^2 \qquad \text{(VI.1.27)}$$

where μ is a weighting coefficient.

There are two constraints which must be considered while the performance index in Equation VI.1.37 is minimized.[33] The first is called stability constraint, and the second is called control limits constraint.

1. Stability constraint—supposing λ is the absolute value of the largest characteristic root of $A(z^{-1})$, then α. λ is the absolute value of the largest closed-loop characteristic root of $A(\alpha z^{-1})$. To guarantee the stability of the closed-loop system, α ought to satisfy the following inequality:

$$-\frac{1}{\lambda} < \alpha < \frac{1}{\lambda} \qquad \text{(VI.1.28)}$$

2. Control limits constraint—the control limits should be taken into account in the stabilizer design to avoid servo system saturation or equipment damage. If u_{min} and u_{max} are the lower and upper hard limits of the system, a dynamic control limits range $[u_{limit}^-, u_{limit}^+]$ can be calculated by the method proposed by Chen et al.[33] While minimizing the performance index in Equation VI.1.27, the control signal $u(t)$ should be limited between u_{limit}^- and u_{limit}^+ instead of between u_{min} and u_{max}:

$$u_{limit}^- < u(t) < u_{limit}^+ \qquad \text{(VI.1.29)}$$

The performance index in Equation VI.1.27 with the constraints in Equations VI.1.28 and VI.1.29 constitute the optimization problem to be solved to obtain the control signal $u(t)$. Some special techniques have been proposed to solve this optimization problem effectively.[34]

1.3.1.6 Performance Studies with Pole-Shifting Control PSS

The performance of the self-tuning adaptive controller based on the pole-shifting control algorithm has been investigated by conducting simulation studies on single machine[25, 28] and multi-machine

TABLE VI.1.2   Dynamic Stability Margin Results

	Without PSS	With CPSS	With APSS
Maximum Power	1.95 pu	2.65 pu	2.90 pu
Maximum Rotor Angle	1.1828 rad.	1.4733 rad.	1.5730 rad.

TABLE VI.1.3   Transient Stability Margin Results

	Without PSS	With CPSS	With APSS
Maximum Clearance Time	120 ms	150 ms	165 ms

systems,[22] on a single machine,[30] and on a multi-machine physical model[32] in the laboratory and on a 400 MW thermal machine under fully loaded conditions connected to the system.[31]

Results of two simulation studies are given to demonstrate the effect of the self-tuning adaptive controller on the dynamic and transient stability margins. For both tests, the initial operating conditions were 0.95 pu power, 0.9 pf lag.

For the dynamic stability margin test, the input torque reference was increased gradually from the initial value. Since the voltage reference setting for AVR was kept constant, the generator terminal voltage remained at the specified value as long as the system was stable. The dynamic stability margin is described by the maximum power output at which the system loses synchronism. This test was repeated for the system without a stabilizer, with CPSS, and with adaptive PSS (APSS), respectively. The results are given in Table VI.1.2. It can be seen that the proposed APSS can improve the system dynamic stability margin much further than the CPSS.

For the transient stability margin test, with the system operating at the same initial point as above, a three phase to ground fault was applied near the sending end of one transmission line. The maximum clearance time for the power system without PSS, with CPSS, and with APSS is shown in Table VI.1.3. The APSS provides the largest maximum clearance time, which indicates that the transient stability margin of the system is improved most by the APSS.

This APSS was implemented on a microprocessor and tested in real-time on a physical model of a single-machine infinite bus system. With the system operating at a stable operating point, the APSS was applied to the system. The torque reference was then increased gradually to the level

$$P = 1.307\,\text{pu, pf} = 0.95\,\text{lead,}\ V_t = 0.950\,\text{pu}$$

At this load, the system was still stable with the APSS. At 5 s, the APSS was replaced by the CPSS. After the switch-over, the system began to oscillate and diverge, which means that the CPSS is unable to keep the system stable at this load level. At about 25 s, the APSS was switched back to control the unstable system and the system came under control very quickly. The whole process is shown in Figure VI.1.16. This test demonstrates that the APSS can provide a larger dynamic stability margin than the CPSS. Also, more power can be transmitted with the help of the APSS if an overload operation is necessary under certain circumstances.

1.3.2   Adaptive PSS with ANN Identifier and Pole-Shift Control

A self-tuning adaptive PSS as described above can improve the dynamic performance of the synchronous generator by allowing the parameters of the PSS to adjust as the operating conditions change. However, unless proper care is taken, problems such as the identifier using the RLS technique going to sleep and unstable parameter identification, i.e., $P(t)$ matrix blow-up, can arise.[35]

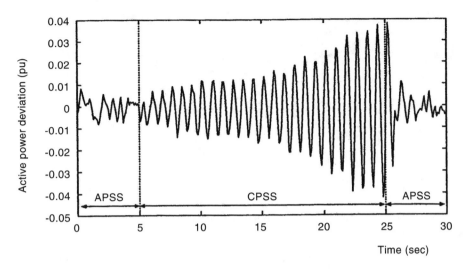

FIGURE VI.1.16 Dynamic stability improvement by the APSS.

The success of ANNs in controlling unknown systems under significant uncertainties makes them very attractive. Using the online learning features of neural networks, the time-varying power plant can be tracked and the control signal can be computed accordingly. Because of their inherent features, ANNs do appear to be able to implement many functions essential to control systems with a high degree of autonomy.[16]

It is possible to overcome the problems of RLS identification by using an ANN for identifying the system model parameters. An analytical technique, such as the pole-shift control, can be retained to compute the control signal. Two versions of this approach, one using an ADALINE and the other using a radial basis function (RBF) network for model parameter identification, have been developed. The APSS shown in Figure VI.1.14 now consists of an ANN identifier and the pole-shifting control algorithm described above.

1.3.2.1 ADALINE Identifier

Using the third-order discrete ARMA model, Equation VI.1.2, to describe the generating unit, an ADALINE network is modeled so that its weights have a one-to-one relationship with the regression coefficients of the ARMA model. This overcomes the black-box-like description of the ADALINE and the information is analyzed at each sampling interval to track changes in the controlled system.

The ADALINE model shown in Figure VI.1.17 is used as the identifier. It consists of a single layer and has one output linear neuron. $\Delta P_e(t)$ is the active power deviation and $u(t)$ is the APSS control signal, both at time step t. The linear transformation (*lin*) calculates the neuron's output by simply returning the value passed to it as given in Equation VI.1.30 below:

$$\Delta \hat{P}_e(t) = lin(WV) = WV \qquad \text{(VI.1.30)}$$

The output of the identifier is the predicted active power deviation at time step t, $\Delta P_e(t)$. W is the weight vector. The input vector of the ADALINE Identifier is

$$V(t) = \lfloor \Delta P_e(t-1), \Delta P_e(t-2), \Delta P_e(t-3), u(t-1), u(t-2), u(t-3) \rfloor \qquad \text{(VI.1.31)}$$

The cost function for the identifier is:

$$J(t) = \frac{1}{2}[\Delta P_e(t) - \Delta \hat{P}_e(t)]^2 \qquad \text{(VI.1.32)}$$

where $\Delta P_e(k)$ is the desired active power deviation.

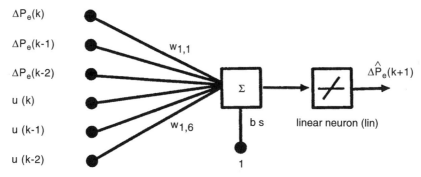

FIGURE VI.1.17   ADALINE network model.

The Widrow–Hoff learning rule[36] is used to train the ADALINE. The weights are updated by the following equation:

$$W(t+1) = W(t) + 2\gamma e(t)V^T(t) \qquad (VI.1.33)$$

where γ is the learning rate and $e(t)$ is the identification error. $e(t)$ is given by

$$e(t) = \Delta P_e(t) - \Delta \hat{P}_e(t) \qquad (VI.1.34)$$

The learning algorithm adjusts the weights of the ADALINE so as to minimize this mean square error. γ decides the speed of convergence of the iterative procedure. If γ is large, learning occurs quickly, but if it is too large, it leads to instability and errors increase. To ensure stable learning and to ensure that the weights converge to optimum W^*, γ is chosen as:[37]

$$0 < \gamma < 1/\lambda_{\max} \qquad (VI.1.35)$$

where λ_{max} is the largest eigenvalue of the correlation matrix $V^T V$ of the input vectors.

The ADALINE identifier has a linear structure. From Equations VI.1.4, VI.1.5, VI.1.32, and VI.1.33, it can be seen that the weight vector W has a one-to-one relationship with the system parameters θ which can be used by the adaptive controller in computing the control signal. This provides insight into the operation of the ADALINE model, and the Widrow–Hoff learning procedure can be used to find the optimum set of regression coefficients. The process of obtaining ARMA parameters from ADALINE is an important point, as it is possible to apply linear analysis control methods such as the pole-shift control technique to obtain the control signal.

For off-line training of the ADALINE identifier, data were collected at a number of operating conditions for the generator and various disturbances. In Equation VI.1.35, 2γ was chosen equal to $0.999V^T V$.

After the off-line training, the network is further trained online every sampling period, making this an adaptive approach. The online updating of weights allows the APSS to track the operating conditions of the power system and any changes in the ARMA parameters.

Results of one study on a five machine, two area interconnected system (Figure VI.1.18) that exhibits multi-modal oscillations are given. To simulate the dynamic behavior of this system, each generator is represented by a non-linear fifth-order model. An IEEE Standard 421.5, Type STIA AVR and exciter model and a PSSIA type CPSS were used in the simulation studies. The input to the PSS was the active power deviation, ΔP_e. A sampling rate of 20 Hz was used for both parameter identification and control using the pole-shift algorithm.

A three phase to ground fault was applied at the middle of one transmission line between buses 3 and 6 at 1 s and cleared 50 ms later by removing the faulted line. The faulted transmission line was

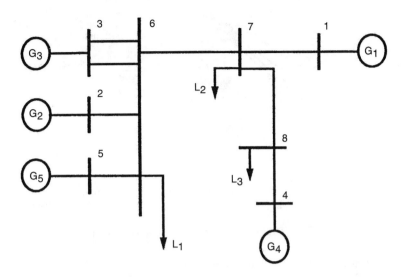

FIGURE VI.1.18 A five machine power system configuration.

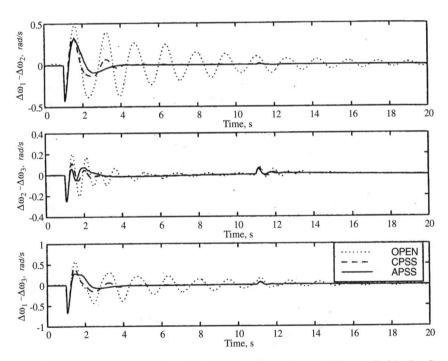

FIGURE VI.1.19 System response to a three phase to ground fault with PSSs installed in G_1, G_2, and G_3.

restored successfully at 11 s. Response of the system with the adaptive PSSs only and with CPSSs only installed on G_1, G_2, and G_3 given in Figure VI.1.19 shows that both the local mode and the interarea mode oscillations are damped effectively.

1.3.2.2 Radial Basis Function Network Identifier

The RBF model shown in Figure VI.1.20 is used to identify the system model parameters. It consists of three layers: the input, hidden, and output layers. Each of the six input variables is assigned to

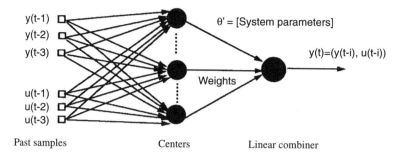

Centers: Adjusted using n-means clustering (offline)
Weights: Adjusted using recursive least-squares algorithm (online)

FIGURE VI.1.20   Radial basis function network model.

an individual node in the input layer and passes directly to the hidden layer without weights. The hidden nodes are called the RBF centers and calculate the Euclidean distance between the centers and the network input vector. The result is passed through the most widely used Gaussian function,[38]

$$\phi(x) = e^{-x^2/\sigma^2} \tag{VI.1.36}$$

This is characterized by a response which has a maximum value of 1 when the distance between the input vector and the center is zero. Thus, a radial basis neuron acts as a detector which produces 1 whenever the input vector $p = y(t-1), \dots, u(t-1), \dots$ is identical to the center (active neuron). The other neurons with centers quite different from the input vector will have outputs near zero (non-active neurons).

The connections between the hidden neurons and the output node are linear weighted sums, as described by Equation VI.1.37.

$$y = \sum_{i=1}^{nh} \theta' \exp\left(-\frac{\|p - c_i\|^2}{\sigma^2}\right) \tag{VI.1.37}$$

where c_i, σ, and θ' are the centers, widths, and weights respectively. nh is the number of hidden layer neurons.

To make the proposed RBF identifier faster for online applications, the hidden layer is created as a competitive layer wherein the center closest to the input vector becomes the winner and all the other non-active neurons are deactivated. Also, the scalar weights w are modified as a vector θ' whose size equals the size of the input vector. The weight vector θ' is given by

$$\theta'(t) = [a_1'\ a_2'\ a_3'\ b_1'\ b_2'\ b_3'] \tag{VI.1.38}$$

To obtain the linear parameters of the standard ARMA model, the output of the RBF, $y(t) = f(y(t-1), u(t-1))$ is linearized as follows at each sampling instant:

$$\Delta y = \frac{\partial y}{\partial y(t-1)} \Delta y(t-1) + \cdots + \frac{\partial y}{\partial u(t-1)} \Delta u(t-1) + \cdots \tag{VI.1.39}$$

It is seen from Equation VI.1.39 that the weight vector θ' has a one-to-one relationship with the system parameters θ which can be used in computing the control signal.

In a procedure similar to the ADALINE identifier, for off-line training of the RBF identifier to choose appropriate centers, data were collected at a number of operating points for various

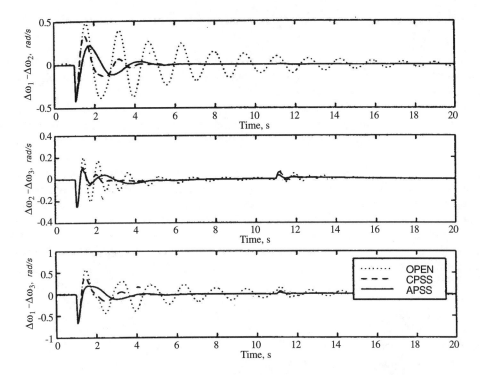

FIGURE VI.1.21 System response to a three phase to ground fault with PSSs installed on G_1, G_2, and G_3. (From Ramakrishna, G. and Malik, O.P., *Conference Proceedings*, Vol. 1, IEEE Power Engineering Society, July 2000, Seattle, 116–121. With permission.)

disturbances. The n-means clustering algorithm[39] yielded 15 centers for the RBF model. With the off-line training finished, the weights (system parameters) were updated online to obtain the appropriate control signal using the pole-shifting controller. A 100 ms sampling period was chosen for digital implementation.

Results of a simulation study, similar to that given in Figure VI.1.19 for the ADALINE identifier, are shown in Figure VI.1.21.

1.3.2.3 Comparison of ADALINE and RBF Identifiers

A comparison of the results obtained in various studies using the ADALINE identifier with the results obtained by the RBF identifier showed that the performance of the two APSSs is similar even though the RBF identifier performs better when tracking the system dynamics. The reason for this is discussed below:

1. The identification results obtained show that both procedures track the low frequency oscillations quite accurately. The only difference is that the RBF performs better compared to the ADALINE at the peaks where the system exhibits high frequency dynamics. Since the main point of interest for an APSS is to control the low-frequency variations (0.5–2.0 Hz), the inputs to the PS control are filtered using a low-pass filter. Thus, the PS control sees similar identification by both procedures, and, hence, results in a similar control action.
2. Both identifiers are first trained off-line to reduce their MSEs. They are further trained online to track the varying dynamics. The ARMA coefficients for use in the PS control are obtained from the identifiers online. Since both techniques result in an optimum set of ARMA coefficients in the MSE sense, even though the ARMA coefficients obtained by the identifiers can be different

individually at any given instant of time, the PS control results in a similar performance with both identifiers.

This raises an important question as to which of the two APSSs, ADALINE identifier/PS-control or RBF identifier/PS-control, is suitable for practical use. At first glance, the choice would be ADALINE which, using a simple neuron, involves the simple Widrow–Hoff learning rule and needs only 6 weights that have to be updated on-line, whereas RBF uses a non-linear Gaussian function in the hidden nodes (9) and uses an elaborate learning procedure (n-means clustering to find RBF centers and RLS to find the 9 weights at the output layer). However, from a theoretical standpoint discussed below, the balance weighs in favor of the RBF.

Chen et al.[38] point out that a simple linear network such as ADALINE can be used as a linear approximation to a non-linear problem as long as it can reduce the MSE within the performance desired. However, if it cannot achieve the above objective, then feedforward ANNs, RBF networks, etc. are the alternatives.

In an RBF network, the transformation from the input space to the hidden-unit space is non-linear, whereas the transformation from the hidden-unit space to the output space is linear. Mathematical justification for this rationale may be traced back to Cover.[40] The important point that emerges from this chapter is that a complex pattern-classification problem cast in high-dimensional space non-linearly is more likely to be linearly separable than in a low-dimensional space.

Basically, the non-linear mapping in RBF is used to transform the non-linearly separable classification problem into a linearly separable one. In a similar fashion, the RBF identifier can be used to transform the difficult non-linear power system problem into an easier one that involves a linear regression model. Thus, there is, in general, practical benefit to be gained in using the RBF identifier compared to the ADALINE identifier. The RBF identifier is more likely to find a set of ARMA coefficients at all operating conditions than an ADALINE identifier, which can reduce its MSE only to a certain level for any given problem because of its limitation in size (limited to the number of inputs).

In summary, for a given power system, the ADALINE identifier may give adequate results. However, because of the mathematical limitations discussed above, it is not necessarily true that it will work in all systems. An RBF identifier is more likely to achieve the desired results for different systems.

1.3.3   Adaptive PSS with ANN Identifier and ANN Controller

Identification of the power plant model using an online recursive identification technique is a computationally extensive task. Neural networks offer the alternative of a model-free method. An adaptive neural-network based controller using an indirect adaptive control method has been developed. It combines the advantages of neural networks with the good performance of adaptive control. This controller employs the learning ability of neural networks in adaptation process and is trained in each sampling period.

The controller consists of two subnetworks, as shown in Figure VI.1.22. The first one is an adaptive neuro-identifier (ANI) which identifies the power plant in terms of its internal weights and predicts the dynamic characteristics of the plant. The identifier is based on the inputs and outputs of the plant and does not need knowledge of the states of the plant. The second subnetwork is an adaptive neuro-controller (ANC) which provides the necessary control action to damp the oscillations of the power plant.

The success of the control algorithm depends on the accuracy of the identifier in tracking the dynamic plant. For this reason, the ANI is initially trained off-line before being hooked up in the final configuration. The training is performed over a wide range of operating conditions and a wide spectrum of possible disturbances for the generating unit. After the off-line training stage, the ANI is hooked up in the system. Further training of the ANI and ANC is done online every sampling

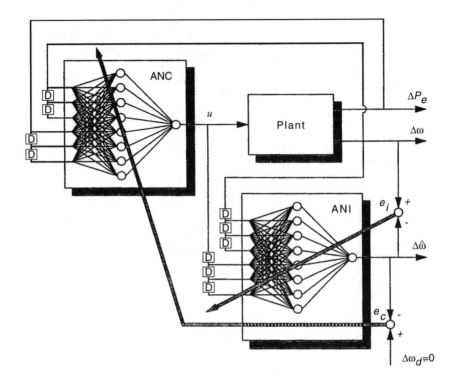

FIGURE VI.1.22 Controller structure for single machine study. (From Shamsollahi, P. and Malik, O.P., *IEEE Trans. Energy Convers.*, 12(4), 382, 1997. With permission.)

period. Online training enables the controller to track the plant variations as they occur and to provide control signals accordingly. It also considers the non-linear nature of the plant.

Two versions of this controller have been developed and studied.

1.3.3.1 Multi-Layer Network Based[41]

In this case, the feedforward multi-layer network is employed in each of the two subnetworks to build the adaptive neural network PSS. It is trained in each sampling period using an online version[42] of the back propagation algorithm. The errors used to train the ANI and ANC are both scalar, and the learning is done only once in each sampling period for each of the two subnetworks. This simplifies the training algorithm in terms of computation time.

System performance in response to a three phase to ground fault on one circuit of the double-circuit transmission line in the five machine interconnected system (Figure VI.1.18) is shown in Figure VI.1.23. In this case, the adaptive neural network-based PSSs were installed on two generators, and CPSSs were installed on the other three generators. It can be seen that both the local mode and the interarea mode oscillations are damped effectively.

1.3.3.2 Recurrent-Network Based[43]

In this case, a recurrent network is employed in each of the two subnetworks to build the adaptive neural network. The main difference with respect to the feedforward network is that a recurrent network has at least one feedback loop. Feedback has a profound impact on the learning ability of the network and on its performance. The feedback loop involves the use of a unit delay element, which results in non-linear dynamic behavior of the network.

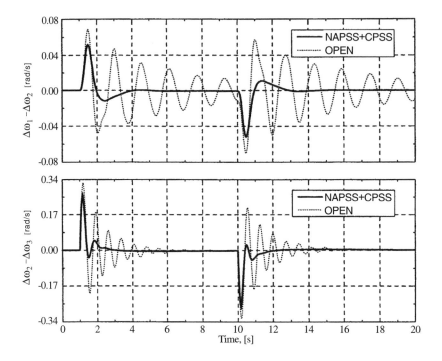

FIGURE VI.1.23   System response with NAPSS installed on generators G_1 and G_3 and CPSS on G_2, G_4, and G_5.

In all other respects, the two versions are similar. The errors used in training are scalar, and the learning is done only once in each sampling period for each of the two subnetworks.

Both simulation and experimental studies showed that the performance of this PSS with two recurrent-network based subnetworks was very similar to that of the multi-layer network-based structure. Therefore, no additional results are shown here.

1.3.4   Adaptive-Network Based Fuzzy Logic Controller

The characteristics of fuzzy logic and neural networks complement each other with respect to their pros and cons. That offers the possibility of using a hybrid neuro-fuzzy approach in the form of an adaptive-network based fuzzy logic controller whereby it is possible to take advantage of the positive features of both fuzzy logic and neural networks. Such a system can automatically find an appropriate set of rules and membership functions.[44,45]

1.3.4.1   Architecture

In the neuro-fuzzy controller, the fuzzy system is implemented in the framework of a network architecture. Considering the functional form of the fuzzy logic controller, FLC (Figure VI.1.3), it becomes apparent that the FLC can be represented as a five-layer feedforward network, as shown in Figure VI.1.24. With this network representation of the fuzzy logic system, it is straightforward to apply the back propagation or a similar method to adjust the parameters of membership functions and inference rules.

Each layer in Figure VI.1.24 corresponds to one specific function, with the node functions in each layer being of the same type. Nodes in layer 1 perform the membership function, and every node in layer 2 represents the firing strength of the rule. The normalized firing strength of each rule is calculated by the nodes in layer 3, and the output of each node in layer 4 is the weighted consequent

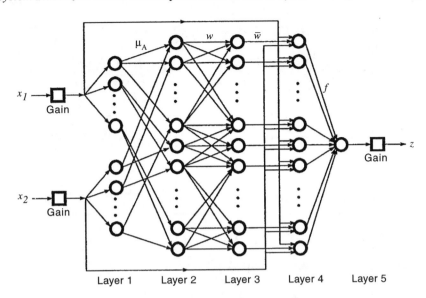

FIGURE VI.1.24   Architecture of an adaptive fuzzy controller.

part of the rule table. The single node in layer 5 computes the overall output as the summation of all incoming signals.

In this network, the links between the nodes from one layer to the next layer only indicate the direction of flow of signals, and part or all of the nodes contain the adjustable parameters. These parameters are specified by the learning algorithm and should be updated to achieve a desired input/output (I/O) mapping.

This FLC has a learning capability. In order to achieve a desired I/O mapping, its parameters are updated according to given training data and a gradient-based learning procedure. It can be used as an identifier for non-linear dynamic systems or as a non-linear controller with adjustable parameters.

1.3.4.2   Training and Performance

Various approaches are possible to train a neuro-fuzzy controller. Because it has the property of learning, fuzzy rules and membership functions of the controller can be tuned automatically by the learning algorithm. Learning is based on the error in the controller output. Thus, in this approach, it is necessary to know the error, which can be evaluated by comparing the output of the neuro-fuzzy controller and a desired controller.

This controller was used as an adaptive-network based fuzzy PSS (ANF PSS). For operation as a PSS, a self-optimizing pole-shifting APSS[29] was chosen as the desired controller. The training was performed over a wide range of operating conditions of the generating unit including various types of disturbances. A total of 18,000 input–output data pairs were obtained for training the ANF PSS.

The number of membership functions of each input variable is determined by the complexity of the training data and by trial and error, similar to choosing the number of neurons in the hidden layers of the ANN. Based on earlier experience, seven linguistic variables for each input variable were used to get the desired performance. The membership functions for two inputs, $\Delta\omega$ and $\Delta\dot{\omega}$, before and after training are shown in Figure VI.1.25. The universe of discourse for both input and output variables is normalized and the gain parameters are chosen based on the input–output space.

Extensive simulation[46] and experimental studies with the ANF PSS show that it can provide good performance over a wide operating range and can significantly improve the dynamic performance of the system. It was also able to provide system performance very close to that of the adaptive PSS.

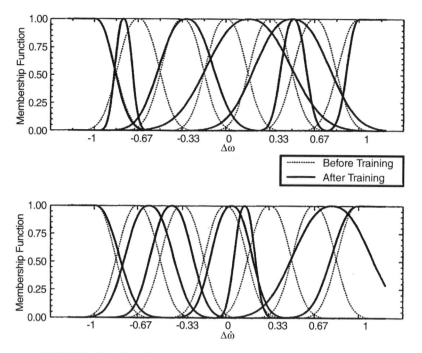

FIGURE VI.1.25   Membership functions before and after training.

1.3.4.3   Self-Learning ANF PSS

In the above case, the ANF PSS was trained by data obtained from a desired controller. However, in a general situation, the desired controller may not be available. In that case, the neuro-fuzzy controller can be trained using a self-learning approach.[47]

In the self-learning approach, a separate neuro-fuzzy identifier is first trained to behave like the plant. This plant identifier can compute the derivative of the plant's output with respect to the plant's input by means of the back propagation process. The final output error of the plant is back propagated through the neuro-fuzzy identifier to obtain the equivalent error for the controller output. It thus uses two neuro-fuzzy systems, as shown in Figure VI.1.9, one acting as the controller and the other acting as the plant identifier. In Figure VI.1.9, the back propagation process is illustrated by the dashed line passing through the forward identifier and continuing back through the neuro-fuzzy controller that uses it to learn the control rule.

The self-learning neuro-fuzzy controller used as an ANF PSS was trained for both the controller and the plant model (identifier) over a wide range of generator operating conditions and a wide spectrum of possible disturbances. A number of studies were performed based on a detailed model of a generating unit connected to a constant voltage bus through two parallel transmission lines. System performance with this ANF PSS was compared to that obtained with a CPSS.

Response of the power system to a three phase to ground short-circuit at the middle of one transmission line, cleared 200 ms later by the disconnection of the faulted line and successful reclosure after 4 s, is shown in Figure VI.1.26. The results show that the ANF PSS significantly improves the dynamic performance of the system.

The ANF PSS was implemented on a DSP mounted on a PC, and its performance was investigated on a physical model of a power system in the laboratory environment. A digital CPSS was also implemented in the same environment on the DSP board for comparative studies.

The self-learning ANF PSS was initially trained off-line on a power system simulation model over a wide range of operating conditions and disturbances. In this case, electric power deviation and its

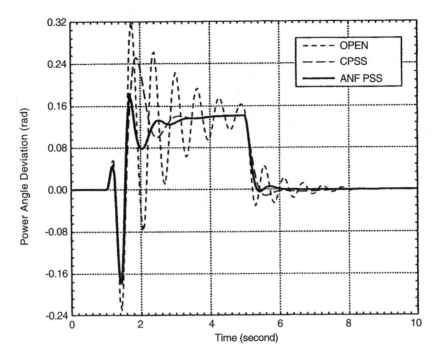

FIGURE VI.1.26   Response to a three phase to ground fault at the middle of a transmission line.

integral were used as the inputs to the stabilizer. After a complete off-line training procedure, the parameters of the fuzzy controller, membership functions, and inference rules, were transferred to the PC to build the ANF PSS on the DSP board.

Out of the various tests, results of a dynamic stability test are given here. With the ANF PSS connected to the system, the system input torque reference was increased gradually to the level:

$$\text{Power output} = 1.20 \, \text{pu}, \text{power factor} = 0.90 \, \text{lag}, \text{terminal voltage} = 1.05 \, \text{pu}$$

and the system remained stable. At 4 s, the ANF PSS was switched off and the CPSS was switched on. After switch-over, the system started to oscillate without any external disturbances. This shows that the CPSS was unable to maintain stability at this load level. The ANF PSS was switched back at 17 s, and the system very quickly regained stability, as shown in Figure VI.1.27.

This test demonstrates that the ANF PSS can provide a larger dynamic stability margin, thereby allowing the generating unit to operate at heavier load conditions.

Simulation studies on a single machine connected to a constant voltage bus[48] and on a multi-machine power system[49] and experimental studies on a physical model of a power system have demonstrated the effectiveness of the ANF PSS in improving the performance of a power system over a wide operating range and a broad spectrum of disturbances.

1.3.4.4 Neuro-Fuzzy Controller Architecture Optimization

Adaptive fuzzy systems offer a potential solution to the knowledge elicitation problem; however, the structure of the fuzzy system needs to be set in advance. The structure, expressed in terms of the number of membership functions and the number of inference rules, is usually derived by trial and error. The number of inference rules has to be determined from the standpoint of overall learning capability and generalization capability.

The above problem can be resolved by employing a genetic algorithm to determine the structure of the adaptive fuzzy controller.[50] By employing both genetic algorithm and adaptive fuzzy controller

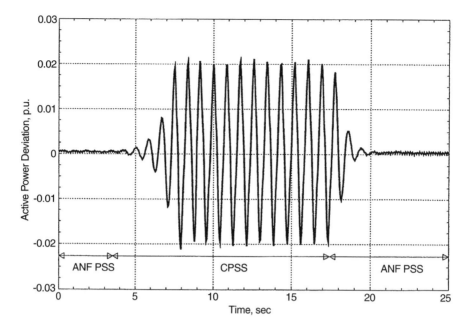

FIGURE VI.1.27   Dynamic stability improvement by ANF PSS. (From Malik, O.P. and Hariri, A., *Trans. Inst. Meas. Control*, 24(2), 87–107, 2002. With permission.)

approaches, the inference rules' parameters can be tuned and the number of membership functions can be optimized at the same time. This optimization contains two major processes:

- Search for the optimum number of rules and shape of membership functions by using a genetic algorithm.
- Train the network to determine the consequent parts of the rule base by the gradient descent algorithm.

1.4   Concluding Remarks

Power system stabilizers based on all the control algorithms described above have been studied extensively in simulation on a single machine infinite bus system and on a multi-machine system. They have also been implemented and tested in real-time on a physical model of a single-machine infinite bus system in the laboratory with very encouraging results. The pole shifting control-algorithm based adaptive PSS has also been tested on a multi-machine physical model[32] and on a 400 MW thermal machine under fully loaded conditions connected to the system,[31] and it is now in regular service in a power station after extensive testing in the field.[52] These studies have quite clearly shown the advantages of the advanced control techniques and intelligent systems.

Acknowledgments

A summary and salient features of the work of a number of graduate students and colleagues in our research group is given in this article. It in no way does justice to the many years of hard work and efforts of so many individuals, whose contributions are acknowledged collectively.

References

1. Larsen, E.W. and Swann, D.A., Applying power system stabilizer: Parts 1-3, *IEEE Trans. Power Apparatus Syst.*, PAS-100(6), 3017, 1981.
2. Concordia, C. and deMello, F.P., Concepts of synchronous machine stability as affected by excitation control, *IEEE Trans. Power Apparatus Syst.*, PAS-88, 316, 1969.
3. Roy, S., Influence of amortisseurs on stabilizer design requirements for damping local oscillations of a generator, *IEEE Trans. Power Syst.*, 14(3), 935, 1999.
4. Schlief, F.R. et al., Excitation control to improve power line stability, *IEEE Trans. Power Apparatus Syst.*, PAS-87, 1426, 1968.
5. Kundur, P., Lee, D.C., and Zein el-Din, H.M., Power system stabilizers for thermal units: analytical techniques and on-site validation, *IEEE Trans. Power Apparatus Syst.*, PAS-100, 81, 1981.
6. Farmer, R.G., State-of-the-art technique for system stabilizer tuning, *IEEE Trans. Power Apparatus Syst.*, PAS-102, 699, 1983.
7. El-Metwally, M.M., Rao, N.D., and Malik, O.P., Experimental results on the implementation of an optimal control of synchronous machines, *IEEE Trans. Power Apparatus Syst.*, PAS-94(4), 1192, 1975.
8. Chen, S. and Malik, O.P., An H_∞ optimization based power system stabilizer design, *IEE Proc. Generation Transmission Distribution*, 142(2), 179, 1995.
9. Chen, S. and Malik, O.P., Power system stabilizer design using μ-synthesis, *IEEE Trans. Energy Convers.*, 10(1), 175, 1995.
10. Chan, W.C. and Hsu, Y.Y., An optimal variable structure stabilizer for power system stabilizer, *IEEE Trans. Power Apparatus Syst.*, PAS-102, 1738, 1983.
11. Hiyama, T., Application of rule-based stabilizer controller to electric power system, *IEE Proc. C*, 136(3), 175, 1989.
12. Takagi, T. and Sugeno, M., Derivation of fuzzy control rules from human operator's control action, *Proceedings of the IFAC Symposium on Fuzzy Information, Knowledge Representation, and Decision Analysis*, July 1983, 55.
13. Zadeh, L.A. et al., *Calculus of Fuzzy Restriction in Fuzzy Sets and their Application to Cognitive and Decision Process*, Academic Press, New York 1975.
14. El-Metwally, K.A. and Malik, O.P., Application of fuzzy logic stabilizers in a multi-machine power system environment, *IEE Proc. Generation Transmission and Distribution*, 143(3), 263, 1996.
15. El-Metwally, K.A., Hancock, G.C., and Malik, O.P., Implementation of a fuzzy logic PSS using a micro-controller and experimental test results, *IEEE Trans. Energy Convers.*, 11(1), 91, 1996.
16. Antsaklis, P.J., Neural networks in control systems, *IEEE Control Syst. Mag.*, 8, 1992.
17. Zhang, Y. et al., An artificial neural network based adaptive power system stabilizer, *IEEE Trans. Energy Convers.*, 8(1), 71, 1993.
18. Zhang, Y. et al., Application of an inverse input/output mapped ANN as a power system stabilizer, *IEEE Trans. Energy Convers.*, 9(3), 433, 1994.
19. Zhang, Y. and Malik, O.P., Experimental studies with a neural network based power system stabilizer, *Proceedings of the International Conference on Intelligent Systems Applications to Power Systems*, Orlando, Florida, Jan. 28–Feb. 2, 1996, 104.
20. Zhang, Y., Malik, O.P., and Chen, G.P., Artificial neural network power system stabilizer in multi-machine power system environment, *IEEE Trans. Energy Convers.*, 10(1), 147, 1955.
21. Eykhoff, P., *System Identification*, Wiley, London, 1974.
22. Cheng, S.J., Malik, O.P., and Hope, G.S., Damping of multi-modal oscillations in power systems using a dual-rate adaptive stabilizer, *IEEE Trans. Power Syst.*, PWRS-3(1), 101, 1998.
23. Kuo, B.C., *Discrete-Data Control Systems*, Prentice-Hall, London, 1970.

24. Malik, O.P. and Mao, C., An adaptive optimal controller and its application to an electric generating unit, *Int. J. Electrical Power Energy Syst.*, 15(3), 169, 1993.

25. Cheng, S.J. et al., An adaptive synchronous machine stabilizer, *IEEE Trans. Power Syst.*, PWRS-1(3), 101, 1986.

26. Chandra, A., Malik, O.P., and Hope, G.S., A self-tuning controller for the control of multi-machine power systems, *IEEE Trans. Power Syst.*, 3(3), 1065, 1988.

27. Pahalawaththa, N.C., Hope, G.S., and Malik, O.P., A MIMO self-tuning power system stabilizer, *Int. J. Control*, 54(4), 815, 1991.

28. Malik, O.P. et al., Adaptive self-optimizing pole-shifting control algorithm, *IEE Proc. D*, 139(5), 429, 1992.

29. Chen, G.P. et al., An adaptive power system stabilizer based on the self-optimizing pole shifting control strategy, *IEEE Trans. Energy Convers.*, 8(4), 639, 1993.

30. Chen, G.P., Malik, O.P., and Hancock, G.C., Implementation and experimental studies of an adaptive self-optimizing power system stabilizer, *Control Eng. Pract.*, 2(6), 969, 1994.

31. Malik, O.P. et al., Tests with a microcomputer based adaptive synchronous machine stabilizer on a 400 MW thermal unit, *IEEE Trans. Energy Convers.*, 8(1), 6, 1993.

32. Malik, O.P. et al., Experimental studies with power system stabilizers on a physical model of a multi-machine power system, *IEEE Trans. Power Syst.*, 11(2), 807, 1996.

33. Chen, G.P., Malik, O.P., and Hope, G.S., Control limit considerations in discrete control system design, *IEE Proc. D*, 140(6), 413, 1993.

34. Chen, G.P., Malik, O.P., and Hope, G.S., Generalized discrete control system design method with control limit considerations, *IEE Proc. D*, 141(1), 39, 1994.

35. Malik, O.P., Hope, G.S., and Cheng, S.J., Some issues on the practical use of the recursive least squares identification in self-tuning control, *Int. J. Control*, 35(5), 1021, 1991.

36. Etxebarria, E., Adaptive control of discrete systems using neural networks, *IEE Proc. Control Theor. Appl.*, 141, 209, 1994.

37. Hagan, M.T., Demuth, H.B., and Beale, M.H., *Neural Network Design*, PWS Publishing, Boston, MA, 1996.

38. Demuth, H. and Beale, M., *Neural Network Toolbox User's Guide*, The Math Works, Inc., MA, 1998.

39. Chen, S., Billings, S.A., and Grant, P.M., Recursive hybrid algorithm for non-linear system identification using radial basis function networks, *Int. J. Control*, 55(5), 1051, 1992.

40. Cover, T.M., Geometrical and statistical properties of systems of linear inequalities with applications in pattern recognition, *IEEE Trans. Electroni. Comput.*, EC-14, 326, 1965.

41. Shamsollahi, P. and Malik, O.P., An adaptive power system stabilizer using on-line trained neural networks, *IEEE Trans. Energy Convers.*, 12(4), 382, 1997.

42. Haykin, S., *Neural Networks: A Comprehensive Foundation*, Macmillan, New York, 1994.

43. He, J. and Malik, O.P., An adaptive power system stabilizer based on recurrent neural networks, *IEEE Trans. Energy Convers.*, 12(4), 413, 1997.

44. Jang, J.G., ANFIS: adaptive-network-based fuzzy inference system, *IEEE Trans. Syst. Man Cybernet.*, 23(3), 665, 1993.

45. Lin, C. and Lee, G.S.G., Neural-network based fuzzy logic control and decision systems, *IEEE Trans. Comput.*, 40(12), 1320, 1991.

46. Hariri, A. and Malik, O.P., A fuzzy logic based power system stabilizer with learning ability, *IEEE Trans. Energy Convers.*, 11(4), 721, 1996.

47. Jang, J.G., Self-learning fuzzy controllers based on temporal backpropagation, *IEEE Trans. Neural Networks*, 3(5), 714, 1992.

48. Hariri, A. and Malik, O.P., A self-learning fuzzy stabilizer for a synchronous machine, *Eng. Intelligent Syst.*, 5(3), 157, 1997.

49. Hariri, A. and Malik, O.P., A self-learning adaptive-network-based fuzzy logic power system stabilizer in multi-machine power system, *Eng. Intelligent Syst.*, 2001, in press.

50. Hariri, A. and Malik, O.P., Fuzzy logic power system stabilizer based on genetically optimized adaptive network, *Fuzzy Sets and Syst.*, 102(1), 31, 1999.

51. Eichmann, A. et al., A prototype self-tuning adaptive power system stabilizer for damping active power swings, *Conference Proceedings*, Vol. 1, IEEE Power Engineering Society 2000 Summer Meeting, July 16–20, Seattle, WA, 122.

52. Hariri, A. and Malik, O.P., *Excitation Control in Power Systems, Encyclopedia of Electrical and Electronic Engineering*, Vol. 7, John Wiley and Sons, New York, 1995.

53. Malik, O.P. and Hariri, A., Power system stablizer based on self-learning adaptive network fuzzy inference system, *Trans. Inst. Meas. Control*, 24(2), 87–107, 2002.

54. Ramakrishna, G. and Malik, O.P., Radial basis function based identifiers for adaptive PSSs in a multimachine power system, *Conference Proceedings*, 1, IEEE Power Engineering Society, July 2000, Seattle, 116–121.

2

Machine Learning Applied to Power System Dynamic Security Assessment

2.1 Introduction VI-36
2.2 Dynamic Security Assessment VI-37
2.3 Case Study—The Crete Power System VI-39
2.4 Learning Set Generation VI-42
    Identification of Security Problems • Generation of the
    Learning Set (LS)
2.5 Application of Decision Trees VI-46
    Theoretical Background • Application of Decision Trees
    for Online Dynamic Security Assessment of Crete
2.6 Application of Feedforward Neural Networks VI-53
    Theoretical Background • Application of Artificial Neu-
    ral Networks for Online Dynamic Security Assessment of
    Crete
2.7 Application of Entropy Networks VI-56
2.8 Online Implementation and Evaluation VI-58
2.9 Conclusions VI-59
Acknowledgments VI-60
References VI-60

N.D. Hatziargyriou
National Technical University of Athens

E.S. Karapidakis
National Technical University of Athens

Abstract. In this chapter, the application of machine learning techniques to the dynamic security assessment (DSA) of power systems is described. The online dynamic security of power systems is a very demanding task dictated by the needs of increased efficiency under stressed operating conditions. The task assumes the evaluation of the power system's ability to face a number of critical contingencies, requiring short computational time and high reliability. Online DSA functions based on advanced inductive inference and statistical methods as well as artificial neural networks have been developed and integrated within an advanced control system installed on the island of Crete in Greece. These functions are used to assess the dynamic behavior of the system regarding its frequency excursions when subject to some prespecified disturbances. This application is used as a practical example that helps expose the relevant issues, mainly from a practical point of view. Results from the implementation and evaluation of the relevant software are also provided. It should be noted however, that the machine learning techniques described are not limited to this application or type of power system, but are equally applicable to larger, interconnected systems for the assessment of various types of instability.

2.1 Introduction

The recent developments in electric energy systems, namely, the on-going liberalization of the energy markets, the pressing demands for power system efficiency and power quality, and the increase of renewable and dispersed generation and power exchanges among utilities, dictate the need for improvements in power system planning, operation, and control. Machine learning techniques together with the traditional analytical techniques can significantly contribute to the solution of the related problems.

This chapter focuses on isolated power systems like the ones operating on large islands. These systems provide interesting case studies because they face greater problems with respect to their operation and control. Some of those problems are related to economic operation, since electric power production, mainly by diesel units and gas turbines, requires high costs due to fuel imports and transportation. In these systems, the production of electric energy from renewable energy sources, mainly wind, is of particular interest, especially when important wind energy potential exists, allowing significant displacement of conventional fuels by high wind power penetration. Regarding security, however, mismatches in generation and load or unstable system frequency control might lead to system failures much more easily than interconnected systems. The control of frequency and the management of system generation reserves are of primary importance.[1,2] The introduction of a high penetration from wind energy causes additional difficulties, i.e., fast wind power changes and very high wind speeds, resulting in sudden loss of wind generator production, can cause frequency excursions and dynamically unstable situations. Moreover, frequency oscillations may easily trigger the frequency protection relays of the wind parks, thus causing further imbalance in the system generation/load.

In order to guard isolated power systems against these disturbances and retain acceptable security levels, online dynamic security assessment functions need to be provided, which means assessment of the dynamic performance of the system for each critical contingency online. For this purpose, dynamic simulation of the system after each dispatch, taking into account the dynamic response of the generators, the load, and the protective devices, is required in order to calculate the frequency and voltage excursions in the system. The long execution times associated with full dynamic simulations, however, make the online application of this method unsuitable for practical systems. In this chapter, the application of artificial intelligence techniques for online dynamic security assessment of power systems is described. Such functions based on advanced inductive inference and statistical methods, as well as artificial neural networks, have been developed and integrated within an advanced control system installed on the island of Crete in Greece.[3-5] More specifically, decision trees are used in order to identify the most important system parameters and to classify dynamic security regarding frequency, while artificial neural networks emulate the degree of security, evaluated by predicting the expected minimum value of system frequency and the maximum rate of frequency change for each critical disturbance specified.[6-8] In the control center software, the relevant security evaluation functions can be activated "on call" by the operator providing dynamic security monitoring. A combination of previous structures and analytical mathematical methods leads to a more robust operation of the system, while application of these techniques in the power system of Crete and comparison with actual events has shown that security can be assessed accurately online.[3,4] In addition, the application of entropy networks, a hybrid neural network approach, has been investigated. This approach aims at combining the attractive features of the two techniques, namely, the simplicity and transparency of decision trees and the information accuracy of multilayer perceptions. It should be noted that the application of these methods is not limited to the examined frequency stability problems nor to isolated systems, which are merely used as examples of a successful application.

In the following section, the terms and notions of dynamic security and the related classes of stability problems are discussed. Section 2.3 describes the case study of the power system of Crete in order to present the relevant security issues from a practical point of view. The application of machine learning techniques requires three general steps, namely, application of (1) random sampling techniques to scan all relevant operating states of the system and simulation of these states under various predetermined disturbances to build the data set, (2) various techniques to extract and synthesize relevant information and to reformulate it in a way suitable for decision-making, and (3) extracted synthetic information, either in real-time for fast and effective decision making, or in an off-line study environment, so as to gain new physical insight and to derive better system operation and planning strategies.[3] In Section 2.4, the derivation of the data set is described. This is followed by a description of the application of the decision trees and ANNs and the main results obtained in Sections 2.5 and 2.6, respectively. The application of entropy networks is explored in Section 2.7. These techniques, as integrated into the advanced energy management system (EMS) of Crete and applied to online dynamic security assessment and monitoring, are described in Section 2.8. Evaluation of these functions compared to actual events shows that the timely and accurate assessment of frequency deviations during the dynamic disturbances recorded is provided. Finally, general conclusions regarding the applications of machine learning techniques to power system security problems are provided in Section 2.9.

2.2 Dynamic Security Assessment

Power system security is the "art and science of ensuring the survival of power systems."[9] This definition describes what system planners and operators might intuitively understand as security. There are many more formal definitions of security. It is defined by the North American Electric Reliability Council (NERC) as, "prevention of cascading outages when the bulk power supply is subjected to severe disturbances,"[10] and by CIGRE as, "the ability of the system to cope with incidents without the operator being compelled to suffer uncontrolled loss of load."[11] Dr. Dy Liacco, in his pioneering reports and papers,[12–14] originally defined security in terms of satisfying a set of inequality constraints over a subset of possible disturbances called the next contingency set. Figure VI.2.1.[15] shows the different operating modes of a power system together with their interactions. A power system in its normal operating state can be led to an alert state, which is still a fully operational system state, but some further contingencies would send it to an emergency state. In case of a temporary emergency condition, operator action can relieve unacceptable operating conditions (line overloads or low voltages) and bring the system back to an alert or normal state. In a controlled emergency state, load shedding is necessary, while in an extreme emergency state, the stability and integrity of the system are threatened.

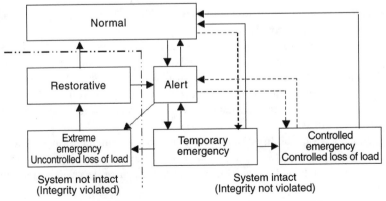

FIGURE VI.2.1   Operating states and transitions.[15]

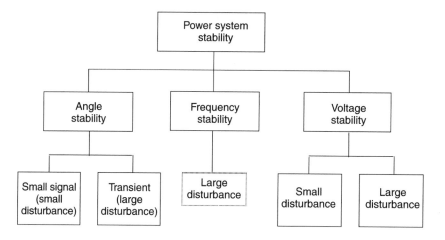

FIGURE VI.2.2   Types of power system stability.[16]

Security assessment[27] means the evaluation of the power system's ability to face various disturbances and the proposal of remedial actions when necessary. Preventive security assessment examines if a system in its normal state is able to withstand every plausible disturbance. If not, preventive control aims at bringing the system into a secure operating state. Steady state security, in contrast with dynamic security, is concerned with voltages and line currents being within acceptable operating limits after the transient voltage and angle variations following a disturbance extinguish and the system reaches an equilibrium point. The following sections focus on preventive dynamic security assessment.

Figure VI.2.2 adapted from Kundur and Morison,[16] provides an overview of the various types of system instability that threaten the system security. These are related to:

- Rotor angle stability, expressing the ability of interconnected synchronous machines to remain in synchronism under normal operating conditions and after being subjected to a disturbance. This depends on the ability to maintain/restore equilibrium between electromagnetic torque and mechanical torque of each synchronous machine in the system. The instability that may result occurs in the form of increasing angular swings of some generators, leading to their loss of synchronism with other generators. Small signal stability is concerned with the ability of the power system to maintain synchronism under small disturbances, while transient stability is when the system is subjected to a severe transient disturbance.
- Voltage stability, expressing the ability of a power system to maintain steady, acceptable voltages at all buses in the system under normal operating conditions and after being subjected to a disturbance. The main factor causing voltage instability may occur in the form of a progressive fall or rise of voltage of some buses. Another factor causing voltage instability is the inability of the power system to maintain a proper balance of reactive power throughout the system. Large disturbance voltage stability is concerned with a system's ability to maintain steady voltages following large disturbances, while small disturbance voltage stability is concerned with a system's ability to control voltages following small perturbations such as incremental changes in system load.
- Frequency stability is the ability of a power system to maintain the frequency within a nominal range following a severe system upset, which may or may not result in the system being divided into subsystems. It depends on the ability to restore balance between system generation and load with minimum loss of load.

The latter type of instability is typical of isolated systems like the ones operating on islands.

Traditionally, frequency stability has been accounted for by ensuring sufficient reserve capacity in the system so that it can withstand various disturbances. The determination of this reserve capacity was based on off-line stability analysis of credible contingencies under a variety of operating conditions. This was implemented as a fixed emergency reserve (security margin) in the economic dispatch algorithms, allocating generation to the online units, so that the system load was supplied most economically. Carrying unnecessary reserve capacity, however, might significantly increase the cost of operation. This is inevitable if a fixed security margin is assumed since the dynamic performance of the system depends critically on its operating conditions and the dynamic characteristics of the generation response, as shown in Hatziargyriou et al.[2]

The new operating and planning policies imposed by the current demands for increased efficiency and economy dictate the need to operate the power system in a more stressed way and to consider more effective means of providing a given level of security. Online dynamic security assessment (DSA) techniques have been developed with the goal of selecting critical contingencies and assessing the dynamic performance of the system for each critical contingency online.[17] Only indicative references of a very dynamic activity in this area are provided in this section. These include techniques based on energy functions[18–20] and artificial intelligence.[21–27] Moreover, in order to operate optimally within the new market conditions, the price of providing a given level of security has to be accounted for. This is directly linked to the provision of remedial actions in case of insecure situations. For dynamic security, unlike steady state security, remedial actions can only be preventive, leading to load shedding or generation rescheduling.[28,29] For isolated systems, this means that the cost of load shedding has to be balanced with the cost of providing adequate spinning reserves to avoid it. In O'Sullivan and O'Malley,[29] a successful methodology for the determination of reserves in isolated power systems is presented, which enables assessment of the costs for the loss of any given unit. This methodology is based on the dynamic simulation of the system after each dispatch, taking into account the dynamic response of the generators and the demand, and calculates minimum frequency so that load shedding is avoided. The long execution times associated with full dynamic simulations however, make the online application of this method unsuitable for larger systems.

In the following sections, the development of online DSA functions based on decision trees (DTs) and artificial neural networks (ANNs) are described. These functions have been developed and integrated within an advanced control system tailored for the needs of isolated power systems with increased renewable power penetration. The control system has been installed on the Greek island, Crete, and is currently used to assist its operators by proposing optimal operating set points for the various power units, both conventional and renewable. The DSA module has the task of evaluating if a given dispatch scenario is dynamically secure for a number of predetermined disturbances. The related problems can be better explained by focusing on the case study of the power system of Crete that is briefly described next.

2.3 Case Study—The Crete Power System

The power system of the island of Crete is the largest autonomous power system in Greece with the highest rate of increase in energy and power demand nationwide. The one-line diagram of the system is shown in Figure VI.2.3. The transmission network consists mainly of 150 kV lines and some 66 kV lines. In 1998, the conventional generation system consisted of two major power plants, one in Linoperamata and one in Chania, with several types of oil-fired units located near the major load points of the island. These are 18 thermal generating units with a total capacity of 524 MW installed, including 6 steam units of total capacity 112 MW, 4 diesel units with 50 MW, 7 gas turbines with 227.5 MW, and one combined cycle plant with 134.5 MW. The annual peak load demand occurs on a winter day, and overnight loads can be assumed to be approximately equal to 25% of the corresponding daily peak loads. One characteristic of the load profile is the large variations (low night valleys–high

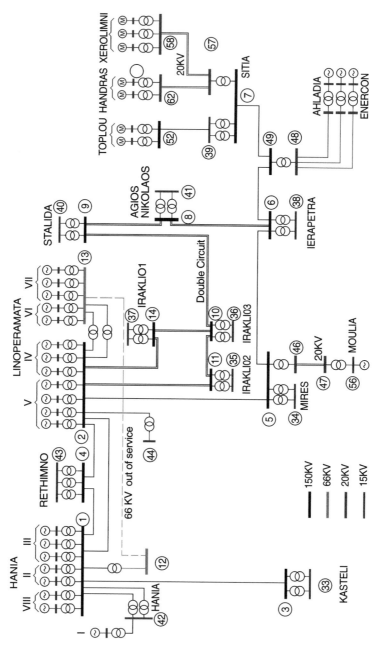

FIGURE VI 2.3   One-line diagram of Crete power system.

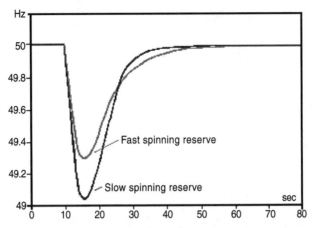

FIGURE VI.2.4　Frequency change during disconnection of three wind parks.

evening peaks). In 1998, the peak load exceeded 380 MW, while the lowest load was about 100 MW. The combined cycle and the steam units mainly supply the base load, while the gas turbine units supply the peak load at a high running cost that significantly increases the average cost of the electricity. Eleven wind parks (WPs) of a total capacity exceeding 60 MW have already been installed, and many more will be installed in Crete in the near future, operated mainly by Independent Power Producers.[10] The majority of these WPs (TOPLU, HANDRAS, XIROLIMNI, ACHLADIA) are located at the eastern part of the island that presents the most favorable wind conditions. As a result, in case of faults on the lines connecting the eastern part of the island, the majority of the wind power production might be disconnected. Furthermore, protection of the wind turbines might be activated in case of frequency variations, additionally decreasing the dynamic stability of the system. Extensive simulations on the power system model using EUROSTAG software[30] have shown that for the most common wind power variations, the system remains satisfactorily stable if sufficient spinning reserve is provided.[2] On the other hand, for various short circuits and conventional unit outages, the system frequency might undergo fast changes and reach low values that can activate load shedding. In any case, the dynamic security of the system depends critically on the amount of spinning reserve provided by the conventional machines and the response of their speed governors.

For example, Figure VI.2.4 shows the change of the system frequency in two different operating conditions following the disconnection of three wind parks producing approximately 30 MW. First, the system is considered to operate with 28% wind power, equal to 46 MW, and with the fast thermal units such as the diesel machines and gas turbines to provide the spinning reserve (fast spinning reserve). The lower value of the frequency is 49.31 Hz. Second, the system is again considered to operate with the same high penetration of wind power but with the slower machines, such as the steam turbines, to cover the main spinning reserve plus some diesel machines (slow spinning reserve). In this case, the lower frequency value, close to 49 Hz, could cause the operation of the protection devices of the rest of the wind parks. The total wind power disconnection would lead the system to collapse.

The effect of the wind power penetration, as well as of the response of the available spinning reserve, is further shown by simulating the disconnection of a gas turbine producing 20 MW. In Figure VI.2.5, the change of frequency and the diesel machine power in three different operating conditions are shown. First, the system is considered to operate without wind turbines, and it seems to be quite stable. Second, the system is considered to operate with 28% wind power, equal to 46 MW, and with the fast conventional units such as diesel machines and gas turbines (fast spinning reserve). In this case, the lower value of the frequency is almost the same as in the previous case. Third, the system is again considered to operate with the same percent wind power but with the slow machines

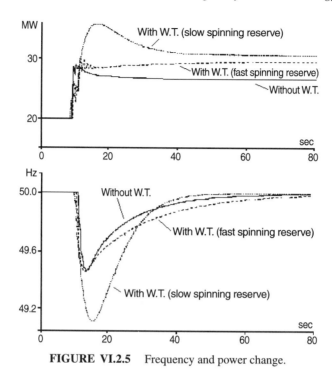

FIGURE VI.2.5   Frequency and power change.

such as steam turbines to cover the main spinning reserve (slow spinning reserve). In this case, the lower frequency value is equal to 49.14 Hz and might cause the operation of the underfrequency load shedding protection, disconnecting a number of consumers. This shows that spinning reserve needs to be optimized both in quantity and speed of reaction.

2.4   Learning Set Generation

The application of machine learning techniques in the field of dynamic security assessment requires a four-step approach:

1. Identification of the security problems, where a physical evaluation of the phenomena is performed in order to select the most critical disturbances and the attributes that characterize the operating conditions relative to the security problems
2. Generation of a data set consisting of records of operating points with information regarding the behavior of the system in several operating conditions (this data set is split into a learning set (LS), used to derive security evaluation structures, and a test set (TS), used for testing the developed structures)
3. Design of security evaluation structures that will be used online providing information on the system robustness for the disturbances under consideration
4. Performance evaluation, necessary to assess the quality of the security evaluation structures obtained in step 3.

Steps 1 and 2 are common to both learning from example techniques applied, i.e., decision trees and neural networks, and are described in this Section. Steps 3 and 4 depend on the security evaluation structure and are described in Sections 2.5 and 2.6.

2.4.1 Identification of Security Problems

2.4.1.1 Modeling of the System

A detailed model of the power system of Crete including the 150 kV and 66 kV lines has been developed. It consists of 64 buses, 20 generator (PV) buses, and 33 load (PQ) buses plus the 11 wind park connection buses. The 20 kV lines where the wind parks are connected are represented in detail, while aggregate load models have been used for the other distribution 20 kV and 15 kV lines.

The dynamic models,[31] used for the presentation of the system components, are suitable for the duration of the transient phenomena under study, that is, between 0.1 and 10s approximately.

2.4.1.1.1 *Diesel Engines and Gas Turbines*

The generic model, illustrated in Figure VI.2.6, is used for the simulation of the diesel engines and the gas turbines speed governors.

Δf is the per unit frequency change ($f_o = 50\,\text{Hz}$).
Pm is the mechanical power of the diesel motor.
R is the droop of the speed governor.

Input is the frequency change and output is the produced mechanical power, while the diesel engine or the gas turbine is represented by a first order lag with a time constant T_D. T_G is the time constant of the hydraulic actuator of the governor mechanism.

2.4.1.1.2 *Steam Unit*

The block diagram of Figure VI.2.7 represents the speed governor system considered for the steam units.

Δf is the per unit frequency change ($f_o = 50\,\text{Hz}$).
Pm is the mechanical power of the steam turbine.

The transfer function for the governor includes speed relay. The steam turbine is represented as single type whose transfer function is:

$$G(s) = F_{HP}/1 + sT_R$$

F_{HP}, is the fraction of total turbine power generated by the high pressure section. It is taken equal to one because only a high-pressure state is considered. An integrating control block parallel

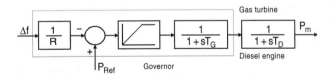

FIGURE VI.2.6   Diesel-gas speed control system.

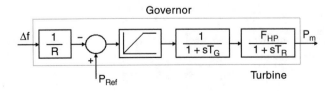

FIGURE VI.2.7   Steam speed control system.

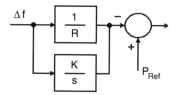

FIGURE VI.2.8   Addition of integrating control block.

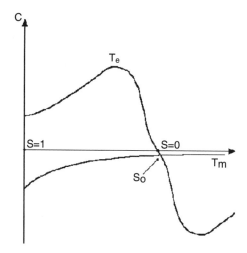

FIGURE VI.2.9   Electrical and mechanical torques.

to machine droop is added to the speed controllers of Figures VI.2.6 and VI.2.7, as shown in Figure VI.2.8.

2.4.1.1.3 *Voltage Regulator*
The standard IEEE DC1 model is used to represent the voltage regulator of each generator.

2.4.1.1.4 *Synchronous Generators*
The typical fifth-order model with two armature, two damper, and one field winding, 3 for the D-axis and 2 for the Q-axis, are used for the synchronous machines.

2.4.1.1.5 *Asynchronous Generators*
Each wind park is modeled as an aggregate asynchronous or induction generator with a short-circuited double cage rotor. The initial slip corresponds to the intersection of the electrical torque curve and the opposing mechanical torque, as shown in Figure VI.2.9. The mechanical torque is a linear function of the asynchronous wind generator speed:

$$P_m = T_m \cdot \omega_r$$

In steady state conditions and in case of a disturbance where the wind remains stable, the mechanical power is assumed to be constant:

$$T_m \cdot \omega_r = \text{constant}$$

2.4.1.1.6 *Loads*

Power system loads are composed of a variety of electrical devices. For resistive loads, such as lighting and heating loads, the electrical power is independent of frequency. In case of motor loads, the electrical power changes with the frequency due to changes in motor speed. The overall frequency dependence of a composite load may be expressed as:

$$\Delta P_e = \Delta P_L + D\Delta f$$

where ΔP_L is the non frequency sensitive load change.
$D\Delta f$ is the frequency sensitive load change.
D is the load damping constant.

2.4.1.2   Selection of Critical Disturbances

Based on the previous model and for various operating conditions, a number of disturbances were simulated using EUROSTAG.[30] Two disturbances were finally selected after extensively studying the behavior of the network. These are presented in decreasing order of severity:

1. Three-phase short circuit—transient short circuit in a 15 kV busbar of Ag. Nikolaos. Total disconnection from the grid takes 200 ms.
2. Outage of the largest gas turbine—Outage of a unit generating 30 MW maximum.

It should be noted that a gas turbine disconnection is a frequent event, and a three-phase fault, although rare, is a severe event that commonly occurs during stormy conditions in autumn.

2.4.2   Generation of the Learning Set (LS)

The learning set consists of a large number of operating points (OP) of the system characterized by their degree of security with respect to each selected disturbance. These operating points should be representative of the actual system operation. Each OP comprises a vector of variables, called attributes.[8] The attributes are properly selected predisturbance variables, which are considered important for the security of operation of the system and can be directly measured (powers, voltages, etc.) or indirectly calculated quantities (wind penetration, spinning reserve, etc.). In addition, each OP carries a class label denoting the degree of security (secure or insecure for a two-class partition) for the applied disturbance (contingency), or it includes values of post-disturbance variables characterizing the system security obtained by simulation.

It would be desirable for the OPs in the LS to represent real states of the system, recorded and stored in a data base, on which the critical disturbances would be imposed (simulated). If such a database is unavailable, it is necessary to generate one by simulation. The basic consideration in this process is to ensure the representativity of the LS, i.e., that it is a good statistical sample of the population of the possible operating states of the system. The considered disturbances are then simulated for each OP of the LS. This task is by far the most time consuming since it involves step-by-step simulations of several contingencies for several thousand OPs; however, it is completed off-line without any impact on the online performance.

For Crete, OPs were obtained by randomly varying the load for each load busbar and the wind power for each wind park. These variables are assumed to follow normal distributions around four starting operating profiles.

- Low-load with a total load $P_L = 120$ MW
- Medium-load with a total load $P_L = 180$ MW
- High-load with a total load $P_L = 240$ MW
- Peak-load with a total load $P_L = 300$ MW

The total load demand follows a ± 30 MW change around the four above operating profiles (final load range: 90–330 MW). A dispatch algorithm approximating actual operating practices followed in the control system of Crete is applied next in order to complete the predisturbance OPs. This means that the steam units and the combined cycle plant cover the base load, while the gas turbines mainly supply the peak.

For a given load demand P_L and wind power P_W, the conventional unit generation is approximately:

$$P_C = P_L - P_W$$

For a required wind margin (WM), which is defined as the ratio of the spinning reserve of the operating conventional units to the total wind power, the minimum necessary nominal capacity $P_{C\min}$ is given by:

$$P_{C\min} = P_C + WM \cdot P_W$$

Considering that the load varies from 90 to 330 MW and the total wind power varies from 10 to 60 MW, the wind power penetration varies from 3% to 66%.

The selection of candidate attributes, i.e., of the predisturbance variables which probably carry significant information with respect to the system post-disturbance state, is an important step that determines the reliability of the security evaluation structures and depends on each application, e.g., frequency and/or voltage stability, preventive assessment, or preventive control, etc. Selection relies upon engineering judgment and understanding of the dynamics of the system. It might also be necessary to test several alternative candidate attribute lists in order to determine the most efficient. Due to the relatively small size of the Crete power system, it is possible to concentrate practically all information concerning its state in a 60-candidate attribute list. These variables comprise active and reactive powers of all system components, bus voltages at major buses, and a number of further attributes such as spinning reserve, wind margin, and wind penetration. According to the desired application and based on engineering judgement, a more limited set can be used, as shown in Sections 2.5 and 2.6

Commonly used indicators for the security evaluation of each OP are the critical clearing time (CCT) for a particular fault (transient stability), transient excursions of frequency (frequency stability), the available transfer capacity (ATC), etc. In the case study presented, the frequency deviation is used as a security index. In particular, the minimum value of system frequency and the maximum rate of frequency change are recorded for each OP. For security classification, these parameters are checked against the values that activate the frequency protection relays of the system leading to load shedding, and the OPs are labeled accordingly as secure/insecure (two-class partition). This corresponds to the prevention of the system from entering the emergency state according to the terminology of Section 2.2. The selected security criterion is expressed as follows:

If *fmin* < 49 Hz **or** *df/dt* > 0.4 Hz/sec **then** the OP is insecure

      else it is secure

The development of LS and selection of the appropriate attributes is discussed further in Section 2.5.

2.5 Application of Decision Trees

2.5.1 Theoretical Background

The decision tree (DT) method is a non-parametric learning technique, independent of the statistical distribution of the LS. DTs are constructed off-line by an inductive inference procedure (similar to the ID3) and have the hierarchical form of a tree of rules built upside-down. Their structure permits both forward and backward chaining of the rules, which makes hypothetical reasoning possible. Their construction is based on the use of a learning set (LS), which is comprised of a number of preclassified operating states or points (OPs) of the system. Using the available information in the LS, the developed DTs are able to produce classifiers about a given problem in order to deduce information for new, unobserved cases.

A DT is comprised of the following node types:

- Test Nodes: redirect the examined OP to a successor node using a suitable test called the node splitting test.
- Terminal Nodes: classify the state of the system as secure or insecure. There are two types of terminal nodes: (1) leaves, characterized by class-pure OP subsets, these classify deterministically the security of the OP, and (2) deadends, which are characterized by non-uniform OP subsets and provide the probability of the OP being secure or insecure.

Except for the *root node* (or top node), every node of a decision tree is the successor of its parent node. Each of the test nodes has two successor nodes. Nodes that have no successor nodes are called terminal nodes. In order to detect if a node is terminal, i.e., sufficiently class pure, the classification entropy of the node with a minimum preset value H_{min} is compared. If it is lower than H_{min}, then the node is sufficiently class pure and it is not further split. Such nodes are labeled leaves. Otherwise, a suitable test is sought to divide the node by applying the optimal splitting rule. The optimal splitting rule decides the best attribute and its threshold value so that the additional information gained through that test is maximized. For a two-class partition, it has the form of a dichotomic test T, defined as:

$$T : A_i \leq t$$

where t is the optimal threshold value of the chosen attribute A_i. The best attribute and its threshold value are obtained by sequentially testing all attributes and candidate thresholds and comparing their information gain. The selected test is applied to the LS of the node, splitting it into two subsets corresponding to the two successor nodes. The optimal splitting rule is applied recursively to build the corresponding sub-trees. If the node cannot be further split in a statistically significant way, it is termed a deadend, carrying the two class probabilities estimated on the basis of the corresponding OP's subset. The degree of non-uniformity of each node is given by the node security index, defined as the percentage of secure states in the corresponding subset. For example, if the security index of a terminal node of the decision tree is 0.9, then the OPs falling to this node have a 90% probability of being secure. An excellent description of the DT methodology and its application to power system problems is provided in the relevant literature.[3,4]

2.5.1.1 Performance Evaluation of the DTs

The accuracy or reliability of the developed DTs is tested using the test set (TS), which is similar but different than the LS. The DT reclassifies the states of the TS, and the *a posteriori* classification results are compared with the *a priori* classification of the OPs in order to determine the success and error rates.

The most important evaluator of DT reliability and performance is the rate of successful classifications, defined as:

$$\text{Success Rate} = \frac{\text{OPs successfully classified by the DT}}{\text{Total number of OPs in the TS}}$$

For a two-class partition, two types of errors can be distinguished, depending on the actual class of the misclassified OP:

$$\text{False Alarm Rate} = \frac{\text{Secure OPs misclassified as Insecure by the DT}}{\text{Total number of Secure OPs in the TS}}$$

$$\text{Missed Alarm Rate} = \frac{\text{Insecure OPs misclassified as Secure by the DT}}{\text{Total number of Insecure OPs in the TS}}$$

The global error rate is defined, as:

$$\text{Global Error Rate} = 1 - \text{Success Rate}$$

2.5.2 Application of Decision Trees for Online Dynamic Security Assessment of Crete

Decision trees have been applied in various ways to determine online the dynamic security of the Crete power system. The selection of the initial candidate attributes critically determines each particular application. This is particularly the case when preventive control is desired. For example, if it is desired to allow the secure acceptance of all wind power available, leading to maximization of wind power penetration, preventive control would suggest unit redispatch to ensure security. In this case, only active powers and spinning reserves of the conventional units should be included in the initial attributes. If disconnection of wind parks in order to improve system stability in case of insufficient spinning reserve is considered, then wind power production and penetration should be included in the initial attribute list. Alternatively, the selection of interruptible loads as initial attributes and the preparation of the appropriate LS would provide load-shedding advice. These different DT applications reveal the versatility and flexibility of the DT approach. This is better shown in the following applications.

2.5.2.1 Application 1: Security Classification Using Extended List of Candidate Attributes

Based on the procedure described in Section 2.4.2, 2765 OPs of the Crete power system were created. Two-thirds of the OPs are used as the learning set, and the remaining one-third is used as the test set. According to the adopted security criteria, the learning sets and the test sets for each disturbance are shown in Tables VI.2.1 and VI.2.2.

TABLE VI.2.1   Secure and Insecure OPs

	Learning Set—Disturbance (Short Circuit)		
Set	Total OPs	Secure OPs	Insecure OPs
LS	1844	800	1044
TS	921	401	520

TABLE VI.2.2   Secure and Insecure OPs

	Learning Set—Disturbance (Machine Outage)		
Set	Total OPs	Secure OPs	Insecure OPs
LS	1844	1536	310
TS	921	687	222

TABLE VI.2.3　List of Candidate Attributes

AT ID	Description	Units	Symbol
ATTR1	Power Gen.1	MW	Pc1
ATTR2	Spinning Res.1	MW	SR1
ATTR3	Power Gen.2	MW	–
ATTR4	Spinning Res.2	MW	–
ATTR5	Power Gen.3	MW	Pc3
ATTR6	Wind O Total	MVAr	–
ATTR7	Wind Park 2	MW	–
ATTR8	Wind Park 3	MW	–
ATTR9	Wind Park 4	MW	–
ATTR10	Total	MW	–
ATTR11	Wind Power	MW	$\sum P_w$
ATTR12	Spinning Res.3	MW	–
ATTR13	Power Gen.4	MW	–
ATTR14	Spinning Res.4	MW	–
ATTR15	Wind Penetration	%	WP
ATTR16	Wind Margin	–	–
ATTR17	Active Power	MV	–
ATTR18	Reactive Power	MVAr	–
ATTR19	Conv. Gen. Total	MW	$\sum P_c$
ATTR20	Total Active Load	MW	$\sum P_L$
ATTR21	Total React. Load	MVAr	–
ATTR22	Capacitors	MVAr	–

TABLE VI.2.4　Decision Tree – Disturbance (Short Circuit)

Classification Performance Evaluation	
Global Error	2.28%
False Alarm	1.87%
Missed Alarm	2.58%

Based on engineering judgement, the 22-candidate attribute set listed in Table VI.2.3 is specified. For each OP, the disturbances selected in Section 2.4 are simulated and the minimum frequency and the maximum rate of frequency variation are stored. A description of the terminology is provided in Figure VI.2.10. In Figure VI.2.11, a representative DT for the short-circuit disturbance is shown. In the root node (node number 1) and in the non-terminal nodes, information related to the number of OPs, the safety index, and the dichotomic test is included. In the terminal nodes, information about the number of OPs included, their security index, and the type of the node is given.

The most important attributes, which are shown in the developed DTs, are:

- The percentage of the wind power according to the generation of conventional units, denoted as WP, which appears many times in the DT structure
- The amount of spinning reserve, denoted as SR, that should be available by the various types of units, taking into account the speed of response of their governors for various loading conditions and wind power penetration levels.

A similar DT is obtained for the gas turbine outage disturbance. The classification performance evaluations for both disturbances are shown in Tables VI.2.4 and VI.2.5.

Nonterminal node

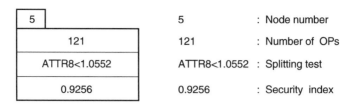

5	: Node number
121	: Number of OPs
ATTR8<1.0552	: Splitting test
0.9256	: Security index

Terminal node

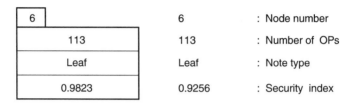

6	: Node number
113	: Number of OPs
Leaf	: Note type
0.9256	: Security index

FIGURE VI.2.10   DT terminology.

TABLE VI.2.5   Decision Tree – Disturbance (Machine Loss)

Classification Performance Evaluation	
Global Error	2.06%
False Alarm	1.31%
Missed Alarm	4.4%

The DT of Figure VI.2.11 can be readily converted to a number of if ... then ... else rules relevant to the secure operation of the system in the event of the simulated disturbance. The number of rules is equal to the number of terminal nodes. For example, in case of the three-phase fault, the system is totally insecure if wind penetration exceeds 37.85%. If it is less, the active power production of GEN1 should be examined. If it is higher than 37.2 MW, the system is secure. Such rules can be applied for corrective actions; it should be noted, however, that they are based solely on security and they do not consider economical operation of the system. Further application of DTs for corrective security applications are described next.

2.5.2.2 Application 2: Security Classification Using Only Active Powers

The aim of this application is to obtain operating rules relevant to the scheduling of the conventional units. These rules can be incorporated in the constrained optimization algorithms used for the economic scheduling of the system. It is considered that all available wind power should be exploited.

Using the procedure of Section 2.4.2, 5735 acceptable OPs were obtained, which were divided into the LS, comprised of 3748 Ops, and the TS, comprised of 1987 OPs. For each OP, the maximum frequency deviation and the rate of change of frequency are recorded.

Ten attributes characterizing each OP were selected, five attributes corresponding to the active production of the conventional unit groups, and five corresponding to the spinning reserves. These

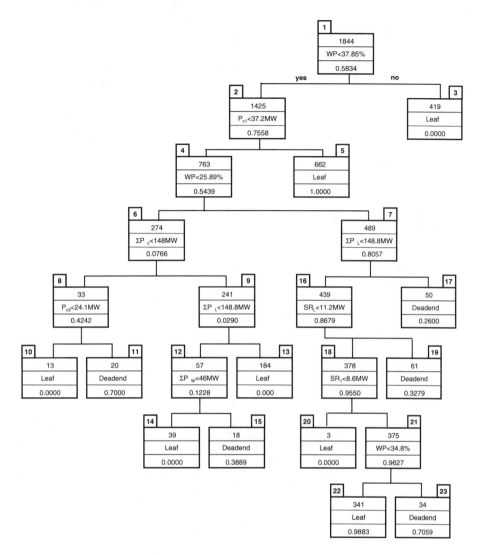

FIGURE VI.2.11   Decision tree for short-circuit contingency.

TABLE VI.2.6   System Generating Unit Groups

Unit Groups	Code	Active Power	R
Combined Cycle (1, 2)	Comb	45–133.4 MW	0.0764
Diesel Machines (3, 4)	Diesel	12.3–49.2 MW	0.0538
New Steam Units (5, 6)	NStm	25–75 MW	0.1421
Old Steam Units (7, 8)	Ostm	7–37 MW	0.1602
Gas Turbines (9, 10)	NGas	16.2–228.6 MW	0.0550

are coded, as shown in Table VI.2.6. *R* is the average droop of each group, indicating the speed of governor reaction. The spinning reserve of each group is denoted by the extension *Res* (e.g., CombRes means the spinning reserve of the combined cycle plant).

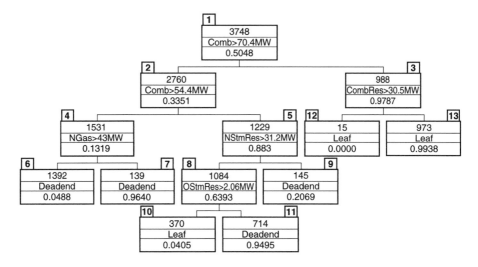

FIGURE VI.2.12   Decision tree for short-circuit disturbance.

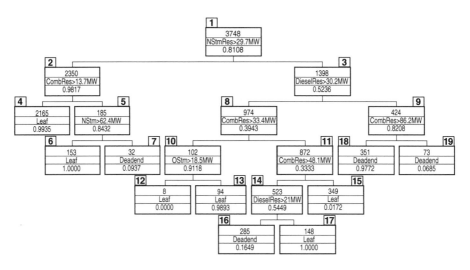

FIGURE VI.2.13   Decision tree for machine-outage disturbance.

After several studies, based on the previous learning set, the DTs shown in Figures VI.2.12 and VI.2.13, are developed. It can be seen that operating rules relevant to the active production and spinning reserve of the conventional unit groups can be readily obtained. For example, from the first DT it can be observed that the system is classified as secure if the combined cycle generation exceeds 70.4 MW and has over 30.5 MW spinning reserve. Similarly, the system is classified as insecure if the combined cycle production is less than 54.4 MW and the gas turbines provide less than 43 MW. This is apparently related to a lower spinning reserve provided by online gas turbines. In general, it is shown that the combined cycle attributes appear many times in both DTs, indicating the importance of this base unit in the secure operation of the power system. The classification performance evaluation for the developed DTs are shown in Tables VI.2.7 and VI.2.8.

It can be seen from the above examples that the DTs can provide a fast and reliable classification of dynamic security suitable for online applications.

TABLE VI.2.7　　Performance Evaluation of DT of Figure VI.2.12

Numerical Performance Evaluation	
Global Error	4.681%
False Alarm	3.329%
Missed Alarm	5.871%

TABLE VI.2.8　　Performance Evaluation of DT of Figure VI.2.13

Numerical Performance Evaluation	
Global Error	3.473%
False Alarm	4.684%
Missed Alarm	3.141%

2.6　Application of Feedforward Neural Networks

2.6.1　Theoretical Background

There are many excellent publications describing the theoretical foundations of artificial neural networks and their several applications. Indicatively, the relevant textbooks[27,32,33] are included in the references. In this work, the most popular class of artificial neural networks, called multi-layer pereceptrons (MLP), are applied for DSA. Multilayer perceptrons have proven most useful in regression problems. They are composed of several layers of neurons (perceptrons) character-ized by a smooth (nonlinear) activation function followed by an output layer of linear neurons. Their interconnection allows the network to learn non-linear and linear relationships between input and output vectors. The transfer function of the hidden layer, in our networks, is the Tan-Sigmoid function that generates outputs values between −1 and +1. The transfer function of the output layer is the linear function in order to give the appropriate range of output values for our study.

The back propagation algorithm is commonly used for training MLPs. This is an iterative, gradient search, learning algorithm, which adjusts each weight in a multilayer network so as to reduce the error in the outputs. Its name stems from the fact that the gradient of the MLP output error with respect to the weights is computed by propagating errors backward from the output layer. The method can be used either by adapting the weights after each presentation of an input attribute vector (OP) or by cumulating derivatives over the whole LS before adapting the weights. In this case, the global error with a given set of weights is the sum over all output units of the sum over all training cases of the squared distances between the actual and desired outputs of a unit.

Mismatches between the predicted value y' and the actual value y are quantified by the mean absolute error (MAE) and the root mean square error (RMSE), defined as:

$$MAE = \frac{\sum_{OP_i} |y_i' - y_i|}{Total\_Number\_of\_OPs(TS)}$$

$$RMSE = \sqrt{\frac{\sum_{OP_i} (y_i' - y_i)^2}{Total\_Number\_of\_OPs(TS)}}$$

2.6.2 Application of Artificial Neural Networks for Online Dynamic Security Assessment of Crete

2.6.2.1 Application 1: Frequency Prediction Using Extended List of Attributes

Using the learning set and the attribute list of Section 2.5.2.1, two multilayer ANNs were trained (one for each disturbance) using the adaptive back propagation algorithm. The networks were trained with constant learning rate of 0.1, momentum of 0.06, and maximum error tolerance of 0.001. For the two ANNs, the following structure was selected: one input layer with 22 attributes as inputs, one hidden layer with 8 neurons, and one output layer with the two security indices as outputs (Figure VI.2.14). The 22 inputs are the attributes presented in Table VI.2.3. The testing set performance evaluation results obtained for the designed security structures are presented in Tables VI.2.9 and VI.2.10.

In Figure VI.2.15, the regression capabilities of the ANN structures to the fmin due to short circuit are illustrated. In this graph, each point represents one TS sample, where its vertical distance to the diagonal presents the predicting error of applying the ANN to that OP.

2.6.2.2 Application 2: Frequency Prediction Using Only Active Powers

Using the learning set of application 2 of Section 2.5 and the active powers of the unit groups of Table VI.2.6 as inputs, two multilayer ANNs have been selected for both disturbances. For each case, the same ANN architecture was used, as shown in Figure VI.2.16, but with different weight matrixes and bias vectors. The networks were trained with constant learning rate of 0.1, momentum of 0.06, and maximum error tolerance of 0.001. The neural network toolbox of the Matlab package was used for this study.[34]

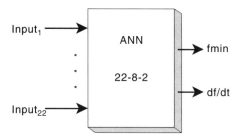

FIGURE VI.2.14   Structure selected for training the ANNs.

TABLE VI.2.9   Performance Evaluation of ANN1 for Short-Circuit Disturbance

	fmin (Hz)	df/dt (Hz/sec)
MAE	0.0330	0.0085
RMSE	0.0654	0.0177

TABLE VI.2.10   Performance Evaluation of ANN2 for Machine-Outage Disturbance

	fmin (Hz)	df/dt (Hz/sec)
MAE	0.0215	0.0047
RMSE	0.0432	0.0094

Using the corresponding TS, the accuracy of the results of the developed ANNs were evaluated. The comparison of the real values of the test set and the network outputs for 212 OPs are shown in Figure VI.2.17.

The absolute error for each OP of the first disturbance does not exceed the 0.18 Hz, as shown in Figure VI.2.18.

The quality of the results is evaluated by the mean relative error, the mean absolute error, and the mean square error relative to the target output values y, in this case the minimum frequency deviation of the power system, fmin. The performance evaluations for both disturbances are shown in Tables VI.2.11 and VI.2.12. These results are indicative of the fact that, similar to the DTs,

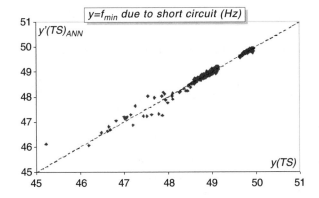

FIGURE VI.2.15   Performance evaluation of the ANN.

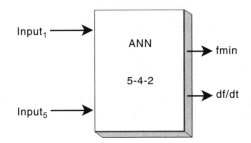

FIGURE VI.2.16   A feedforward neural network.

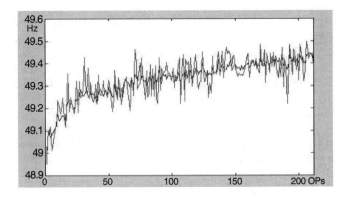

FIGURE VI.2.17   Comparison of frequency results.

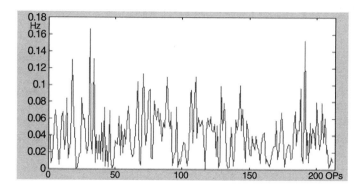

FIGURE VI.2.18   Absolute frequency error of a test set.

TABLE VI.2.11   Performance Evaluation of ANN1 for Short-Circuit Disturbance

	fmin (Hz)
MAE	0.0262
RMSE	0.0554

TABLE VI.2.12   Performance Evaluation of ANN2 for Machine-Outage Disturbance

	fmin (Hz)
MAE	0.0175
RMSE	0.0412

ANN-based methods can provide timely and reliable online dynamic security assessment. In addition, ANNs give a fairly accurate estimation of minimum frequency values in the case of the selected disturbances.

2.7   Application of Entropy Networks

In feedforward neural networks with one input layer, two hidden layers, and one output layer, the first hidden layer is the partitioning layer or test layer (TL), which divides the entire feature space into several regions. The second hidden layer is the AND layer (AL), which performs adding of the partitioned regions. The output layer is the OR layer (OL), which combines the previous layer results to produce a disjoint region of arbitrary shape. It can be seen that in this case, DTs and ANNs are equivalent in terms of input/output relations.[23,27,35] Thus, a DT can be reformulated into a neural network, following the next steps for the formation of:

1. The input layer (IL): It contains one neuron per attribute selected and tested by the corresponding DT. In our application, Section 2.5.2.1, there are 7 system attributes selected in the DT of Figure VI.2.11.

2. The first hidden layer (TL): It contains one neuron per DT test node. Each TL neuron is linked to the IL neuron corresponding to the tested attribute. The test nodes for the respective DT are 11.
3. The second hidden layer (AL): It contains one neuron per DT terminal node. Each AL neuron is connected to the TL neurons corresponding to the test nodes located on the path from the top node toward the terminal node. There are 12 terminal nodes for the respective DT.
4. The output layer (OL): It contains one neuron per DT class, connected to the AL neurons corresponding to the DT terminal nodes, where the class is the majority class. Its activation is high if at least one of these is active. Thus, all weights arriving at an OL neuron are equal to 1, and its threshold is equal to its number of inputs minus one.

The entropy networks (ENs) may be used to approximate a continuous security margin rather than to merely classify. In this case, the above described output layer would be replaced by a single output neuron, fully connected to all neurons of the AL, while the weights would be recalculated through retraining. The developed EN is illustrated in Figure VI.2.19. The neurons are represented by a circle, within which the corresponding DT node number is indicated.

Once the network structures are defined, they are trained by adapting their weights and thresholds to the input/output pairs observed in the LS. Each layer has a weight matrix W, a bias vector b, and an output vector that is input for the next layer. The total number of weights for the above example is 76, ([11] + [53] + [12]). The back propagation algorithm was used for the learning procedure, while the networks were trained with constant learning rate of 0.1, momentum of 0.06, and maximum error tolerance of 0.001.

As in the case of the DTs, the classification success is evaluated by quantifying the success rate, the false alarms, and the missed alarms using the test set of 921 OPs. The results are shown in Table VI.2.13. These rates are compared to the corresponding rates of the DTs' performance evaluation in Table VI.2.4.

It is shown that the performance of the ENs is slightly improved compared with the DTs' performance. This means that the application of the ENs increases the reliability of the approach, while fast computational time, which is very important for online applications, is maintained.

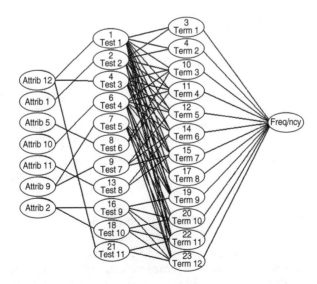

FIGURE VI.2.19   Entropy network corresponding to the DT of Figure VI.2.11.

TABLE VI.2.13   Performance Evaluation of EN

Success Rate	1.99%
False Alarm	1.73%
Missed Alarm	2.14%

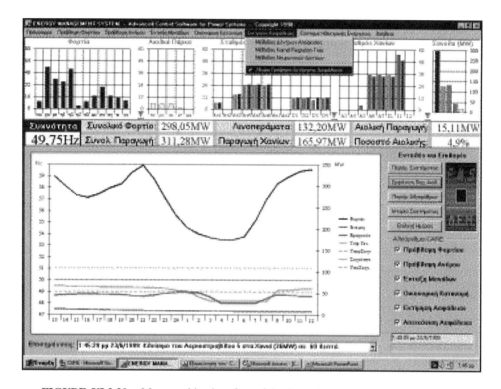

FIGURE VI.2.20   Man–machine interface of the dynamic security assessment module.

2.8   Online Implementation and Evaluation

The security evaluation structures described in Sections 2.6 and 2.7 were integrated as modules in the CARE software, developed within the European R& D program, "Advanced Control Advice for Power Systems with Large Scale Integration of Renewable Energy Sources" within the JOULE III framework. The software comprises various modules for short-term load and wind power forecasting, unit commitment, economic dispatch, and online security assessment oriented to the needs of isolated power systems with increased renewable power penetration.[25] In the execution cycle, security assessment, activated "on call" by the operators, follows the unit commitment and dispatch modules, leaving to the operator the decision to activate the module for validation of the proposed dispatch scenario. In Figure VI.2.20, the display of results produced by the dynamic security assessment module are shown.

On top of the screen, the load at critical buses, the current production of the wind parks, and the production of the various thermal units in the two thermal stations are displayed in the form of bar charts. The dynamic security assessment results for 48 hours ahead are displayed on the main screen under the forecasted load curve in the form of lines representing the expected frequency in case of the considered disturbances. The expected frequency is provided by ANNs. It is shown that the expected

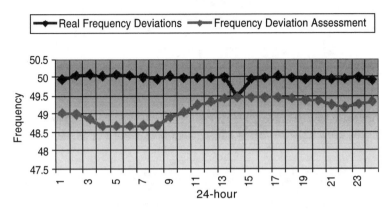

FIGURE VI.2.21   Actual frequency compared to frequency deviation assessment.

frequency deviation in the short circuit is unacceptable for most of the time. This contingency has a low probability of occurrence unless the weather conditions are bad, and the operator might select to ignore it. Between 1 and 8 o'clock however, the expected frequency deviation is shown to be below the 49 Hz threshold for both simulated contingencies. It is characteristic that this period corresponds to the low load period, when a significant wind power penetration can lead to poor dynamic security.

A pilot installation of CARE has been operating in Crete since May 1999. Figure VI.2.21 shows the information recorded during the trip of a 20 MW gas turbine at Chania on the 23rd of June (taking place between the 14th and 15th hours). The minimum frequency recorded by the SCADA system was 49.46 Hz, while the frequency drop predicted by the DSA functions was 49.4 Hz, very close to reality.

It is clear that enrichment of the security structures with more disturbances can provide online security assessment of very satisfactory accuracy.

2.9   Conclusions

Online dynamic security assessment is one of the most challenging problems faced by modern power systems. The task assumes the evaluation of the power system's ability to face a number of critical contingencies, requiring short computational time and high reliability. This can be achieved by the application of machine learning techniques in cooperation with traditional analytical techniques. This chapter describes the practical application of advanced inductive inference methods and of artificial neural networks to online dynamic security assessment of isolated power systems. A main concern in these systems is frequency instability. In the application described, decision trees are used in order to classify dynamic security, while artificial neural networks emulate the degree of security, evaluated by predicting the expected minimum value of system frequency and the maximum rate of frequency change for each critical disturbance specified. These functions have been integrated within the control system of Crete and can be activated "on call" by the operator providing dynamic security monitoring. The results obtained so far are considered highly satisfactory, both in accuracy, which increases the reliability of the method, and in computational time, which is a necessity for real time applications. The application of these techniques to larger interconnected power systems to assess online their dynamic security against various types of instability can provide equally satisfactory results leading to a more robust operation.

Acknowledgments

The authors wish to thank the European Commission for financing the R&D projects: "CARE: Advanced Control Advice for Power Systems with Large Scale Integration of Renewable Energy Sources" and "MORE CARE: More Advanced Control for Secure Operation of Isolated Power Systems with Increased Renewable Energy Penetration and Storage." They also thank all partners in the projects, especially Professor J.A. Pecas Lopes from INESC, Portugal for his valuable contribution.

References

1. Dialynas, E.N. et al., Effect of high wind power penetration on the reliability and security of isolated power systems, paper 38–302, 37th Session, CIGRE, Aug. 30–Sept. 5, 1998.
2. Hatziargyriou, N., Karapidakis, E., and Hatzifotis, D., Frequency stability of power systems in large islands with high wind power penetration, Bulk Power Systems Dynamics and Control Symposium—IV, Restructuring, Santorini, August 24–28, 1998.
3. Hatziargyriou, N.D., Ed., Special issue, *Wind Eng.*, 23(2), 1999.
4. CARE: advanced control advice for power systems with large scale integration of renewable energy sources, contract JOR3-CT96-0119, final report, August 1999.
5. Hatziargyriou, N.D. et al., Operation and control of island systems—the Crete case, IEEE PES WM, Singapore, January 23–37, 2000.
6. Hatziargyriou, N.D. et al., On line dynamic security assessment of power systems in large islands with high wind power penetration, 13th Power Systems Computation Conference (PSCC), Trondheim, Norway, June 28–July 2, 1999.
7. Hatziargyriou, N.D. et al., Artificial intelligence techniques applied to dynamic security assessment of isolated systems with high wind power penetration, 2000 Session of CIGRE, Paris, August 2000.
8. Karapidakis, E.S. and Hatziargyriou, N.D., Application of artificial neural networks for the security assessment of medium size power systems, MELECON, Limasol, Cyprus, May 29–June 1, 2000.
9. Marceau, R.J. et al., Power system security assessment: a position paper, CIGRE TF 38.03.12, June 30, 1997.
10. Balu, N., On-line power security analysis , *Proc. IEEE*, 80(2), 280, 1992.
11. Haubrich, H.J. and Nick, W.R., Adequacy and security of power systems at planning stage, *Electra*, No. 149, August 1993.
12. Dy Liacco, T.E., The adaptive reliability control system, *IEEE Trans. Power Apparatus Syst.*, PAS-86(5), 517, 1967.
13. Dy Liacco, T.E., Real-time computer control of power systems, *Proc. IEEE*, 62, 884, 1974.
14. Dy Liacco, T.E., System security: the computer's role, *IEEE Spectrum*, 15, 43, 1978.
15. Fink, L.H. and Carlsen, K., Operating under stress and strain, *IEEE Spectrum*, 15, 48, 1978. Real-Time Computer Control of Power Systems, Proceedings of the IEEE, Vol. 62, pp. 884–891, July 1974.
16. Kundur, P. and Morison, G.K., A review of definitions and classification of stability problems in today's power systems, Panel session on stability terms and definitions, IEEE PES Meeting, New York, Feb. 2–6, 1997.
17. Meyer, B. et al., New trends and requirements for dynamic security assessment, CIGRE TF 38.02.13, 1997.

18. Vaahedi, E., Mansour, Y., and Tse, E., A general purpose method for on-line dynamic security assessment, *IEEE Trans. PWRS*, 13(1), 243, 1998.
19. Li, W. and Bose, A., A coherency based rescheduling method for dynamic security, *IEEE Trans. PWRS*, 13(3), 810, 1998.
20. Bettiol, A.L., Wehenkel, L., and Pavella, M., Transient stability-constrained maximum allowable transfer, *IEEE Trans. PWRS*, 14(2), 654, 1999.
21. El-Sharkawi, M. and Neibur, D., Artificial neural networks with applications to power systems, *IEEE-PES*, special publication 96TP-112-0, 1996.
22. Wehenkel, L. and Pavella, M., Eds., Special issue on automatic learning application to power systems, *Revue-E*, SRBE-Belgium, Dec. 1996.
23. Wehenkel, L. and Akella, V.B., A hybrid decision tree neural network approach for power system dynamic security assessment, in *Proceedings of the 4th International Symposium on Expert Systems Application to Power Systems*, Elsevier—North Holland, Amsterdam, 1993, 397.
24. Wehenkel, L. and Pavella, M., Decision tree approach to power system security assessment, *Int. J. Electrical Power Energy Syst.*, 15(1), 1993.
25. Hatziargyriou, N.D., Papathanassiou, S., and Papadopoulos, M., Decision trees for fast security assessment of autonomous power systems with large penetration from renewables, *IEEE Trans. Energy Convers.*, 10(2), 1995.
26. Miranda, V. et al., Real time preventive actions for transient stability enhancement with a hybrid neural network—Optimization Approach, *IEEE Trans. PWRS*, 10, 1995.
27. Wehenkel, L., *Automatic Learning Techniques in Power Systems*, Kluwer Academic, Boston, MA, 1998.
28. La Scala, M., Trovato, M., and Antonelli, C., On-line dynamic preventive control: an algorithm for transient security dispatch, *IEEE Trans. PWRS*, 13(2), 601, 1998.
29. O'Sullivan, J.W. and O'Malley, M.J., A new methodology for the provision of reserve in an isolated power system, *IEEE Trans. PWRS*, 14(2), 519, 1999.
30. TRACTEBEL, User's Guide, Eurostag 3.2 package.
31. Kundur, P., *Power System Stability and Control*, McGraw-Hill, New York, 1994.
32. Wasserman, P., *Neural Computing, Theory and Practice*, Van Nostrand Reinhold, New York, 1989.
33. Haykin, S., *Neural Network*, Comprehensive Foundation, Ontario, Canada, 1994.
34. Neural Network Toolbox, User's Guide Version 3.0, Matlab 5.3 package.
35. Karapidakis, E.S. and Hatziargyriou, N.D., Entropy networks for security assessment of isolated medium size power systems, 1st RiMAPS Conference, Madeira, Portugal, Sept. 25–27, 2000.

3

Knowledge-Based Systems Techniques and Applications in Power System Control Centers

3.1   Introduction VI-63
3.2   Power System Operation VI-66
3.3   Power System Control Centers VI-67
3.4   Artificial Intelligence Applications in Power System
      Control Centers VI-69
      Overview • Areas of Artificial Intelligence Application in
      Power Systems • Knowledge-Based Systems Applications
3.5   SPARSE—A Case Study VI-76
      SPARSE Overview • Knowledge Base • Inference Engine •
      SPARSE Explanations • Machine Learning Module • User
      Interface • Verification and Validation • Tutor • Example
3.6   Conclusions and Future Directions................   VI-107
References VI-109

Zita A. Vale
Polytechnic Institute of Porto (IPP)

3.1   Introduction

Control centers (CC) play a very important role in power system operation and control. They receive real-time information on the power system and are responsible for most decisions concerning the power system operation. This chapter deals with the use of knowledge-based techniques and their applications in these control centers. Reasons for this use will be explained and the problems raised will be clearly stated. Solutions to these problems will be pointed out whenever the author believes they exist.

A successful knowledge-based system (KBS) application in a power system control center can have a large impact, enhancing the already achieved reputation of real practical use of this technology. Power system control centers are a very challenging domain for knowledge-based systems because they can provide solutions for a large set of complex problems for which traditional software techniques are not suitable. In fact, power systems are complex and dynamically changing environments, comprised of a lot of complex plants and equipment. These characteristics of power systems require knowledge-based applications in control centers to deal with non-monotonic and temporal reasoning. On the other hand, the analysis of these situations is event-driven and requires each piece of information to be analyzed in its context and not independently from the other available information. For instance, when analyzing lists of alarm messages, CC operators must have in mind the group of messages that describes each type of fault. That same group of messages can show up in the reports of different types of faults. So, CC operators have to analyze the arrival of additional information, whose presence or absence determines the final diagnosis.

As there may be errors during information transmission or even missing information, CC operators also have to deal with uncertain, incomplete, and inconsistent information. The introduction of new plants and new equipment in the power systems requires efficient updating of knowledge. Moreover, power systems require real-time control that considers a huge amount of information, which is also very demanding for knowledge-based applications.

Let us consider a small example: a simplified situation that may occur in a power system that helps to understand the importance of temporal reasoning when dealing with power system operation. Whenever a fault occurs in a power system, its protection system should respond to that fault, giving automatic opening orders to one or more breakers. The opening of these breakers ensures the isolation of the fault because the protection system is designed in such a way that only an area as small as possible is affected. Protection systems are very important for the performance and security of power systems and can be rather complex, especially in the case of transmission networks, involving a lot of different protection devices. In our example, we will consider that a fault occurs in a line connecting two substations of a power system, and, in consequence, the protection system gives opening orders to two breakers installed in the two ends of this line (Figure VI.3.1).

Figure VI.3.2 considers one of these breakers and a possible sequence of operations after the occurrence of the fault.

Let us consider that, as shown in Figure VI.3.2 this breaker opens at instant T1, closes at instant T2, and opens again at instant T3. The interpretation of this fault depends not only on this sequence of events but also on the time intervals between the events. In fact, the breaker closing at T2, after the first opening at T1, is likely to be due to the automatic reclosure procedure of the protection. Fast automatic reclosures are widely used in power systems in order to minimize the impact of faults. In this case, the time interval between T1 and T2 would depend on the type of fault and on the regulation of the automatic reclosure in the protection. Let us consider that, for instance, for a fault involving only one of the three phases (single phase fault), this time would be 900 milliseconds, whereas for a fault involving the three phases (three-phase fault), it would be 300 milliseconds. Apart from considering these times, we have to allow for some tolerance in the dating and transmission of the information from the plant to the control center. For this reason, let us say that in the case of a three-phase fault, the time interval between T1 and T2 should not exceed 500 milliseconds.

FIGURE VI.3.1   Power system line.

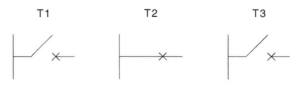

T1<T2<T3

FIGURE VI.3.2   Sequence of breaker's operations.

So, if T2−T1 is less than or equal to 500 milliseconds, we can interpret the first two messages as a consequence of a three-phase fault. After this, we have to consider the third message reporting a new opening of the breaker at T3. Assuming that this is a consequence of a tripping order sent by the protection system, it is due to an incident situation. Once more, the time T3 is crucial for the interpretation of this part of the incident. If this tripping takes place in a short interval of time (say, within 5 seconds) after the reclosure of the breaker, it is considered that it is caused by the same fault that originated the first opening of the breaker considered in this example. Under these circumstances, with T3 − T2 equal to or less than 5 seconds, the whole incident would be seen as a three-phase fault with unsuccessful reclosure in this end of the line. If T3−T2 was greater than 5 seconds, the third message would be considered as reporting a fault independent from that already considered.

This small example shows the complexity of the analysis of the messages that CC operators have to interpret. Notice that the same sequence of messages can be interpreted in different ways, depending on the time intervals between messages. If a knowledge-based system is used to assist this interpretation, its inference engine must be prepared to deal with the temporal nature of the problem. For instance, after receiving the second message considered in this example, the incident could be described as a three-phase fault with successful reclosure, but the inference engine will have to wait at least 5 seconds for the possible arrival of a message reporting another opening of the breaker. If this message arrives, the incident is described as a three-phase fault with unsuccessful reclosure.

In fact, if we consider all the messages that are generated during the period of the incident, including not only the messages originated in the plants involved in the incident but also in other plants of the power system, operators may have to consider as many as several hundred messages in just a few minutes. It is important to note that, usually, the occurrence of an incident, besides the messages that are important for this particular analysis of the incident, causes the generation of more messages, which are not important for the analysis in the context of the incident but that increase the total number of generated messages. However, in other contexts, these messages could be important, a fact that stresses the need for a contextual interpretation of the information. On the other hand, several incidents can take place almost at the same time, and one incident can have consequences in more than two plants, resulting in a much more complex interpretation of the situation. If we also take into account the need to consider missing information, we can have an idea of the difficulties CC operators face and also of the complexity of a knowledge-based application for this area.

In the remainder of this chapter, we will present in more detail the added value of knowledge-based systems and techniques for incident analysis, restoration support, and other tasks in control centers.

Section 3.2 provides a general view of power system operation and justifies the importance of control centers. Section 3.3 deals with control centers, referring to their functions and the most important issues to understand the following sections. In fact, Sections 3.2 and 3.3 intend to help readers understand the field of application treated in this chapter and do not provide detailed knowledge of power system operation or power system control centers.

Section 3.4 is concerned with the application of artificial intelligence (AI) techniques in control centers. Some of the situations where artificial intelligence can give important contributions are presented. Several examples of knowledge-based applications are included in this section. Section 3.5 provides a case study of a knowledge-based system developed for a transmission control center. More than presenting this specific system, this section aims to identify problems that usually appear in the application of knowledge-based systems in this field, to discuss those problems, and to point out some solutions.

Finally, Section 3.6 presents some conclusions and possible scenarios for the future. The restructuring of the electricity market is bringing new challenges and needs to the power industry; those are considered in that section.

3.2   Power System Operation

The use of electrical energy in a disseminated way, as we are familiar with today, is indeed very recent. After some early attempts to use direct current, namely, in electric dynamos and motors, it was only with the use of alternating current that the transmission of electrical power over great distances became technically and economically possible. At the early stage, generation of electrical energy served the needs of small groups of consumers, and power plants were operated locally and individually.

People enjoyed the benefits of the use of electrical energy, not only because of the improvement it made possible in industrial activities, but also due to the comfort that it added to their daily lifes. Electrical lighting, achieved in the early days of electrical energy only by means of incandescent lamps, which are still widely used today, was a basic but very important item for this comfort.

The demand for electrical energy started to rise, and this led to a flourishing new industry: the power industry. With the spread of electrical energy use, reliability and economy issues gained more importance in the scope of the power industry. The interconnection of generating plants increased the reliability of the electrical networks and also allowed optimization of the operation of those plants in economic terms. This situation led to the present state where electrical networks are interconnected whenever geographical conditions do not make this impossible, crossing frontiers and resulting in very large and complex power systems.[1]

The wide availability of electrical energy changed the human way of life in such a deep way that it is now very difficult to imagine how humankind could live without electrical energy in an organized and developed society. This fact made quality of electrical power become more and more important as clients became more and more demanding.

The operation of power systems had to cope with all the difficulties that arose during this evolution. In fact, to operate and efficiently control modern power systems requires huge technical support. This support is important both in terms of the equipment as well as methods used in power system planning and operation.

In this evolution process that led power systems from small isolated systems to large and complex interconnected systems, these methods played a very important role to ensure the reliability of the systems but also to enable their optimization. These two key issues can be better addressed if the power system is seen as a whole, with information about all existing resources and needs. In power system operation, this is implemented through the use of SCADA (supervisory control and data acquisition) systems, which provide real-time information about the power system to control centers (CC) where most operation decisions are made. High quality redundant telecommunication systems are used in the power industry for this purpose. On the other hand, these centers are equipped with the capability of sending orders through the telecommunication system, which allows for remote control of the power system in most situations. The use of computers in the power industry started very early in the area of accounting, and was then extended to the planning and design of the power system and only later, in the early 1960s, to the most demanding area of power system operation.[2] The use of computer systems made the use of telephone lines for communications between power plants, substations, and CCs completely obsolete. Nowadays, control centers are based on computerized systems and use complex software for control and operation purposes. In fact, a lot of computational algorithms have been developed for power systems and have become essential aids in power system operation. Although this software is vital, the human operators who work in these centers still play the most important role in the decision process. As these centers are, in fact, decision centers, artificial intelligence techniques can provide an important contribution to some of the most difficult tasks these operators have to face.

The application of artificial intelligence techniques in this field is not, however, always very easy. In fact, power systems are dynamically changing environments, involving a huge amount of data and requiring fast decisions, especially in critical situations. These characteristics make the use of artificial intelligence techniques rather challenging, resulting in many experiments and contributions but only a few successful real applications.

3.3 Power System Control Centers

In the early days of the power industry, most of the plants had technical staff who worked in shifts, guaranteeing the operation of the plant twenty-four hours a day. The development of reliable telecommunication systems changed this situation. Nowadays, most power system plants, such as generating plants and substations, are unmanned. Operating and control actions are concentrated in centers especially developed for this purpose, generally known as control centers.

These centers receive real-time information about the power system through a SCADA system. Moreover, control center operators can send orders using the telecommunication system. This way, the most important apparatus of the Power System are remotely supervised and operated. This is the case with the most important breakers and isolators. The equipment for which remote operation is not implemented is operated by sending technical staff to its location. Usually, this is the case with equipment that rarely needs to be operated.

This operation philosophy allows coordination of all the existing resources and the optimization of the operation of the system as a whole, considering all relevant issues, namely, economy, reliability, and security.

Power system performance is dependent on the performance of control centers. This makes these centers very important for the quality of the whole system. Efficient performance of these centers requires high quality telecommunication equipment for receiving reliable real-time information about power system state. On the other hand, control center operators must have the means to send orders to remotely operate the power system equipment. The complexity of the tasks that must be accomplished in these centers requires top level computational equipment and complex software.

Presently, control centers normally include not only SCADA functions but also EMS (energy management system) functions. Control centers can include a large set of functions, from which we consider the following:

- Monitoring — Acquisition and processing of information about power system components through the SCADA system, presenting, for instance, cases of limit violations
- Network operation — Operation, from the CC, of network equipment such as breakers, isolators, reactive voltage compensators, and control automation equipment installed in network plants
- Economic dispatch — Scheduling the production of generators, considering the forecast demand and minimizing operating costs
- Automatic generation control — Scheduling the production of generators, with a temporal anticipation of some seconds, considering the most updated demand forecast and considering the control of the system (namely the regulation of the frequency value) and an initial operating point for each generating unit (which is usually obtained by the economic dispatch function)
- Power flow — Real-time power flow applications use real-time data obtained by the SCADA system to calculate the values of operating variables such as voltages, currents, and active and reactive power flows
- Security assessment — Using a large set of available data, security assessment applications analyze the operating conditions of the power system, including contingency analysis

There are a lot of other possible functions which have not been listed, such as reactive power and voltage control, state estimation, load forecasting, emergency control, restoration, and training simulation.

Power systems aim to consistently respond to the demands of their clients with high levels of quality at minimum costs. These costs should include several aspects, namely, the ecological impact of the techniques being used.

The identification of the most important problems in power system operation is very important for power system performance. The current operational problems working group has been established under the scope of the system operation subcommittee of IEEE (The Institute of Electrical and Electronics Engineers) to bring attention to power system operators' problems. This working group has undertaken several surveys that led to some interesting conclusions. In 1989, this working group published the results of a survey involving 130 replies from utilities from all regions of the United States and Canada.[3] Some of the problems in power system operation are directly related to control center issues. Let us consider the following:

- Alarms in normal and emergency system operation
- Dispatcher and restoration training simulators
- Evaluation of training program
- Dispatcher performance

It is important to note that regarding the question of alarms in emergency situations, it was reported that "artificial intelligence systems are needed to observe, make conclusions and report the nature of trouble to the dispatcher." Moreover, the working group analysis divided the problems into three categories: people problems, analysis and control problems, and computer system problems. The first class includes the last three problems referenced and deals generally with the quality, training, and performance of control center operators. The problems pointed out in alarms are classified as computer system problems. In fact, software applications were not able to produce a list of alarms that could be easily understood by CC operators, especially in emergency situations. This problem was not only referred to in a high number of responses, but was also referred to as a high priority one.

Although an increasing set of sophisticated means is used for assisting CC operators, their tasks are still complex and utilities tend to be increasingly demanding with them. The information that arrives to the CC is processed by computer applications and presented to operators in several forms, including graphics, tables, and lists of alarm messages. These lists intend to be highly informative and to provide CC operators with a means to evaluate in detail the situation of the power system.

An alarm message can be generated in one of the following situations:

- An analogue value has under-passed or over-passed a pre-established limit (e.g., in the case of an undervoltage or an overcurrent situation)
- A digital value has been modified (e.g., a breaker has opened)

Some software applications can also generate alarm messages in some situations, which are included in the alarm list presented to CC operators.

The importance of alarms for power system operation justifies the need for efforts in order to improve the quality of alarms presented to control center operators.

Since pioneer power system control centers have been installed, the great evolution in computer hardware made it possible to process more and more information in the same period of time. As human capabilities have not increased as much, this led to an important gap between computer applications and their users because the former do not respect human limitations. This gap is especially dangerous under incident conditions when control center operators receive a large amount of information that they must quickly interpret in order to take the required actions to prevent the incident from extending to a wider area and perform power restoration.

The IEEE power system control center's joint working group has undertaken a survey about the excessive number of alarms in CCs. The results of this survey were published in 1989[4] and showed the need to improve CC alarm processors. This survey included a total of 65 questions, including questions for which a yes or no response was required, questions that required a numerical response, questions with multiple answers, and questions that required respondents to express their opinions. Companies that responded to this survey operated several types of power systems, ranging from large, interconnected systems to small, regional systems. In order to be able to take the scope and dimension of the companies into account, a classification based on the number of points being monitored was used. Companies with less than 2000 points were considered small, companies with 2000 to 8000 points were considered medium, and companies with more than 8000 points were considered large. A total of 87 companies were considered in the survey, 24 small, 45 medium, and 18 large. A total of 87 responses were analyzed. Among the large set of survey conclusions, some are very interesting. For instance, in what concerns operator opinions, only 38 (43.7%) and 21 (24.1%) responses say that operators are satisfied with alarm processing during busy operation and emergency operation, respectively. On the other hand, most of the responses (66, corresponding to 75.9%) say that some alarms are a nuisance rather than helpful. Only in 42 responses (48.3%) did operators say that they felt that the "state" of the system could be automatically determined from information they had in the computer.

The conclusions of this survey, together with several other studies, showed, without any doubt, that there was in fact a problem with the alarms in power system CCs and with an excessive number of alarms during emergency situations. Artificial intelligence applications have been appointed as a promising approach to solve this kind of problem.[5]

As alarm processing has been identified as one of the most important problems of power system control centers that is not efficiently addressed by traditional computer applications, it has been one of the most important areas for the application of artificial intelligence in control centers. In 1991, CIGRÉ (International Council on Large Electric Systems) task force 38.06.02 published the results of a survey about expert system use in alarm handling.[6] A total of 30 responses were considered, including power companies as well as other entities involved in alarm handling such as SCADA and energy management systems (EMS) manufacturers and research groups.

The responses to this survey point out as major difficulties with alarm handling the existence of too many alarms, the production of multiple alarms by one event, and the lack of an alarm priority order. This survey also concludes that both inexperienced and experienced operators have problems in the interpretation of alarms when an excessive number of alarms is presented. This shows a real need for the development of decision-support applications to assist control center operators. The responses show that expert operators are the main source of knowledge (50%) for the expert systems developed for alarm handling, while manuals (20%) and computer procedures (20%) are also important. At the time this survey was undertaken, most expert systems were in early phases of development or in the prototype phase, and only a few were in an early stage of use. After completion of the survey results, more expert systems were deployed, with a total of 6 systems deployed in May 1991.[6]

In 1999, CIGRÉ task force 35.13.02 published a brochure titled, "Knowledge Based Applications in SCADA / EMS—A Practical Approach."[7] This text recognized that intelligent applications were at a stage at which it was already possible to use them as practical applications in control centers. The aim of that reference was to distribute the lessons learned with previous experiences in order to help future developments. This brochure includes information about actual installations and experiences of some utilities. This information will be analyzed in Section 3.4 in more detail.

3.4   Artificial Intelligence Applications in Power System Control Centers

3.4.1   Overview

The application of artificial intelligence to power systems[8] has been proposed since the 1970s, but only much later were real applications considered in a more serious way. In the last decade of the

20th century, a significant number of utilities had already had important experiences using artificial intelligence applications.[9–14] Most of these applications are developed for control centers because, being important centers of decision, they present problems whose nature makes them addressable by artificial intelligence techniques. In control centers, artificial intelligence applications are mainly intended to assist control center operators in their complex tasks.

The efficiency of power system operation has been improved by automating many operating procedures that used to be performed manually and also by using computer applications. However, traditional software applications do not provide adequate decision support, especially in emergency situations.[15] In fact, control center operators have to make decisions that are mainly qualitative in nature, for which their past experience is very important. The methods used by traditional computer applications such as the ones based on operational research are not suitable in these situations.

As mentioned in Section 3.3, human operators have always been and still are the most important players in power system operation. CC operators have a lot of responsibility in power quality, and the tasks they have to accomplish are of very diverse and difficult natures.[2] Some of these tasks are routine activities, but most of them involve difficult decisions and have a huge impact on the performance of the power system.

The reasoning involved in the accomplishment of these tasks is of several different types, namely, deductive, inductive, intuitive, and a combination of these.[2] Most of the time, operators use deductive reasoning because they follow *a priori* established policies and practices. Inductive reasoning is used when operators use their past experience to make decisions based on the anticipation of their consequences.

Intuitive reasoning, however, is the most important ability of an experienced operator who is capable of making efficient and rapid decisions for unexpected situations for which there are no *a priori* defined plans or solutions. This kind of reasoning allows a quick analysis of complex situations without being conscious of the complete reasoning process. Intuitive reasoning is especially valuable for high level administrative decisions and emergency situations.[2] At present, this is the only of the three referenced types of reasoning that traditional computer applications have serious difficulties in efficiently accomplishing. The capabilities of intuitive reasoning are very close to what we generally consider specific to human intelligence. This is why artificial intelligence applications can mean a significant difference in CC operators' assistance. CIGRÉ task force 39.03 analyzes some practical aspects of using artificial intelligence in control centers and the future role of these applications.[16] This is a very interesting reference to understand the utilities' point of view. This task force recognizes the difficulties of CC operators, who are assisted by information systems that produce a large volume of raw data but do not provide them with adequate support and assistance when they need it the most: at incident situations.

The training of CC operators is also considered a very important issue because well-trained operators can deal with difficult situations in a more efficient way, improving the performance of the whole power system. However, nowadays, power systems are so reliable that even operators with several years of experience may not have experienced even one severe incident in their career. Operator training through the use of dispatcher training simulators (DTS) is the usual method, but this method has some important drawbacks, primarily because the preparation of training scenarios is very time consuming. Artificial intelligence can also provide an interesting contribution to this field by developing intelligent tutoring systems for operator training.[17,18]

In order to be successfully used in practice, artificial intelligence applications must consider the need of the power system operation but also the needs of CC operators. Concerning this last issue, these applications should not demand too much attention from CC operators, stressing them even more, especially in critical situations. AI applications should provide adequate decision support and enhance the interface between the users, i.e., CC operators, and the system. This interface cannot be seen simply as a graphical user interface but as the complex and crucial process of communication between the system and human operators. If this interface is not adequate, even a high quality system can become completely useless and, even worse, damaging to power system operation. In fact, the user interface must be seen as a crucial part of the system, and artificial intelligence techniques can also provide extra value to the system.[19,20]

3.4.2 Areas of Artificial Intelligence Application in Power Systems

There have been several instances of applying artificial intelligence techniques in a wide range of problems in the power system area.[21] Some areas of application gained more importance, mainly because the problems they raise are more suitable for artificial intelligence techniques but also because some of them are crucial for power system operation.

Artificial intelligence has been applied to diagnosis in general, involving the consideration of sensor measurements in order to analyze the state of the system or a part of it. Diagnosis problems addressed include transmission networks, generating plants, substations, distribution networks, and several kinds of equipment such as generating units and transformers.

Because of their importance in the scope of power system operation and their suitability to be treated by artificial intelligence techniques, fault analysis and diagnosis[22–30] in transmission and distribution networks is an important area of application for artificial intelligence in power systems. CIGRÉ has undertaken a survey of this issue[31] that considers expert systems as well as artificial neural networks, fuzzy logic, and genetic algorithms. Fault analysis and diagnosis is related to alarm processing and the protection system. Alarm processing[32–34] is one of the most important fields of artificial intelligence in CCs. In Section 3.5, we will consider a case study in this field.

Remedial actions, especially restoration after the occurrence of an incident, are also an important field in artificial intelligence applications in power systems.[35,36] Applications in restoration concern transmission and distribution networks. Security assessment, forecasting problems, especially load forecasting, and voltage/reactive power control are other areas where artificial intelligence applications have been able to produce some good results.

Application of artificial intelligence techniques to power systems also includes the area of planning and scheduling problems, including several design problems and the determination of which components should be used by the system and when maintenance operations should take place. These problems include, for instance, unit commitment, hydrothermal scheduling, network optimization, and maintenance scheduling.

Knowledge-based and expert systems are among the most important areas of application of artificial intelligence techniques in power systems. In Section 3.4.3, knowledge-based applications in control centers are considered in more detail. In addition to knowledge-based systems, other artificial intelligence techniques have been used for addressing several problems in power systems,[37] such as neural networks,[38,39] fuzzy systems, and genetic algorithms. A combination of these techniques is also considered, in an attempt to overcome the limitations of each technique. Some of the most important combinations are neuro-expert systems,[40] fuzzy expert systems,[41] and the combination of knowledge-based systems, artificial neural networks, and fuzzy systems.[42]

3.4.3 Knowledge-Based Systems Applications

Knowledge-based systems (KBSs), particularly expert systems, are suitable to address a large set of problems in power system operation. In 1990, CIGRÉ formed its task force 38.06.03, titled "Practical Use of Expert Systems in Planning and Operation of Power Systems." This task force published the final report of its Phase I work in 1993.[43] This report was based on the responses of 26 organizations in the power industry to a survey conducted by the task force. The motivations for the development of the expert systems considered in this survey included, among others, the following: human-error prevention, knowledge maintenance, and inefficiency of the traditional methods. The areas of application included fault diagnosis, restoration, alarm processing, operation assistance, remedial control, and security assessment. The survey covered a wide range of aspects related to expert system development and use, namely, the evaluation of system performance. Expert systems were already gaining acceptance by utilities and were recognized as being able to address several problems in power systems.

The brochure published by CIGRÉ task force 35.13.02 in 1999[7] considers several references in the area and indicates the following proportion of knowledge-based systems developed for electric utilities:

- Fault diagnosis: 46%
- Restoration: 37%
- Security assessment: 21%
- Alarm processing: 12%

Notice that these percentages equal more than 100% because the considered areas of application are not completely independent. For instance, applications that perform fault diagnosis can also consider alarm processing, restoration, and operation assistance. On the other hand, it is still very difficult to make a complete list of deployed systems and even more difficult to follow their stages of use. The referenced areas of application can be, at least, indicative of the most important control center problems being treated with knowledge-based techniques.

The brochure considers nine examples of knowledge-based applications. As one of them is installed in a substation and not in a control center, we will list only the remaining eight here:

- SPARSE, used by REN (Portugal) for alarm handling
- Langage, used by Hydro Quebec (Canada) for alarm handling
- SRS, used by Iberdrola (Spain) as a restoration assistant
- An intelligent alarm processing system, used by Northern States Power Company (USA)
- A restoration assistant system, used by Northern States Power Company (USA)
- A restoration assistant system, used by The National Grid Company PLC (United Kingdom)
- A restoration assistant system, used by Kansai Electric Power Company (Japan) in its ShinSonezaki control center
- A restoration assistant system, used by Chubu Electric Power Company Inc. (Japan)

This list includes systems developed by utilities, vendors, and by mixed teams of academic and utility people and of utility and vendor staff. Although most of the utilities believed the main objectives of their systems had been met, the period of use was still very short to achieve reliable conclusions.

SPARSE[12–14] will be used as a case study in Section 3.5.

3.4.3.1   Problems of Knowledge-Based Applications in Control Centers

Although the advantages of knowledge-based applications are very appealing, we must also consider the difficulties they present, especially in the case of control center applications.[43]

Some of the most important difficulties of their use[44] are the following:

- Difficulties in knowledge acquisition — the knowledge acquisition phase is usually very time consuming and poses some problems related to the communication between experts and knowledge engineers. The participation of several experts, which is usually required in applications for power system control centers, is also problematic.
- Difficulties in rule updating — when the rule base requires some changes, this process is considered rather complex and the changes may question the consistency of the rule base.
- Lack of methods to validate and verify knowledge-based systems — as methodologies to validate and verify knowledge-based systems are not yet commonly accepted, informal (and usually incomplete) procedures are used. Therefore, it is difficult to guarantee the correctness and efficiency of the knowledge-based system. Verification methods are even more difficult to

achieve for power system control center applications requiring non-monotonic and temporal reasoning. The need for reliable validation and verification methods is even more acute for a knowledge-based system already deployed after undertaking changes in its knowledge base.

- Problems in knowledge maintenance — even when modification of the rules is not necessary, the remaining knowledge and information encoded in the knowledge-based system has to be updated. Knowledge-based systems usually require more information than other applications that exist in the control center, and their maintenance cannot be fully automatic. This represents extra work, which is sometimes forgotten. Lack of undertaking adequate maintenance is a rather common reason for some unsuccessful experiences with knowledge-based systems in power systems.

It is interesting to note that the two disadvantages presented first are directly related to the most appreciated characteristics of knowledge-based systems, namely, their ability to encode explicit knowledge. In fact, knowledge acquisition has always been one of the serious weak points of knowledge-based systems and especially of expert systems. Although there are some tools available to support knowledge acquisition, it is not usually considered useful to use a generic tool for a specific application.

The establishment of good relations between the experts and the knowledge engineers is extremely important. When the experts are strongly involved, believe that the project is important, and trust knowledge engineers, knowledge acquisition problems are considerably reduced. However, usually, the development of knowledge-based applications for power system control centers requires the involvement of experts in power system operation. As these people are responsible for tasks concerning power system operation, they are usually very busy with their regular activities and, therefore, hardly available.

The second disadvantage is the negation of a commonly accepted advantage of knowledge-based systems. In fact, providing a representation of explicit knowledge which is easy to change is one of the typical advantages of knowledge-based systems. Is it really easy to include modifications in the rule base of a KBS or not? The change itself poses no problem, in most of the cases even for people not familiar with artificial intelligence. However, there are several items that must be taken into account. The following may be considered the most important:

- The knowledge required to undertake the changes poses a knowledge acquisition problem. The use of machine learning techniques for suggesting changes in the rule base can be useful for this purpose.[45]
- It must be guaranteed that the modified rule base remains consistent and that its use will not make the KBS operate in an incorrect or inefficient way.

This last point leads to the third disadvantage: problems of validation and verification (V&V). Although it is difficult to find total agreement about the meaning of these terms in the specialized literature, let us consider the following definitions:[46]

- Validation — ensures that the KBS provides solutions that present a confidence level similar to the ones provided by the experts. Validation is based on tests, desirably undertaken in the real environment and under real circumstances. During these tests, the KBS is considered to be a black box and only the input and the output are considered important.
- Verification — ensures that the KBS has been correctly conceived and implemented and does not contain technical errors. Verification is intended to examine the interior of the KBS to find any possible errors.

For a great number of KBSs, only validation is undertaken. Although this can guarantee, when the system is deployed, that its performance is correct, the existing problems may show when there is a need to change the rule base.

Undertaking verification is more difficult because it relies on formal methods. Although there are some tools available in the market, specific needs of power system applications usually require the development of specific tools.

As formal methods of verification rely on mathematical and logic foundations, they are able to detect a large number of possible problems. This way, it is possible to guarantee that a KBS that has passed through a verification phase is correct and efficient. Moreover, it is possible to ensure that it will provide correct performance on examples that have not been considered at the validation phase.

The next disadvantage of knowledge-based applications refers to the difficulties found at KBS knowledge maintenance concerning the knowledge that is not part of the rule base, namely, the fact base. One can divide the information included in the fact base of a KBS used in a power system control center into two main categories: (1) real-time information, and (2) static information. Real-time information is acquired through the SCADA system and does not pose any maintenance problems. On the other hand, what we refer to as static information includes information about electrical network equipment, methods of operation, and possibly other kinds of information depending on the KBS considered. In fact, this information also has a dynamic nature because the methods of operation evolve over time, which is also the case of the relevant characteristics of the electrical equipment considered.

A large part of this information is also used by other applications and is part of the SCADA database. This information must be necessarily updated, but this is a general need of the control center and not a special need of the KBS. The problem is that KBSs usually require additional information related with their area of application. For instance, a KBS that deals with power restoration will need information about the methods used for power restoration in the electrical network considered. This kind of information must be kept updated in order to keep the KBS operational. As this means work for the control center, the KBS should provide as much support as possible for people who undertake this task.

Some of the disadvantages of KBSs are really difficult to solve, as is the case of problems with knowledge acquisition. However, most of the problems can be completely solved or, at least, become less serious if all the required tools are provided. Figure VI.3.3 shows what can be considered the

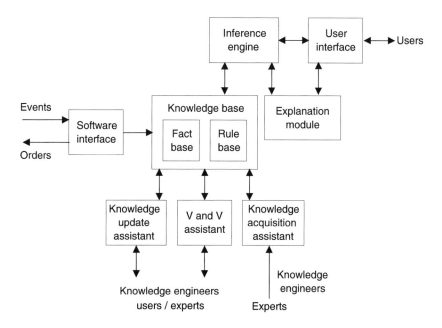

FIGURE VI.3.3   Architecture of a KBS for a control center.

complete architecture of a KBS for a control center. This architecture includes, besides the modules that are always present in KBSs, the following modules:

- Knowledge acquisition assistant
- Knowledge update assistant
- V&V assistant

The inclusion of these modules provides users of the KBS with assistance to solve the most commonly indicated problems. In fact, complaints about KBSs are due to the use of what can be considered an incomplete KBS, one that lacks one or more of these modules.

Most KBSs developed for control centers can be considered expert systems. In fact, expert systems (ES) are knowledge-based systems where the involvement of one or more experts in the field of application of the knowledge acquisition process is especially relevant. In other words, in expert systems, the knowledge originated by these experts is an important part of the knowledge embedded in the system. Expert systems have been developed for addressing several problems in power system operation, alarm processing, fault diagnosis, security assessment, power restoration, and substation monitoring and control being some of the most significant areas of application. Experts in power system operation are the most important source of knowledge considered by these expert systems, but they also consider knowledge of a technical and organizational nature.

Presently, the performance of knowledge-based and expert systems in control centers is considered almost satisfactory, but there still remain some important difficulties in their development and use. The most important of these difficulties are the following:

- Integration of the expert systems in the CC
- Maintenance of the knowledge base to keep it up to date and reliable
- Development and implementation of the knowledge base which is a very time-consuming task
- Testing of the expert systems to ensure a high degree of accuracy
- Handling of maintenance and testing tasks by the utility personnel

The integration of KBSs in control centers can be a complex problem and may endanger the success of the practical use of knowledge-based systems. The last decade has shown a tendency of control center software to be open, so that integration would be easier than in the past. However, this openness is still rather limited, and to integrate an application not provided by the same vendor is still, in practice, either complex or expensive, if not both. Easy integration using standard application program interfaces (API) is highly desirable, but is still a dream at the present time.

This last point is especially important. In fact, it seems very important for utilities to be able to undertake the maintenance and testing tasks of their software, including expert systems. However, these tasks for expert systems cannot be undertaken in the same way as for traditional software applications as the process of ES development requires much deeper involvement of the utility staff than for traditional software.

Utilities desire to profit from the benefits of the use of expert systems which are recognized by them, but, on the other hand, they expect to have off-the-shelf expert systems available from the vendors. From the utilities' point of view, customization should be minimized and knowledge bases must be accurate and robust for untested and unexpected scenarios so that expert systems can be widely used in power system operation.

However, is it possible that a knowledge-based system or expert system can fully play its role without being adequately customized? In fact, the success of knowledge-based systems is directly connected with the knowledge they have about the situations they are dealing with. In the case of power systems, most of the decisions that must be made require knowledge about power systems in general but also knowledge about the specific power system under consideration. This kind of knowledge includes not only knowledge about power system plants, equipment, and topology, but also knowledge about methods and philosophies of operation, rather specific for each utility, and knowledge resulting from experience at that particular power system operation. This kind of knowledge is only obtainable by deep involvement of the power utility in the development of the KBS. Without this involvement, the capabilities of knowledge-based systems are not fully used. Minimum customization would lead to knowledge-based systems that consider only knowledge of very general nature, which can be used for any power system. This would result in a lower contribution to the solution of most problems in power system operation.

3.4.3.2 Considerations about the Success of Knowledge-Based Applications in Control Centers

The success of knowledge-based applications in control centers depends enormously on the consideration of user requirements[47] and on maintenance once the applications are already deployed.[48]

The inclusion of CC operators in a new CC applications design team is a key issue for the success of these applications.[2] When dealing with knowledge-based systems applications, this involvement should not only be considered just to acquire operators' knowledge. On the contrary, their participation should have the primary goal of understanding their needs, the implications of the new applications on their tasks, and, consequently, on the performance of the power system. The cooperation of CC operators during the early phases of application conception and development increases the level of satisfaction and acceptance of the new methods and paradigms used. Moreover, it eases the training of operators on the use of the new tools. These issues are especially important for the use of knowledge-based systems. Otherwise, the new systems easily become targets of operators' suspicion because the human operators tend to see the technology as a kind of substitute for themselves. The development of knowledge-based systems for power system operation tasks has some side effects that are not negligible. One of these effects is the possibility of acquiring the knowledge of the most experienced operators and having a representation of this knowledge. This is of great value, especially when some of the most experienced operators retire and the utilities feel the risk of losing a great part of the knowledge they have gained during several decades of power system operation.

As, nowadays, most power plants and substations are unmanned, there are only few experienced people involved in power system operation before they assume the role of CC operators. In the early days of power industry, with technical staff in every power plant and substation, there were a lot of individuals with significant experience in power system operation. The best could be chosen to become CC operators. Nowadays, this choice is much more limited and young engineers tend to assume the job of CC operators. They are necessarily more distant from the power system equipment than their predecessors who used to have the daily experience of familiarity with this equipment before assuming the operation of a CC. The transfer of knowledge is always a difficult process, and the possession of knowledge bases can be very valuable for utilities.

On the other hand, the development of a KBS is a good opportunity for utility staff to reflect on their own knowledge, policies, and practices. From the point of view of knowledge-based systems' development efficiency, to have only one expert on the development team is usually the best choice. However, from the point of view of the results of an interesting reflection, the participation of several experts is much more profitable. As these experts are usually very busy with the daily operation of the power system, this can be a good opportunity to exchange ideas and discuss different points of view. The result may be a drastic change in utility practices.

Another important contribution of the knowledge acquisition phase of a knowledge-based system is the fact that the utility may realize the importance of this knowledge. This can lead to the adoption of new attitudes and methods, where knowledge management can play an important and primary role.

3.5 SPARSE—A Case Study

This section considers SPARSE,[12–14] a knowledge-based system for alarm processing and operator assistance in service restoration in the Portuguese transmission network. After providing an overview of SPARSE, several of its modules are presented in more detail: knowledge base, inference engine, explanation module, user interface, verification and validation module, and tutor.

3.5.1 SPARSE Overview

SPARSE[12–14] is a knowledge-based system developed for a control center of the Portuguese transmission network, owned and operated by REN (Rede Eléctrica Nacional, S.A.). SPARSE assists control center operators with incident analysis and power restoration. The REN control center operators' function is to perform maneuvers in the electrical grid and to supervise the power system. The role of these operators assumes major importance during critical situations of incident. In these cases, the operators must diagnose the situation and undertake corrective measures, in a short period of time, in order to minimize the incident impacts and bring the system back to its normal state. The operators perform the diagnosis of the situation through the analysis of the information that arrives at the control center, coming from all over the grid, with the list of alarm messages playing a major role in this process.

SPARSE has been developed using Prolog and C and is installed in the control center in a DECstation® (Digital Equipment Corporation). This machine is connected through a local area network ethernet of duplicate configuration with the machines that support SCADA functions in the control center. Off-line versions of SPARSE are available in DECstations® and ALPHAstations® (Digital Equipment Corporation). Figure VI.3.4 presents the software organization of the online version of SPARSE.

In order to send the required real-time information to SPARSE, a software interface—TTLOGW—has been developed and installed in SCADA machines. TTLOGW acquires the information of a time-tagged message list named TTLOG. This application is developed in FORTRAN and includes the following features:

- As SPARSE is intended to perform incident analysis, the only information sent to it under normal conditions concerns state changes (e.g., changes in the state of breakers and isolators and changes in the mode of operation of substations).
- Under incident conditions, all the information arriving at the control center is sent to SPARSE.
- When TTLOGW is initialized or the machines booted, TTLOGW sends to SPARSE a state protocol. This protocol includes all the states relevant to the network considered, and the corresponding messages are time-tagged with the system present time.

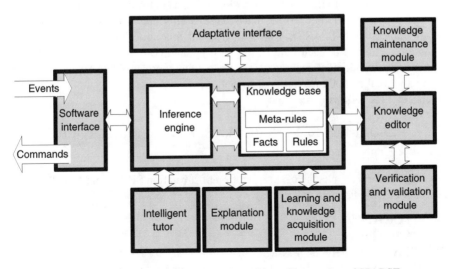

FIGURE VI.3.4   Software organization of the online version of SPARSE.

- In case of communication or remote terminal unit (RTU) faults, a message, "RTU off-line" is generated and time-tagged with the system present time. When the fault is cleared, a message, "RTU online" is generated and the state protocol of the corresponding substation is sent to SPARSE time-tagged with the system present time.
- When there are several RTU faults within a small period of time, the state protocol is not generated.
- When there are repetitive messages, TTLOGW detects this situation and stops sending messages to SPARSE and instead sends a message reporting this fact. When the repetition disappears, TTLOGW sends a new time-tagged message with the present system time.

Concerning the use of the information of TTLOG, one must remember that although TTLOG messages are time-tagged in remote terminal units, these messages arrive at the control center with a delay with respect to the current time and not necessarily in chronological order.

The software application that allows interfacing the knowledge-based application with the SCADA system is a major issue in developing knowledge-based applications for control centers. In fact, the development of TTLOGW has been quite time consuming and requires a high level of expertise about the SCADA system. It is important that this kind of application is developed in parallel with the knowledge-based application and not only when the knowledge-based application is almost completely developed. In fact, this interface with the SCADA system may impose several severe constraints that must be considered during the development of the knowledge-based application.

3.5.2   Knowledge Base

The development of SPARSE required several kinds of knowledge:

- Knowledge about power system elements and topology (e.g., power plants, substations, and lines)
- Knowledge about the alarms that are generated at each moment (alarm lists)
- Knowledge about the information transmission system (e.g., how data are transmitted and what is transmitted)
- Knowledge concerning alarm interpretation

SPARSE's knowledge base includes all these kinds of knowledge. Whenever possible, knowledge is automatically converted from information that already exists in the control center. The messages are treated by a preprocessing module written in language C. This module converts messages into Prolog facts that are transmitted to SPARSE, which processes them online. This preprocessor standardizes the information contained in each field of the message and converts the time and the date of the message into a numeric element, making the treatment of temporal problem easier. This preprocessor is able to treat more than 500 messages per second.

Let us consider the following message:

23-OCT-1999 16:24:32:041 SRA SGER GENERAL SERVICES

OPERATION MANUAL

This message reports the change of the operation of substation SRA to manual. The preprocessor identifies the following fields in this message:

- Date (23-OCT-1999)
- Time (16:24:32:041)
- Plant (SRA) from where the message came
- Code (SGER) of the panel that originated the message

- Name (GENERAL SERVICES) of the panel that originated the message
- Event (change in the type of OPERATION)
- State (MANUAL)

This message is converted into the following Prolog fact:

fact(696, message(23-OCT-1999, 16:24:32:041, ['SRA', 'SGER',

['GENERAL SERVICES']. 'OPERATION', 'MANUAL']), t(25547072, 040))

This Prolog fact is composed by the number of the fact (696), the information about the alarm message, and the numeric element t(25547072, 040) that is the number of seconds and milliseconds since a reference date, the 1st of January, 1999 in this case.

The knowledge included in the rule base is of both a technical and an empirical nature. Technical knowledge models the behavior of the power system and data transmission system. Empirical knowledge has been gathered from the most experienced operators working at Portuguese control centers and allows a faster processing of alarm messages.

A major issue about the knowledge included in the rule base concerns temporal aspects. In fact, temporal aspects are the key for alarm interpretation concerning events occurred and also the transmission of information to control centers. According to this, SPARSE rules include the required features in terms of temporal question consideration.[49]

The left-hand side (LHS) of the rules may be a disjunction of conjunctions of several conditions. The right-hand side (RHS) contains the actions/conclusions to be taken if the LHS is verified. Let us consider the following rule:

rule xx : 'EXAMPLE' :

[

[C1 and C2 and C3]

or

[C4 and C5]

]

==>

[A1,A2].

Conditions considered in the LHS (C1 to C5 in this example) may be one of the following types:

- A fact whose truth must be proved
- A temporal condition

The actions/conclusions to be taken (A1 to A2 in this example) may be of one of the following types:

- Generation of facts (conclusions to be asserted to the knowledge base)
- Elimination of facts (conclusions to be deleted from the knowledge base)
- Interaction with the user interface

Let us now consider a simplified version of a rule concerning a three-phase fault with reclosure:

rule $d8$: 'THREE PHASE TRIPPING WITH RECLOSURE'

[[message(Date1,Time1,[Plant1,Panel1,[Plant2,NL]],' >>>>

TRIP ORDER', 'BEGIN') at T1

and

message(Date2,Time2,[Plant1,Panel1,[Plant2,NL,'BREAKER']],
    'BREAKER','OPEN') at T2

and

panel_in_service(Plant1,Panel1)

and

condition(mod_dif_time_less_or_equal (T2,T1,150))

and

message(Date3,Time3,[Plant1,Panel1,[Plant2,NL,'BREAKER']],
    'BREAKER', 'CLOSED') at T3

and

condition(dif_time_less_or_equal(T3, T2, 500))]] ==>
[remove_fact(breaker_closed(_,_,_ , Plant1,Panel1,Plant2,NL),_ , T2),
generate_fact(breaker_open(Date2,Time2,trip,Plant1,Panel1,Plant2,NL),T2),
remove_fact(breaker_open(_,_,_ , Plant1,Panel1,Plant2,NL),_ , T3),
generate_fact(breaker_closed(Date3,Time3,tp_reclosure,Plant1,Panel1,Plant2,NL),T3)].

The messages included in the LHS of this rule report that a specific breaker opened at T1 and closed at T2. The opening of the breaker can be associated with a tripping order reported by the first message considered in the LHS. However, this association is only considered valid when the time tags of these two events do not differ by more than 150 milliseconds. This is expressed by the first temporal condition of the LHS. The second temporal condition in the LHS states that if the breaker is closed in a 500-millisecond period after its opening, it is considered that this closing is due to a fast reclosure. This kind of reclosure is very usual in power systems and allows increasing the reliability of the network by solving most of the faults without the need of human intervention.

3.5.3 Inference Engine

SPARSE's inference engine has been developed in Prolog and C especially for this application. It allows an efficient processing of alarm messages in real time, dealing with more than 1200 alarm messages per minute.

The inference engine was developed taking into account that:

- The amount of information involved is very large
- The processing must be done in real time, according to the rate at which information arrives
- The analysis involves complex temporal reasoning
- The power system is a dynamically changing environment requiring non-monotonic reasoning

SPARSE's inference engine is conceived as an event driven application: it waits for the arrival of new information in order to get active. The fluxogram in Figure VI.3.5 shows a simple description of this engine.

SCADA messages and SPARSE's conclusions are represented as facts. These facts are used to infer new conclusions. The arrival of a new fact causes the scheduling of a set of rules to be triggered.

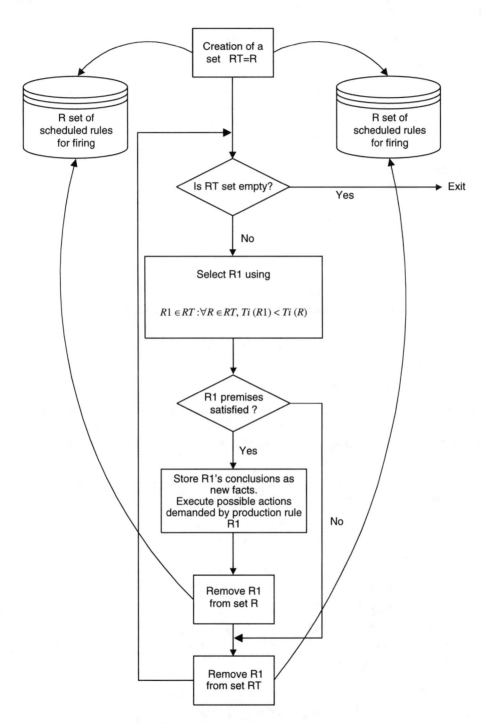

FIGURE VI.3.5   Inference engine fluxogram.

This set of rules is established according to meta-knowledge embedded in SPARSE's rule base. The following fact:

$$\text{trigger (NF,NR,T1,T2)}$$

means that rule NR should be triggered from instant T1 until instant T2 due to the arrival of fact NF. The meta-knowledge is used in order to make the inference process more efficient. It allows dealing with some temporal reasoning requirements, providing means to introduce a delay on rule triggering. The introduction of such a delay is done in an intelligent way, using empirical and technical knowledge concerning temporal interpretation of real-time information. This feature allows consideration of delays in information transmission and also the ability to distinguish situations based on time considerations.

The meta-knowledge involves facts such as the following:

$$\text{fact\_trigger\_rules (F, [(R1, TiR1, TfR1), (R2, TiR2, TFR2), } \dots \text{])}$$

where

F   = type of fact that causes the triggering of rules R1,R2, ...
R1   = first rule to be scheduled for triggering
TiR1 = time (relative to the present time) from which the triggering or rule R1
          should be tried
TfR1 = time (relative to the present time) until which the triggering of rule R1
          should be tried
R2   = second rule to be scheduled for triggering

This information is stored in a structure that supports the scheduling of a variable number of rules.

Let us consider R as the set of all the rules to be triggered at a particular instant in time (all the existing trigger (NF, NR, T1, T2) facts). Considering R, SPARSE's inference engine schedules rule triggering in the following way:

- It creates a set RT = R containing all the rules present in the referred structure
- Until the set RT is empty:

  - It determines which rule belonging to RT has the minimum time for the beginning of triggering, considering only the rules for which this time is equal to or greater than the present time
  - Let R1 be the previously selected rule
  - It tries to trigger rule R1
  - If the triggering of R1 is successful:

    o It stores the conclusions generated by R1 as new facts
    o It executes actions considered in the RHS of R1
    o R1 is deleted from set R

  - R1 is deleted from set RT

This algorithm is shown in Figure VI.3.6 in fluxogram form.

After this, the inference engine verifies if *R* contains rules that have already been triggered at least once and which final time for triggering is greater than the present time. If this is the case, the rules under these conditions are deleted from the structure. Most conclusions are kept only for internal use. Only a small set of conclusions is presented to control center operators.

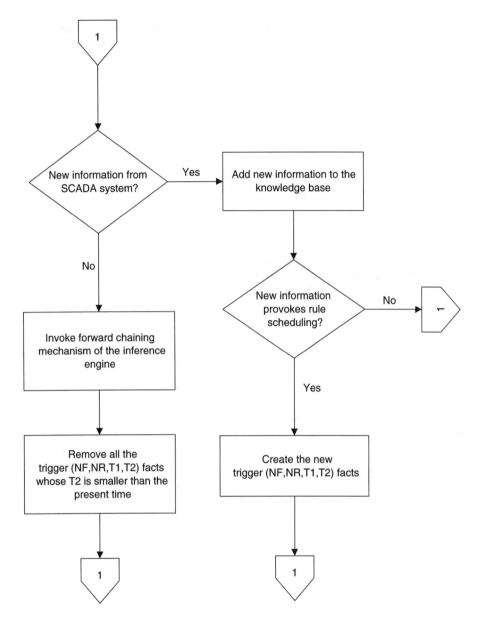

FIGURE VI.3.6   Rule scheduling fluxogram.

It is important to note that this methodology for temporal reasoning allows introducing delays in the triggering of rules without stopping the inference process. Trigger delays are inherent to each rule, and the inference engine continues its task of triggering the other rules scheduled for triggering. Temporal reasoning plays a very important role in the interpretation of events that occur in power systems. The KBS has to be able to understand the temporal evolution of the events. Also, it must be able to reproduce the events that occur based on the arrived SCADA messages, even when these messages do not arrive in the correct chronological order.

SPARSE uses a time-tagged list where each message contains the corresponding date and time. This information is appended to the message at its place of origin (substation or control center). In order to be able to interpret the sequence of events that occurred in the power system, SPARSE must be able to establish relationships between SCADA messages and tolerate some inaccuracy in time relationships among messages and in the order of the arrived messages. The approach to address these issues must take into account that the processing must be done in real-time. This means that the involved reasoning must not be too complex or too heavy in terms of the required computational time.

SPARSE uses technical and empirical knowledge to establish time relationships between SCADA messages in order to characterize incidents. Knowledge about the system used for acquiring and transmitting information is also incorporated in order to consider its influence on the information that arrives to the control center.

SPARSE's inference engine uses two different approaches to treat temporal issues:

- Introduction of delays in the triggering of the rules
- Introduction of time considerations in the LHS of the rules:

  - Association of an instant to a fact
  - Consideration of temporal conditions

In order to analyze temporal relationships, the inference engine includes a set of temporal predicates.

In this intelligent alarm processing system, the knowledge is inherently non-monotonic since the electrical network state is dynamic in time, changing both by commanded operation and actuation of protection devices.

In order for the system to work properly, it must keep all past and present network states. The inference process and recent explanations will use present and recent past states. Older past states will be used for tutoring and explanation purposes. To cope with these specifications, the inference engine has to be able to assert and retract new facts from the system, keeping its coherence. This is done by supplying the knowledge base with methods for both asserting and retracting information. Examples of this have already been seen, such as the predicates generate_fact/2 and remove_fact/3 included in the example rule of Section 3.5.2.

Generation of a new fact is quite straightforward, and the new fact is simply added to the knowledge base. Fact removal is different. As explanation features are very important in any KBS, facts that are already considered false cannot simply be forgotten. It is therefore required that a fact that has become false be kept as an "old fact" so that it can be used for explanation purposes.

The following rule is a simple example of a non-monotonic rule:

> rule g1 : 'MANUAL OPERATION':
>
> [[message(Date,Time,[Inst,_,_], 'OPERATION','MANUAL') at T1
>
> and
>
> automatic(_,_ , Inst)] ==>
>
> [remove_fact(automatic(_,_ , Inst),_ ,T1),
>
> generate_fact(manual(Date,Time,Inst),T1)].

It states that if:

- Premise 1: a message arrives to the system with the information that a particular substation (Inst) has changed to manual operation at time T1, and
- Premise 2: it is known by the system that that same substation is operating in automatic mode, then:
- Conclusion 1: the fact that stated that the substation was operating in automatic mode is unbelieved (it becomes an "old fact") at time T1, and
- Conclusion 2: the fact that the substation in operation in manual mode is asserted at time T1.

Let us consider that, internally, we had a fact such as:

fact(63,automatic(01-AUG-1998,00:35:26.276, 'SPR'), t(207621326, 276)) and a message such as the following arrived to the system:

$$\text{fact}(80,\text{message}(01\text{-AUG-}1998,02\text{:}19\text{:}13.297,[\text{`SPR'},\text{`SGER'},$$

$$[\text{`SGER'},0]],\text{`OPERATION'},\text{`MANUAL'}),\ t(207627553,297))$$

If rule g1 is triggered with facts n. 63 and 80, fact n. 63 will be removed from memory and the following fact will be asserted:

$$\text{old\_fact}(63,\text{automatic}(01\text{-AUG-}1998,00\text{:}35\text{:}26.276,\text{`SPR'}),$$

$$t(207621326,276),\ t(207627553,297))$$

which means that a fact stating that substation SPR is in automatic operation mode became true at instant 207621326, 276 seconds and became false at instant 207627553, 297 seconds.

In the system, there are several rules to handle changes in the state of devices or substations, such as the following:

- Breaker state change (with four possible values: open, closed, moving, or in an error condition)
- Substation under either manual or automatic operation
- Substation under either local or remote operation

The amount of information the system is expected to handle is considerable. Since it is intended for continuous use, its behavior would degrade unless some form of management mechanism was implemented. In order to deal with these knowledge management problems, a special kind of virtual memory has been implemented. It is known that some of the information that the system keeps is for historic, explanation, or tutoring purposes only. An example of this kind of information is an old network state alteration (e.g., referring to the state of a breaker) that no longer corresponds to the present network state.

With this knowledge, an intelligent slot-and-filler mechanism has been implemented in such a way that it minimizes mass data storage accesses, not allowing access to an incident situation. This means that, in an incident situation, explanations about past events will not be available, even though explanations about recent events can be supplied.

In order to fulfill its requirements, this mechanism has to:

- Keep in memory all the believed facts
- Keep in memory a part of the unbelieved facts, large enough to optimize the disk accesses
- Free the system's memory by keeping all the other unbelieved facts on disk

SPARSE has three distinct knowledge bases:

1. The fact knowledge base, which is kept in memory and has a limited memory space to avoid overflow of the system's memory. It keeps the present facts known by the system.
2. The old fact knowledge base, which is also kept in memory and also has a limited memory space to avoid the overflow of the system's memory. It contains the recent past facts known by the system.
3. The old fact knowledge base on disk, which is also kept on disk and also has a virtually unlimited memory space. It contains the remaining past facts known by the system.

Whenever a new fact is created, an old fact is generated as its subproduct. In this case the number of old facts is checked. If this number is greater than or equal to a fixed number of old facts that should remain in memory plus a block of old facts of disk-dependent length, they are removed from memory and stored on disk. The number of unbelieved facts kept in memory is empirically assigned. Forecasting methods to ensure optimization are being studied.

Let us considerer an example to illustrate this behavior, considering the fact:

fact(73,message(01-AUG-1998,02:18:17.142,['SPR',634,['ALFAR.',2,'BREAKER']],

'BREAKER','00'), t(207627497, 142))

which means that on a particular date and time, the breaker of panel 634 of substation SPR, linked to substation ALFAR. by line number 2, was in movement. The third argument is the fact's instant in time. It is a continuous time structure composed of the number of seconds and milliseconds that passed since a reference date, January 1, 1992 in this example.

When this fact is removed, it is replaced by:

old_fact(73,message(01-AUG-1998,02:18:17.142,['SPR',634,['ALFAR.',2,

'BREAKER']], 'BREAKER','00'), t(207627497, 142), t(207627505, 142))

When the number of old facts is greater than a certain value, this old fact is removed from memory and stored on disk.

This particular mechanism is transparent to the rest of the system. It has been designed so that its inclusion in SPARSE brought no need for change. Known facts and inferred conclusions are checked by SPARSE, leaving to this mechanism the task of checking whether they exist and where they are stored, retrieving them by demand. In the case of SPARSE, this mechanism also provides an incident classification in order to ease its later use for postmortem analysis and by a tutor module.

3.5.4 SPARSE Explanations

Explanation ability[50] is one of the most interesting characteristics of knowledge-based systems (KBSs). This ability is very important for its acceptance by the end users because it is a good way to improve confidence in the system. Although providing explanations can be considered a traditional and well-known characteristic of knowledge-based systems, providing explanation features for a real-time application can be very difficult.

Explanations have been used in SPARSE's development since the very beginning. In fact, explanations have been very useful in the knowledge acquisition phase. In an early phase of the project, explanations were internally generated by SPARSE and could be seen in a window included in its user interface. This architecture was due to coexistence in the same process, in that stage of the project, of the knowledge-based system and its graphical interface.

With this kind of architecture, real-time performance of the expert system could be put at risk by the use of the explanation module. In fact, users should not overload the KBS by asking for explanations under incident situations. Changes in SPARSE software architecture led to the separation of the KBS and the user interface into individual processes communicating with each other. Presently, this philosophy is used for most of the components of the system and has proven quite efficient, even under incident conditions.

In a multiagent environment, which is presently SPARSE's case, where more than one process needs the availability of one functionality, a client–server philosophy can be used. In this case, SPARSE, as the server, will be able to supply explanations to any process (which becomes SPARSE's client) that questions it. From this point forward, the SPARSE designation can be used to identify the whole system or just the KBS, depending on the context. The acronym ESES, which means expert system explanation server, will also be used. A certain process becomes an ESES's client by sending it an initial message that contains the following data:

- Client's identification
- Client's means of contact (message queue, socket, file, etc.)
- Client's means of contact identification (message queue key, socket port, file name, etc.)

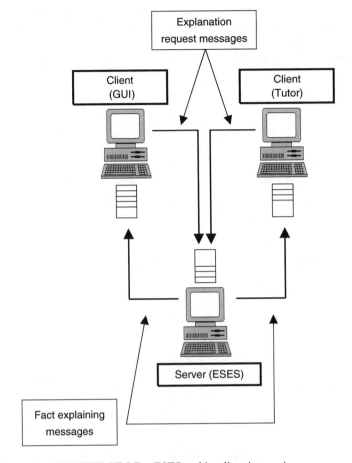

FIGURE VI.3.7   ESES and its client interaction.

This initial message registers the client in the ESES's knowledge base, allowing it to use the explanation production mechanisms. The same process questions the ESES, sending it a message that contains the following data:

- Client's identification
- Explanation chaining method
- Explanation depth
- Fact to explain

The interaction between the ESES and its clients is presented schematically in Figure VI.3.7. These clients can be running either in the same machine or in several machines, connected through a local area network (LAN) or a wide area network (WAN).

This methodology presents the following characteristics:

- It allows more than one process to question ESES at the same time, handling and scheduling these requests
- The explanation production mechanism is independent from the client
- ESES operates with high granularity in explanation production, controlling the time spent in explanation production and delivery
- The explanation server schedules its activity during the periods when the expert system is free

- It makes the explanations available when they are needed by each client
- The requests can be interactive (one level of explanation depth at a time) or full (explanation until the SCADA message level is reached)
- The operation is very flexible, including the form of explanation request and presentation defined by the client

The explanation server has been integrated in SPARSE with minimum modification to the already existing software. The explanation server is presently embedded in the KBS. Other modules, such as the user interface and the intelligent tutor, which exist as independent processes, can use this new functionality. Figure VI.3.8 presents the interface of the explanation module.

SPARSE provides "how" explanations, which provide a description of the way something has been used to infer a certain result. This kind of explanation has several uses:

- To the system builders, in a non-automatic verification and validation perspective
- To help experienced users in knowledge maintenance
- As a tutor for inexperienced users

In the case of SPARSE, the "how" explanation starts in any conclusion drawn by SPARSE, based on the rule which succeeded in obtaining that conclusion and on the facts used as its premises. These premises can, eventually, be explained by rules that fired to create them and their own facts used as premises. This chain will always end in SCADA messages, which are the lowest level of abstraction that the system can reach.

SPARSE also provides inference explanations that show, in natural language, the way that the forward-chaining KBS (SPARSE) works. They start in a SCADA message or in an intermediate conclusion drawn by the system, showing all the information that was inferred. This particular type of explanation is intended to explain the system reasoning in a language that the user can understand. Its obvious use is the validation of the system, supplying the information needed to verify not only whether the system draws the right conclusions, but also that it draws all possible conclusions with the available information. It is quite intuitive that a larger quantity of information is necessary to produce explanations about a certain conclusion than the amount of information needed to infer that conclusion.

In order to provide explanations it is necessary to have:

- The information used to infer
- Additional information about the information used for the inference (e.g., natural language translation)
- Information about the inference process itself (how the information owned by the system is used to infer new information)

It is very important to limit the quantity of the information required to produce any type of explanation in order to make knowledge maintenance effort relatively low. For this purpose, it is desirable to organize the required information in an effective way. In the case of SPARSE, there are internal facts, obtained through SCADA messages, which correspond to the status of devices. These facts can comprehend several devices, each with several possible states. On a first approach, these devices could be considered in an independent form for the translation of the facts to natural language. However, this would be a very redundant form of modeling since grouping their common properties can significantly reduce the knowledge base dimension. This way, it is only necessary to store the essential information: the types of devices that are considered by the system, the possible status for each type, and how a generic fact of this kind can be explained.

Regarding the information concerning the inference process, knowledge about the kind of production rules used must be encoded, including the explanation model for a generic rule. It is also necessary to encode knowledge about the inference engine's characteristics considering its type of

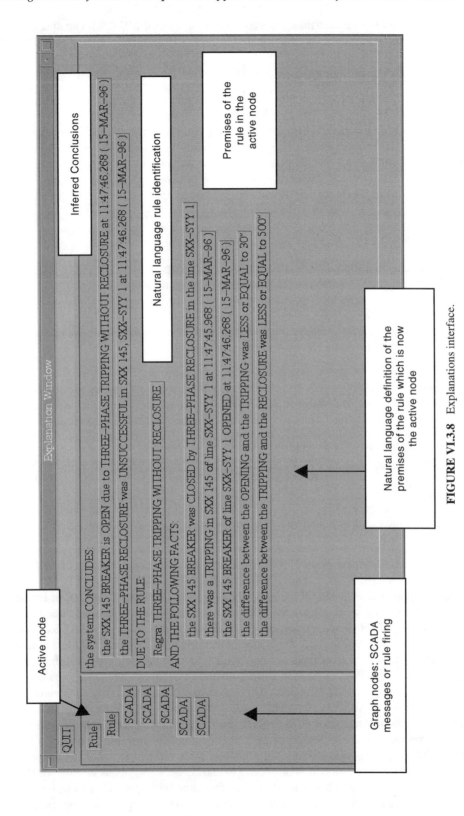

FIGURE VI.3.8 Explanations interface.

inference mechanism (forward-chaining mechanism in the case of SPARSE) and specific features such as temporal and non-monotonic reasoning.

3.5.5 Machine Learning Module

The knowledge acquisition process in SPARSE has been undertaken through a large set of meetings with utility experts. Tests made to the system show that although the general structure of the rules proved to be correct, time parameters used by the rules should be tuned by considering real incidents. In fact, experts tend to assign values to these parameters based on their knowledge about power system operation, control equipment, and automata. Usually, experts only admit changes to these values when faced with real examples. These changes are due to several factors such as data acquisition and communication and particularities of some plants and equipment.

Real-time performance SPARSE experience has shown that most modifications on the rule base are due to the need for tuning the times that are used for temporal reasoning. This tuning requires the analysis of a large set of cases and involves several kinds of times, such as:

- Minimum time that should be verified between two events
- Maximum time that should be verified between two events
- Maximum waiting time allowed for possible related events concerning the two ends of the same line
- Maximum waiting time allowed for possible related events from the same plant
- Maximum time allowed for SCADA information arrival

As this analysis benefits from the consideration of a very large set of examples, machine learning techniques seem to be adequate. For this reason, a machine learning algorithm has been developed and implemented to support the tuning of time in SPARSE's rule base. This machine learning module is called TEMPUS.[45] The goals of machine learning algorithm TEMPUS are as follows:

- Analysis of temporal premises performance of SPARSE's rule base
- For each SPARSE rule, analysis for each plant and within each plant for each panel
- Particular analysis of some rules and plants or a global analysis including all rules and all existing plants
- Elaboration of a report with the obtained results regarding the rules and plants chosen by the user for analysis, and eventual use of the rule editor, in case the expert decides to change SPARSE's rule base

Figure VI.3.9 presents the architecture of SPARSE's machine learning module (TEMPUS) and its integration with SPARSE. SPARSE uses recorded message lists including monthly files containing all the messages generated during the corresponding month. The most interesting incidents are classified according to the work of an incident analysis group. These incidents are classified by type, allowing the use of SPARSE to consider only specific types of incidents. SPARSE automatically analyzes messages contained in those files. This processing work does not require any interaction with the user because, when an analysis covers a large period, it may require a long time. Recorded message lists are fed to SPARSE through a time simulator that allows messages to be processed as if they arrived in real time. This way, there is no need to change SPARSE's architecture. Simply, an off-line version of SPARSE is used. This allows efficient development and maintenance of the system with minimum interference on the online version, which runs at the control center.

After the processing made by SPARSE, the TEMPUS machine learning algorithm is then used. Initially, the TEMPUS algorithm allows the users to choose to analyze either the entire rule base or only a part of it (only one rule, if wished) and to choose either all plants or only some of them. This last possibility is very adequate when treating substations with specific characteristics (e.g., old substations whose equipment has a particular behavior). TEMPUS analyzes the success of the rules

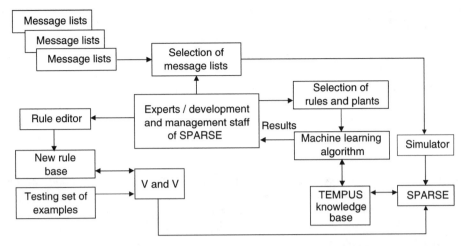

FIGURE VI.3.9   TEMPUS's integration with SPARSE.

not only in terms of "Yes" or "No," but also regarding the "distance to the success." In fact, forward-chaining rule-based systems, like SPARSE, when considering input data, try to find rules that can be successfully fired in order to withdraw conclusions. From this point of view, each considered rule is either successful or unsuccessful for a given data point. However, from the point of view of knowledge enhancement, it is important to have some more information:

- For a successful rule, how far it is from not being successful
- For an unsuccessful rule, how far it is from being successful

This distance is considered either positive, when the rule has been successful, or negative, when the rule has been unsuccessful. This philosophy has the merit of being very close to the process humans usually apply in rule analysis. Also, it can make use of high level knowledge to define the form of evaluating the distance to the success and how to make suggestions for rule modifications based on the philosophy.

TEMPUS makes this evaluation of the distance to the success (Dsuc) for each analyzed rule and comes out with conclusions such as the following:

- For a rule (Rx1), the Dsuc of a specific parameter (Dsucy) is usually positive with a relative value greater than the recommended. In this case, TEMPUS suggests adjusting the value of the corresponding time parameter.
- For a rule (Rx2), the Dsuc of a specific parameter (Dsucz) is usually negative with a relative value that respects time intervals, which suggests adjusting the corresponding time parameter.

TEMPUS has proven very useful in the identification of temporal parameters that can be changed, making knowledge maintenance easier and faster.

3.5.6   User Interface

User interfaces are a key issue in real-time applications, especially on critical applications such as a knowledge-based system assisting control center operators in incident and emergency situations.[19,20] It is important to keep in mind who the end users of the knowledge-based application are and to develop the user interface according to their specific needs. An extensive use of dragging and resizing possibilities is not always well accepted by operators. In emergency situations, operators should have to perform the fewest number of tasks possible, so the most relevant and urgent information should be automatically presented without requiring much effort. A multi-window system where window manipulation is controlled by the application might be the best solution in emergency situations. These are the situations in which it is important to be sure that operators see the most urgent information.

On the other hand, under normal circumstances, the user interface should be more flexible and offer operators a wide range of options. This raises the question of the need for intelligent user interfaces that are able to adapt themselves to each situation. For example, the information presented over the image of the network during an incident should provide the operators with a dynamic view of the evolution of the situation. The success of a user interface for control center applications can be easily increased considering a set of rules that, unfortunately, are almost always forgotten. Some of these rules include:

- Operators should always take part in the process of user interface design and development, including the specification phase. This helps the team understand and attend to operator needs and eases the acceptance of the user interface (and of the whole system) by operators. Operators should also participate in tests performed during user interface development, namely, in the real environment.
- The team involved in designing and developing the user interface should include experts in power system control and operation, computer technology, and human behavior. When human behavioral experts are not present, there is a risk that computer experts will impose the use of the most sophisticated technology even when it is not desirable.
- The operation of the user interface should not require the complete attention of the operator. Even when operating the user interface, operators should be able to continue considering the situation of the electrical network. If operators find the user interface difficult to handle, they will be completely occupied with the user interface and may delay decision making. This is a serious drawback, especially in emergency situations.
- Some important aspects should be kept unchanged. This is the case for colors traditionally used for indicating the state of equipment (e.g., breakers and isolators) and symbols to represent different types of equipment and major commands. Bergstrom[51] compares control center operation to a car, where the way some important controls are performed is standard (e.g., the right foot is still used to brake, in cars with automatic gears as well as in electrical cars).

The main problem with user interfaces is that operators know what they expect from the user interface but the user interface usually has no information about these expectations. This is, in fact, a problem of lack of knowledge. As these expectations are highly dependent on the situation of the network, the operator himself, and the utility methods of operation, artificial intelligence techniques are suitable to solve this problem.

Traditional user interfaces are developed as software applications whose functions are embedded in the code, making most modifications possible only at the programming level. On the other hand, intelligent user interfaces may be provided with knowledge bases that include knowledge about the users, knowledge about the application, and knowledge about the user interface itself. Using this knowledge, intelligent user interfaces are able to adapt their behavior to the specific user and to the situation. In the case of control center applications, user interfaces should adapt themselves to the type of operator (expert, experienced, novice) and to the situation (normal, incident).

This approach of developing user interfaces presents an important set of advantages that include the possibility of changing the knowledge included in the knowledge bases in order to adapt it to particular methods of operation and to new user models. Although the acquisition of knowledge for these knowledge bases represents a difficult and time-consuming task, some of this knowledge is likely to be common to other applications. The knowledge concerning the adaptation of the user interface to the situation requires knowledge about the identification of the situation that should be provided by one or more control center applications. The knowledge concerning the adaptation

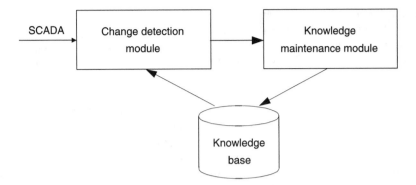

FIGURE VI.3.10   Architecture of the knowledge maintenance module.

to the user requires models of the users that can also be used for training purposes. In this case, the knowledge about user models can be used online by the user interface but also off-line by an intelligent tutor that can be used by operators when the electrical network is in a normal situation, allowing them to follow a training session. An adaptive user interface can be common to both online and off-line applications (e.g., the intelligent training system) that present different requirements. This architecture includes the use of a set of knowledge bases that can be used by several control center applications and also by the user interface. The existence of this set of knowledge bases represents an important step forward for an utility. In fact, these knowledge bases elicit and compile the knowledge that is usually spread among the utility staff and that tends to be at least partially lost with personnel retirement and change of functions.

In the first stage, the development of a SPARSE user interface was based on X-Windows and used Prolog predicates to access X-Windows functions. Although this solution could provide a good user interface, it proved to not be efficient in terms of future development. In fact, this interface was too dependent on the machine and upon the Prolog compiler used. Portability is an important aspect for any system. So, in order to make the migration process of SPARSE to new machines easier, a new user interface, based on X-Windows/Motif and written in C, has been developed. The user interface and the KBS are two processes that communicate through two message queues.

This approach is almost independent of the machine because the user interface uses functions that are common to all X-Windows/Motif environments, and the KBS uses Prolog predicates that are also common to a wide range of Prolog compilers.

The development of this new user interface has required the construction of new knowledge bases. In order to make the maintenance of this knowledge as easy as possible, the system automatically detects changes occurring on the structure of the electrical network. All changes that do not require the intervention of the operator are performed automatically. For the remaining changes, an icon is displayed, alerting operators that there are changes to be performed. The architecture of the knowledge maintenance module is presented in Figure VI.3.10.

The change detection module, developed in C, receives the information coming from SCADA to the KBS. The information received by this module is composed of state information from all transmission network devices. This module has the function of finding changes that have been made on the transmission network. Below, examples of changes that can occur are presented:

- Power plant insertion or removal
- Transmission line insertion or removal
- Panel plant insertion, removal, or change

From the detected changes, the knowledge maintenance module has the function to modify the corresponding knowledge base. However, the changes are accomplished only after asking the control center operator for confirmation. When a change is detected, this module warns the control center operator through an icon displayed on the screen. The operator is not obliged to confirm the change when it is detected; the change can be made later by selecting the icon. Therefore, under incident and emergency situations, when the operator has other urgent tasks to accomplish, he is not overloaded with maintenance tasks of less importance.

The knowledge maintenance module presents a friendly user interface, built in the same environment as the SPARSE user interface, allowing operators to easily confirm the changes detected in the transmission network. This system allows keeping the knowledge bases of the KBS always updated with minimum effort. The system also includes a drawing tool, built in the same environment, that allows operators to build, expand, or change graphic images without any programming effort. This ability may be used when it is necessary to include new plants, change the existing ones, or treat and supervise another network. Figure VI.3.11 shows an application screen used to insert a new plant in the graphical representation of the electrical network

The SPARSE user interface provides an image of the entire electrical network. This image includes information about topology, alarms, and data related to the electrical network, and it is delivered by the transmission system of the control center. The principal aim of this screen is to provide dispatchers with general information about the whole network, identifying the areas that are causing the generation of alarms. Under request, operators can easily access more detailed information about a specific area of the network or a plant diagram without losing the image and the information related to the entire network. Representation of the entire network is provided using the symbols widely known by the utility's staff.

When an incident occurs, the SPARSE interface automatically displays, on a separate window, the affected area. In the case of incidents involving several components, these components are grouped in as many areas as required. The image that is automatically displayed concerns the most recent incident. Operators may select any of the incident zones through the use of the menu, "incident zone." The knowledge concerning the user interface adaptation can be introduced or changed by qualified experts using the "adaptation features" menu. This menu is accessible only to this kind of user. The "user class" menu is used to classify operators in classes according to the introduction of a password. This user class is used to adapt the user interface to the operator.

3.5.7   Verification and Validation

The validation of SPARSE has been undertaken using examples as close as possible to the ones that the application should face in a real environment. According to this, validation should be based mainly on real information about the network. Using TTLOGW, already discussed in Section 3.5.1, files concerning real incidents have been obtained and have been used in order to validate SPARSE conclusions when TTLOGW was still under development and in off-line operation. Experts involved in the project commented on these conclusions, and corrections in the knowledge base were made whenever necessary.

When SPARSE was installed in the control center, receiving real-time information sent by TTLOGW, new techniques of validation had to be applied. The validation of SPARSE considering real-time information was very important for several reasons:

- Temporal reasoning should be tested under real situations in order to ensure its correction
- Consideration of multiple faults is an important aspect of SPARSE performance that is very dependent on the way information flows
- Processing times should be tested in order to guarantee real-time performance, even under incident conditions

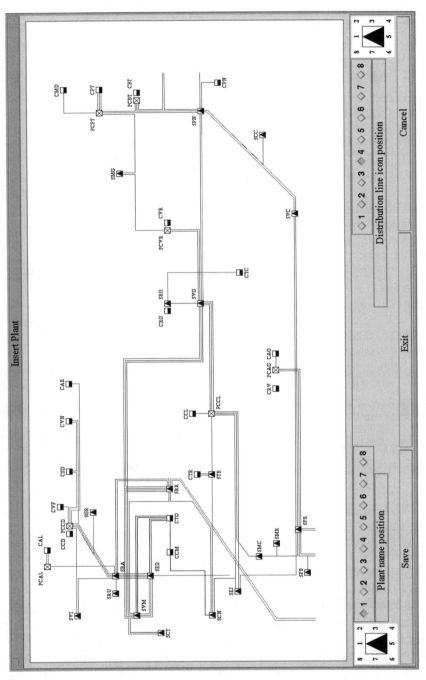

FIGURE VI.3.11 Insertion of a new plant.

Presently, electrical networks are very reliable, so it is not possible to completely validate SPARSE with real incidents. A large number of different types of incidents should be simulated to allow validation. As this simulation should be as accurate as possible, two different techniques have been used:

1. Simulation of incidents by operators located in chosen substations
2. Simulation of incidents using a programmable impulse generator and a remote terminal unit (RTU)

These two techniques complement each other, allowing a complete validation. The simulation of incidents by operators enabled obtaining real-time information presenting exactly the same characteristics as the information generated during a real incident. During these tests, incidents were simulated by operators, allowing the protection equipment to act as if a real incident was taking place. That way, the information used by SPARSE was generated as it would be during a real incident.

Due to the difficulties in coordinating operators at several substations, the simulation was not always correct and the whole process had to be repeated several times in order to obtain a good test. In spite of all the difficulties and costs involved, this kind of test has been considered absolutely necessary for SPARSE's validation, increasing the confidence in its real-time behavior.

Due to the extremely high costs required by this technique, and in order to undertake a complete set of tests, a different technique of testing had to be used. This technique involved the use of a remote terminal unit (RTU) and of a programmable impulse generator (PIG). The alarm messages generated by the SCADA have been simulated by the generated impulses. This test technique may be used on a chosen substation—the most convenient one, namely, a substation close to the control center. This way, the operator who is manipulating the PIG is not far away from the team involved in the SPARSE result analysis and can participate in coordination meetings between two consecutive tests. This technique allowed simulation of a wide set of incidents in order to validate SPARSE's knowledge base as completely as possible.

In order to reduce the validation test time, a time simulator has been developed. This simulator allows using incident files saved by SPARSE and replaying those situations off-line. This way, after carrying on a set of tests, the generated files may be used by the simulator to provide SPARSE with the corresponding information.

These validation methods have been considered sufficient to put SPARSE in service, without the need to undertake formal verification of SPARSE's knowledge base. However, when a knowledge-based system, like SPARSE, is under continuous use, the need to change the rule base arises sooner or later. In the case of the Portuguese transmission network, the introduction of new substations with different types of operation or layout has already imposed some modifications. When these changes happen, it is not possible to undertake complete validation tests like the ones described before. Even if the costs were acceptable, the required time would oblige the knowledge-based system to be either out of service or in service without a validated rule base for a longer time than desirable. This problem must be addressed by using a verification tool that employs formal methods. Verification tools offer an easy and inexpensive way to ensure knowledge quality maintenance.

A specific tool, named VERITAS (Figure VI.3.12), has been developed for use in SPARSE's verification, performing structural analysis in order to detect knowledge anomalies. Notice that an anomaly is considered a symptom of possible errors but cannot, *a priori*, be considered an error.

The detected anomalies can be grouped in four major classes: redundancy, circularity, ambivalence, and deficiency. Considering that the anomalie's definitions depends on KBS characteristics, let us consider some definitions:

- Circularity — a knowledge base contains circularity if and only if it contains a set of rules that allows an infinite loop during rule triggering
- Ambivalence — a knowledge base is ambivalent if and only if, for a permissible set of conditions, it is possible to infer an impermissible set of hypotheses

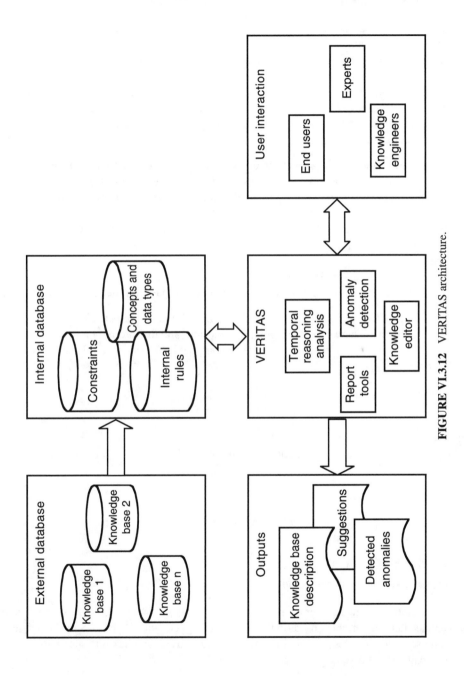

FIGURE VI.3.12 VERITAS architecture.

- Redundancy — a knowledge base is redundant if and only if the set of final hypotheses is the same in the rule/literal presence or absence
- Deficiency — a knowledge base is deficient if and only if there is not a rule that uses a set of conditions representing a possible domain knowledge situation

This classification is based on Preece classification[52] with some modifications.

VERITAS is independent of knowledge domain and of the rule-grammar used. Due to this, theoretically any rule-based system can be analyzed by VERITAS. VERITAS has been developed as an open and modular architecture, allowing user interaction throughout the entire verification process.

SPARSE presents some features that make the verification task more difficult than for most knowledge-based systems. It requires the use of more complex techniques during anomaly detection and introduces significant changes in the number and type of anomalies to detect. The most important features are the following:

- Variable evaluation — the evaluation of variables is crucial, particularly in the case of temporal variables, in order to obtain comprehensive and correct results during the verification process. It is important to note that during anomaly detection (static verification), it is not possible to know the values that a variable will be assigned.
- Rule triggering selection mechanism — this mechanism avoids some run-time errors (such as circular chains) and has to be considered on the verification chains used by the verification process.
- Temporal and non-monotonic reasoning — this kind of reasoning makes the verification process more difficult. It requires considering the evolution of the inference process over time.

In the first step, VERITAS uses one specific filter for each knowledge base it verifies, including the rules for the conversion between the original (or external) and the internal knowledge bases. During this step, a database is constructed for verification purposes. Consider the following example where rule r1 is converted:

Rule r1 : 'EXAMPLE' :

[[msg(A) at T1 and

msg(B) at T2 and

temporal_relation(T1,T2)] or

[msg(A) at T1 and

msg(C) at T2 and ...]] =>

[retract (msg(A)),

retract (msg(B)),

assert(conclusion(X))].

Some of the tasks that are executed during this step are the following:

- Generation of a new rule for each conjunction on the left-hand side of the original rule
- Replacement of variables used on the rules and triggers with VERITAS internal symbols
- Representation of rules in a relational form in order to create indexes to speed up the generation of expansion rules

For the example being considered, we have:

rule(*OldRule,NewRule,Description*)

rule(r1,'r1-l1','EXAMPLE').

rule(r1,'r1-l2','EXAMPLE').

literal*(Label,Type,Functor/Nargs)*

literal(e1,event,msg/1).

literal(t1,temporalOperator,temporal_relation/2).

lit_LHS (Rule,Label,TimeTag,LogValue,Arg,Preds)

lit_LHS(r1-l1,e1, #T1,true, [#A],PREDS)

lit_LHS(r1-l1,e1, #T2,true, [#B],PREDS)

lit_LHS(r1-l1,t1,none,true, [#T1, #T2],PREDS)

lit_LHS(r1-l2,e1, #T1,true, [#A],PREDS)

where PREDS is the list of predecessor rules that allow conclusion of the respective literal, represented by *Label*.

Similar information is stored for the right-hand side of each rule in a structure called lit_RHS. In the second step, VERITAS generates useful information about existing relations between literals (previously obtained). VERITAS considers some type of constraints already described in the literature.[53] The considered constraints can be classified in the following groups:

- Semantic constraints — this type of impermissible set is formed by literals that cannot be present at the same time in the knowledge base. Semantic constraints have to be introduced by the user. If a plant has two distinct modes of operation (remote or local),

$$\{remote(T,P), local(T,P)\}$$

represents a semantic constraint, where T stands for time and P for plant.
- Single value constraints — this type of impermissible set is formed by only one literal, considering different values of its parameters. Notice that those potential constraints are automatically detected. After this, the constraint can be either confirmed or changed by users. Considering that a breaker can be open, closed, or in a changing state between the two states, we have the following single value constraint:

$$\{breaker (T,P,open), breaker (T,P,closed), breaker (T,P,changing)\}$$

where T and P have the same meanings as before.

The anomaly detection module (verification tools) works in an autonomous way with no user interaction and runs in batch mode. Presently, this module can be integrated with a knowledge editing tool (Figure VI.3.12) that, among other functions, allows rule edition. This tool shows the existing relations between the rules that are to be modified and the remaining existing knowledge in the knowledge base. This information is supplied by a graphical interface using a hypergraph type of representation.

The anomaly detection relies on rule expansions and constraint analysis. This method is also used by some well-known verification and validation tools, such as KB-REDUCER[54] and COVER.[55] Considering SPARSE's specific features already discussed, the technique used is a variation of the common assumption-based truth maintenance system (ATMS). Namely, the knowledge represented on the meta-rules has to be considered in rule expansion generation. In the first phase, VERITAS detects anomalies using logic tests. The second phase includes temporal reasoning analysis and filters previously detected anomalies, providing accurate information by preventing false anomaly reporting. When the verified knowledge base is large in respect to the number of rules and inference chains, the information generated during anomaly detection can be huge. The detected anomalies have to be reported using a suitable form to ease analysis. This task has been carefully considered in order to reduce the time needed for the information analysis. So, it is possible to aggregate or select information by type of anomaly, rule number, and literal identification.

3.5.8 Tutor

SPARSE is connected with an intelligent tutoring system (ITS)[17,18] whose main goal is training Portuguese electrical transmission network control center operators in the interpretation of alarm messages in the case of incident situations. Figure VI.3.13 presents the architecture of this intelligent tutoring system.

The domain knowledge used by the intelligent tutor is obtained from SPARSE. The system will be able to produce and select a training scenario according to the training needs of each operator and present it to the operator. An important feature of an ITS is its ability to understand the learner's reasoning to obtain a solution. This reasoning allows the ITS to find out the possible learner's misconceptions and, therefore, to select an appropriated strategy to correct the learner's reasoning. Another system requirement arising at the specification phase is concerned with the ability to support the training activities in the learner's work place. This way, trainees can practice whenever there is not an overflow of other control center tasks.

The domain knowledge used by the intelligent tutor under development is the same as used by SPARSE, which is represented through production rules. However, the way the SPARSE knowledge base is organized is not suitable for use by an intelligent tutor. Moreover, SPARSE's knowledge base does not contain other kinds of knowledge required by operators' learning processes, as well as theoretical knowledge related to the domain.

This intelligent tutor is a "learn-by-doing" system where learners acquire skills to perform diagnosis of the power system's state. Learners obtain these skills by solving problems selected by the ITS. Prediction tables have been used to implement an interaction mechanism between learners and the intelligent tutor. This approach allows a friendly user interface and an accurate way to get the learners' reasoning.

Knowledge representation assumes an important role in the development of knowledge-based systems such as intelligent tutoring systems (ITS). The ITS's ability to interpret the learner reasoning and to transfer the domain knowledge to the learner depends on the domain knowledge representation adequacy.

Another important issue that arises during the development of an intelligent tutor is concerned with the mechanisms used to support the interaction with the learner. This interaction must fit the

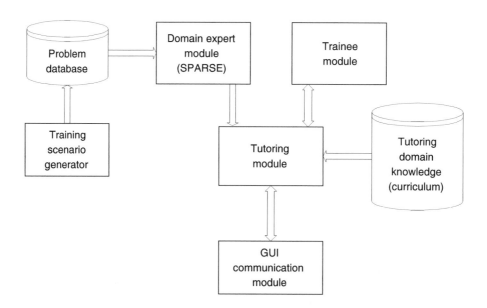

FIGURE VI.3.13   Intelligent tutoring system architecture.

learners' characteristics and supply an environment that invites users to effectively use the system. On the other hand, the interaction mechanism selected must be able to obtain the learners' reasoning in an accurate way, allowing the intelligent tutor to infer the learners' real misconceptions.

Associative or semantic networks are used to represent the domain knowledge. These structures are directed graphs, where the nodes are used to represent concepts or objects and the arcs connecting the nodes represent relationships or associations between the concepts. In the area of knowledge-based systems, associative networks are most commonly used to represent semantic associations. Within engineering applications, they can be used to express both physical and causal structures of systems. When compared with other methods of knowledge representation, such as production rules, associative networks have two advantages:[56]

1. The knowledge can be represented in an explicit and succinct way — connections between nodes can be associated with different kinds of properties such as inheritance
2. Search time is reduced because nodes are directly connected to related nodes rather than being expressed as relationships distributed across a knowledge base (in the case of production rules, this problem can be minimized using meta-rules)

Rodríguez et al.[57] compiled a list of systems on the ITS domain that use associative networks to represent dependencies among concepts, relationships between skills and problems to exercise, the evolutionary nature of knowledge, a curriculum, different levels of abstraction about concepts, and relations among concepts and among concepts and tools. Procedures for solving problems have also been represented using associative networks. In the same systems, procedural networks were used to model skills and erroneous interpretations of these skills performed by the learners.

The Intelligent Tutor domain knowledge base is composed of a knowledge base concerning the electrical grid components and of a hierarchic knowledge units network obtained from the rule base of SPARSE. The knowledge base concerning the electrical grid components (several types of plants, protection devices, etc.) contains information about each component function and about typical values (such as typical protection regulation) and theoretical knowledge about the domain. The elements of this knowledge base include pointers to didactic material in several forms such as text files containing explanations, schematics, images, and animations. The availability of didactic material in several formats allows a better understanding of the domain concepts by the learner.

The main element of the domain knowledge base used by the intelligent tutor is the hierarchic knowledge units network or concepts network, implemented through associative networks. The knowledge units network represents the procedures and the concepts needed to perform any type of diagnosis. This network contains the set of concepts to be understood by the learner, allowing him to get the correct diagnosis for each type of problem proposed by the tutor. The network concepts are related through precedence relations. These precedence relations define the hierarchy of concepts in order to indicate the sequence the learner must observe over the set of concepts and dependencies among concepts.

The knowledge units or concepts included in the network can be of the following types:

- Events
- States
- Relations between events and/or states
- Rules
- Conclusions

Event concepts correspond to SCADA messages that arrive at control centers. State concepts represent hypotheses about power system element states. Events and/or state concepts are related according to the relations established between them by relational concepts. Temporal relations among events and/or states are, for example, relational concepts. Rule concepts relate other concepts forming a set of conditions leading to new concepts which reflect states or hypotheses about states. Finally, conclusion concepts correspond to the highest abstraction level, representing the several types of problem solutions or types of diagnosis.

The concepts included in the concept network correspond to two abstraction levels. The highest abstraction level includes only concepts representing events (concepts of the lowest abstraction level), concepts representing relations between events, and the conclusion concept that represents the diagnostic (concept of the upper abstraction level). The lowest abstraction level includes all the network concepts. In cases where the intelligent tutor presents a problem to an advanced learner, the IT will allow the learner to refer only to the concepts corresponding to the highest abstraction level when the learner exposes his reasoning to obtain the solution. Likewise, during an explanation, the intelligent tutor must present only the concepts belonging to an abstraction level compatible with the learner's skills about that particular problem.

One of the major goals of the intelligent tutor is to provide operators with the knowledge needed to solve problems, namely, situation diagnosis problems. This knowledge transfer needs two kinds of knowledge included in the concept work:

1. Declarative knowledge: knowledge used to select the relevant information from SCADA messages and to relate this information in order to infer new knowledge
2. Procedural knowledge: knowledge needed to split the problem-solving process into separated phases, as usually done by the expert operators

Frame structures were used to implement the knowledge units network. The frames are organized in classes and each knowledge unit corresponds to an instance of one of the frame classes. These classes are event, state, relationship between event and/or state, rule, and conclusion. The frame slots can be active or not, depending on the corresponding type of concept. These slots are the following:

- Concept identification
- Concept overview
- List of keywords related to the concept
- Pointers to didactic material
- Pointers to samples related to the concept
- List of precedent concepts
- List of successor concepts
- Information about a prior visit of the concept by the learner, including the tutor mode used (tutoring or exploratory) and the skill level of the learner (advanced, beginner, etc.)
- The skill level of the learner in the concept, based on the performance in prior training sessions
- Pointer to the beginning of the subnetwork if the concept belongs to the conclusion class
- Criteria used to select the following concepts during a session driven by the intelligent tutor (breadth first, depth first, etc.)
- Set of typical errors related to the concept
- Instant related with the concept in the case of event, state, and conclusion concepts

The information about the learner level and skills regarding each concept is stored in a file for each student. This information is part of the student model and is loaded from the file and updated during each session. The didactic material corresponding to each concept includes the knowledge needed to build up explanations about the contents of the knowledge unit. This framework supports the didactic material in different formats, considering the needs and skills of the learners. The didactic material can be composed of one or more of the following parts: theory underlying the concept, detailed concept description, and application samples.

The set of typical errors related to each concept allows identification of the most common mistakes learners make and formulation of the corresponding explanations. This set of errors can be divided

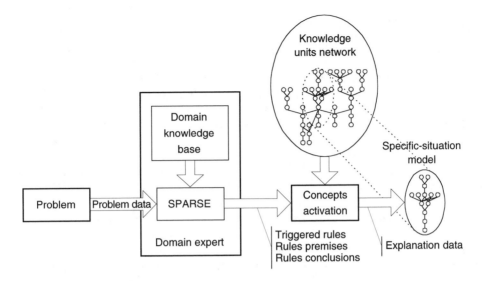

FIGURE VI.3.14   Obtaining the specific-situation model.

into two subsets: a subset of errors common to the concepts belonging to a particular class and a subset of errors specific to a concept class instance.

The adopted knowledge representation structure allows several knowledge operators to use the domain knowledge for different purposes. The same domain knowledge can be shared between training sessions, oriented by the intelligent tutor, and exploratory sessions conducted by either the intelligent tutor or the learner.

During intelligent tutor oriented sessions, the domain knowledge, included in the knowledge units network, will be used not only to evaluate the reasoning performed by the student on the analysis of a problem proposed by the intelligent tutor but also to detect occasional errors in the student's reasoning, including the reasons that led to these errors.

The reasoning undertaken by SPARSE is available to the intelligent tutor so that it can perform the evaluation process. This reasoning contains the triggered rules, the facts used as premises, and the facts obtained as rule conclusions. From this set of particles or knowledge units used by the knowledge-based system to get the diagnosis, the intelligent tutor will activate the corresponding concepts or nodes from the knowledge units network. The activated concepts define the network portion used in the reasoning performed by the expert system. This network portion corresponds to the specific-situation model. As stated by Clancey,[58] this specific-situation model describes some situation in the world, generally an explanation on how the situation came about. The situation-specific model is obtained by SPARSE, applying the inference procedure to its domain model (Figure VI.3.14).

To ensure that an ITS is an effective learning aid, it is required that the ITS can obtain the reasoning performed by the learner when solving a problem. This allows the ITS to detect the learner's misconceptions and to select the most appropriate strategy to guide the learner's reasoning.

To make the diagnosis process easier for the learner, the domain knowledge units are grouped in phases. Each phase corresponds to a step in the diagnosis process. Let us consider the following phases:

- Breaker tripping
- Breaker reclosure
- Breaker tripping after reclosure

- Result of breaker reclosure
- Conclusion about the tripping type
- Conclusion about the number of transmission line ends involved

The number of phases included in the diagnosis of a particular problem depends on the nature of that problem. For instance, in a definitive tripping (a tripping followed by a reclosure and a new tripping), all the phases are present, while in a single tripping (a tripping without reclosure attempting), the diagnosis process includes only phases 1, 5, and 6. Each phase of the diagnosis process is defined by a set of premises leading to a conclusion. The premises and conclusions correspond to concepts of the knowledge units network. The conclusion obtained in a phase is a concept belonging to an abstraction level above the premises' abstraction level. Meanwhile, this conclusion can be used as a premise of an upper phase.

One of the most relevant parts of an intelligent tutor is the user interaction mechanism used to obtain the learner's reasoning. From the learner's reasoning, the tutor can infer the set of reasons that led to the learner's misconceptions and then select the best tutoring strategies. Natural language has been the approach most commonly used to implement such an interaction mechanism. However, this technology is not successful in some aspects such as understanding textual input. Therefore, the implementation of natural language interfaces becomes a hard task. Also, within some environments, a natural language interface does not provide a friendly interaction with the user, leading to ambiguous and boring communication.

In the case of SPARSE, it is important to ensure that the intelligent tutor is easy to use in order to be accepted by the operators and effectively used during less activity periods in the control center. Also, because some end users are undergraduate operators, a natural language interface would be an obstacle to effective use of the intelligent tutor. Prediction tables are used as the main mechanism to support the communication between the learner and the tutor. This interaction mechanism makes the user interface easy to use, fast, appealing, and concise. The major drawback usually pointed out is that this kind of interaction would provide the learner with reminders of issues of knowledge that are often forgotten, which would not be present in the real working situation. This disadvantage can be overcome by adapting the detail level of the information available in the prediction tables according to the learner's expertise.

The interaction between the learner and the intelligent tutor is performed through prediction tables where the user can select the set of premises and the conclusion for each phase. During the resolution of a diagnosis problem, the learner must fill in a prediction table for each phase involved in the case. In each prediction table, the learner must choose the relevant premises and the conclusion from a set of items presented to him. When a premise is a temporal relation, the learner must define the variables and the relations between them.

Using prediction tables, the learner acquires the two types of knowledge included in the concept network which are needed to solve the problems. The learner uses the procedural knowledge to select the phases present in the problem and the corresponding prediction tables. Declarative knowledge is used when the learner fills the slots of the prediction tables.

In order to evaluate the learner's reasoning performed to get the solution of a diagnosis problem, the intelligent tutor will compare the prediction tables' content with the reasoning undertaken by SPARSE. If the matching reveals differences, it means that the learner's reasoning contains some error. If that is the case, the intelligent tutor will present to the learner the reasons behind the error. Figure VI.3.15 presents the process of obtaining the learner's reasoning.

As an example, let us consider a problem whose diagnosis is a breaker single tripping (tripping without an attempt at reclosure). Let us also consider that the learner includes in his reasoning a rule whose conclusion is a breaker reclosure (rule $d4$). In this case, after consulting the knowledge units network, the intelligent tutor must show the learner that the reclosure is not verified because one of the premises of that rule is not present. In fact, the rule mentioned by the learner would need a

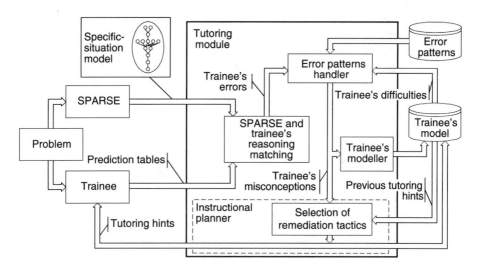

FIGURE VI.3.15   Process of obtaining the learner's reasoning.

premise corresponding to the state "breaker closed." Figure VI.3.16 presents a portion of the lowest abstraction level knowledge units network corresponding to the reasoning leading to the conclusion of single tripping. The concept represented by a dashed circle corresponds to a premise that is not mandatory to trigger the rule.

Figure VI.3.17 presents a prediction table corresponding to phase 1 of the diagnosis process of the problem represented in Figure VI.3.16.

The ITS is able to present the learner with a graphical representation of the knowledge units network. This representation is useful to represent the concepts involved in the reasoning and the relations between concepts. Furthermore, this kind of representation helps the learner acquire the procedural knowledge needed to get the correct diagnosis.

The knowledge units network can also be used in exploratory mode. During the operation in this mode, the learner shall go through each knowledge unit without restrictions or following a path pre-established by the intelligent tutor, making possible the access to the didactic material included in each knowledge unit.

3.5.9   Example

Let us consider a small example based on a real incident. This incident generated a set of messages, from which we have selected the following:

01:58:45.178	SPM	127	F.ALENTEJO	OTHER ALARMS	0 1
01:58:45.218	SPM	127	F.ALENTEJO	>>>TRIP ORDER	0 1
01:58:45.249	SOQ	106	F.ALENTEJO	OTHER ALARMS	0 1
01:58:45.264	SFA	106	PALMELA	OTHER ALARMS	0 1
01:58:45.278	SPM	127	F.ALENTEJO	>>>TRIP ORDER	0 0
01:58:45.278	SPM	127	F.ALENTEJO-BRE	BREAKER	0 0
01:58:45.279	SSN	104	F.ALENTEJO	OTHER ALARMS	0 1
01:58:45.284	SFA	106	PALMELA	URGENT ALARM	0 1
01:58:45.284	SFA	104	SINES	OTHER ALARMS	0 1
01:58:45.284	SFA	102	OURIQUE	OTHER ALARMS	0 1
01:58:45.304	SFA	106	PALMELA	>>>TRIP ORDER	0 1

01:58:45.324	SFA	106	PALMELA-BRE	BREAKER	0 0
01:58:45.378	SSN	104	F.ALENTEJO	OTHER ALARMS	0 0
01:58:45.378	SPM	127	F.ALENTEJO	OTHER ALARMS	0 0
01:58:45.384	SFA	106	PALMELA	>>>TRIP ORDER	0 0
01:58:45.384	SFA	102	OURIQUE	OTHER ALARMS	0 0
01:58:45.389	SOQ	106	F.ALENTEJO	OTHER ALARMS	0 0
01:58:46.178	SPM	127	F.ALENTEJO	OTHER ALARMS	0 1
01:58:46.198	SPM	127	F.ALENTEJO	OTHER ALARMS	0 0
01:58:46.278	SPM	127	F.ALENTEJO-BRE	BREAKER	0 1
01:58:45.404	SFA	106	PALMELA	URGENT ALARM	0 0
01:58:45.404	SFA	104	SINES	OTHER ALARMS	0 0
01:58:45.530	SFA	106	PALMELA	OTHER ALARMS	0 0
01:58:46.290	SFA	106	PALMELA	OTHER ALARMS	0 1
01:58:46.350	SFA	106	PALMELA-BRE	BREAKER	0 1
01:58:46.370	SFA	106	PALMELA-BRE	BREAKER	0 0
01:58:46.390	SFA	106	PALMELA-BRE	BREAKER	1 0
01:58:47.872	SFA	OPA	AUTOMATIC OP.	OPA IN SERV.	0 1
02:01:16.770	SFA	106	PALMELA-BRE	BREAKER	0 0
02:01:16.830	SFA	106	PALMELA-BRE	BREAKER	0 1
02:01:17. 090	SFA	106	PALMELA	NON URG. AL.	0 1
02:01:17. 090	SFA	106	PALMELA	BREAKER AL.	0 1
02:01:17. 092	SFA	OPA	AUTOMATIC OP.	OPA IN SERV.	0 0
02:01:41.430	SFA	106	PALMELA	NON URG. AL.	0 0
02:01:41.430	SFA	106	PALMELA	BREAKER AL.	0 0

These messages correspond to an incident in line Ferreira do Alentejo–Palmela (SFA-SPM) involving only one phase, which implies the tripping of both ends. Automatic reclosure equipment performed the reclosure of the line, successfully in the Palmela substation (SPM) but unsuccessfully in the Ferreira do Alentejo substation (SFA). This end of the line has been closed by the automatic operator (OPA) of the Ferreira do Alentejo substation.

For this incident, SPARSE provides the following conclusions:

01:58:45.218	BEGINNING OF INCIDENT		
01:58:46.390	Line SPM-SFA - DmR in SPM and DmD in SFA		
01:58:46.390	SFA	106	>> Wait closure by OPA
02:01:17.830	SFA	106	Breaker closed by OPA
02:01:17.830	END OF INCIDENT		

SPARSE is able to enormously reduce the number of messages that are presented to operators. Moreover, it gives advice about actions that operators should or should not take. For this purpose, the system takes into account knowledge about characteristics of automata (OPA) and about the procedures usually taken for service restoration.

From the original set of generated messages, almost 50 messages refer to a period of just 130 milliseconds. SPARSE provides operators with a small set of 5 conclusions. The interpretation of this incident is condensed in the first two messages. The third message provides operators with some advice about the restoration procedure, while the fourth message reports the closing of the breaker, which leads to the end of the incident (reported in the fifth message).

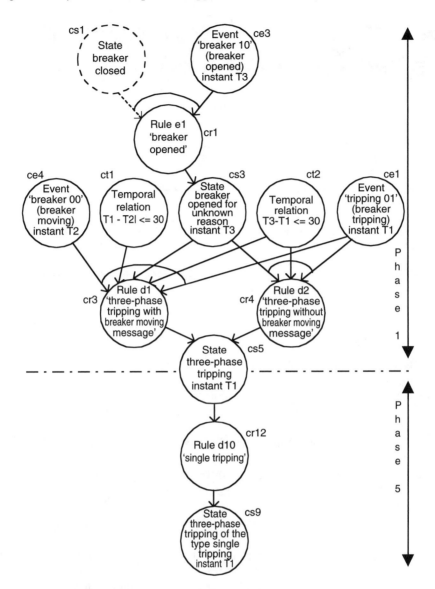

FIGURE VI.3.16   Single tripping representation.

3.6   Conclusions and Future Directions

The development of a KBS for real-time operation in power system control centers can make a significant improvement in the operation of power systems and assist control center operators in their complex tasks. However, a project like this requires rather high human power and a multidisciplinary team in which the deep involvement of utility experts is of prime importance. Artificial intelligence applications for control centers are very demanding because they must deal with enormous amounts of information in real-time and must be integrated with the existing software. This means that the current state of the art that concerns artificial intelligence applications is not usually sufficient for artificial intelligence applications for power system control centers. For this reason, the projects in this area require an important effort that is very time consuming and can take several years to be

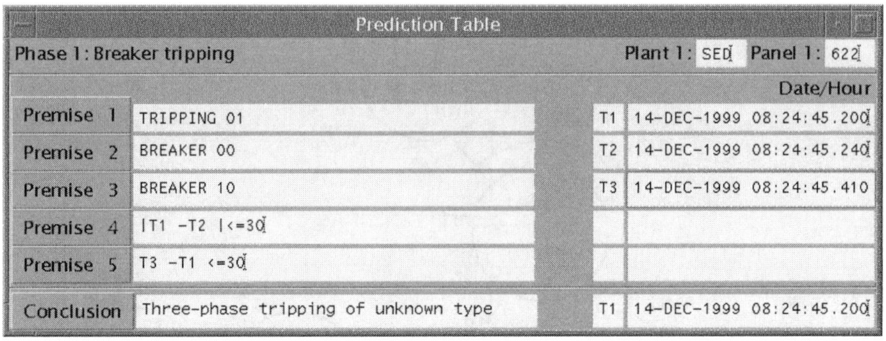

Prediction Table		
Phase 1: Breaker tripping		Plant 1: SED Panel 1: 622
		Date/Hour
Premise 1	TRIPPING 01	T1	14-DEC-1999 08:24:45.200
Premise 2	BREAKER 00	T2	14-DEC-1999 08:24:45.240
Premise 3	BREAKER 10	T3	14-DEC-1999 08:24:45.410
Premise 4	IT1 -T2 I<=30		
Premise 5	T3 -T1 <=30		
Conclusion	Three-phase tripping of unknown type	T1	14-DEC-1999 08:24:45.200

FIGURE VI.3.17   Prediction table of phase 1.

accomplished. However, the integration of these applications in control centers is a key issue for the success of these projects. This integration must consider the existing software and hardware and guarantee that artificial intelligence applications do not interfere with control functions.

There is a need to establish methods that increase the rate of successful projects and make them less time consuming. In fact, presently, this kind of project is not easily justifiable in economic terms mainly because of the huge level of human power that it requires. Discussion of the following points may contribute to this goal:

- The effort that the utilities must consider for this kind of project is much greater than the effort required for the specification of needs in traditional computer applications. Moreover, the experts that must be involved in these projects are usually occupied with the daily problems of network operation and control. However, if the project has a great probability of success, the effort will be rewarding.

- The tools used are very important for the final results. In many cases, in order to reduce the duration of the projects, a shell is used for this purpose. However, it is crucial to be sure that the chosen tool fully meets the needs of the application to be developed, namely, in what concerns real-time performance and integration with other software. It is also very important to be sure that the tool used will not act as an important limitation in further developments that can take place in the future. From our experience, the use of programming languages (such as C and Prolog) is generally preferable from this point of view. These languages are more likely to be portable to other machines and allow developers almost complete freedom. The drawback in using this kind of languages is the higher manpower required (e.g., in user interface development and in the establishment of the rule base). However, this extra effort may be the difference between a successful project and a project that has to be reconsidered after the conclusion that the initial prototype is no longer useful.

- Integration issues should be considered from the very beginning of the projects. In fact, they usually impose important limitations. If they are not considered from the beginning, an important part of the system is very likely to have to be rebuilt when the system is to be integrated in the control center.

- End users should participate in the development process from the beginning. This guarantees that their needs are really considered, and they will feel more motivated to cooperate and not feel excluded.

Once deployed, the knowledge-based system should be supported by good tools to assist in knowledge maintenance and updating. In fact, some of the systems are abandoned after a short period of use because their knowledge bases became obsolete. Knowledge maintenance and updating should be very easy and, whenever possible, guaranteed by the end-users.

Electrical energy markets are presently going through a restructuring process[59] that is changing the traditional methods of power system operation and even power system planning. Privatization, increasing competition, and a complete reorganization of the electricity market are only a few signs of this complex process. In this new environment, the power industry is still ruled by physical laws such as those that govern power flows in the electrical network, those that impose limits on power transmission, and those that require an increasing complex control to ensure the security of the power system. This industry, however, is also ruled by market rules based mainly on the laws of supply and demand and ruled by profit.

This tendency of privatization and liberalization appeared not only in the power industry but also in other crucial areas of business such as insurance, banking, and telecommunications. However, the power industry has some specific features because there is a very strong correlation between the physical laws of power systems and the contracts established between producers and distributors. These contracts have to be technically accepted by an entity such as an independent system operator (ISO). Rules about the choice between contracts when all the proposed contracts cannot be accepted must be defined in a very careful way. For instance, the possibility of a set of producers imposing technical constraints on the power system, either deliberately or not, leading to the impossibility for some producers to get in the business must be checked.

Concerns about the environmental impact of the technologies used is also increasing, imposing new restrictions on power system operation and making it necessary to revise the traditional method of calculating costs and revenues. Pollution quotas may also change the way in which cooperation is established between different actors in the power business.

All these factors are changing the functions and philosophy of the operation of power system control centers and are making power system operation even more complex. Several tasks in this context, such as generation and demand bidding, and the provision of ancillary services, such as regulating power, regulating reserve, blackstart facilities, and voltage/reactive power supply, are not adequately supported by traditional control center technologies. Artificial intelligence techniques can provide a valuable contribution to this process, providing operators with decision support tools.

References

1. Miller, R.H. and Malinowski, J.H., *Power System Operation*, 3rd ed. McGraw-Hill, New York, 1994.
2. Kennedy, T., A system operator's view of evolving applications, *IEEE Comput. Appl. Power*, 8(2), 25, 1995.
3. Prince, W.R., Nielsen, E.K., and McNair, H.D., A survey of current operational problems, *IEEE Trans. Power Syst.*, 4(4), 1492, 1989.
4. Prince, W.R., Wollenberg, B.F., and Bertagnolli, D.B., Survey on excessive alarms, *IEEE Trans. Power Syst.*, 4(3), 950, 1989.
5. Amelink, H., Forte, A.M., and Guberman, R.P., Dispatcher alarm and message processing, *IEEE Trans. Power Syst.*, 1(3), 188, 1986.
6. CIGRÉ Task Force 38.06.02, Survey on expert systems in alarm handling, *Électra*, 139, 132, 1991.
7. CIGRÉ Task Force 35.13.02, Knowledge based applications in SCADA/EMS—a practical approach, CIGRÉ Technical Brochure, 1999.
8. Liu, C.C., Pierce, D.A., and Song, H., Intelligent system applications to power systems, *IEEE Comput. Appl. Power*, 10(4), 21, 1997.

9. Talukar, S.N., Cardozo, E., and Perry, T., The operator's assistant—an intelligent, expandable program for power system trouble analysis, *IEEE Trans. Power Syst.*, 1(3), 182, 1986.

10. Kirschen, D.S. et al., Artificial intelligence applications in an energy management system environment, Proceedings of the Symposium on Expert Systems Application to Power Systems, Stockholm-Helsinki, August 22–26, 1988, 17–1.

11. Miao, H., Marino, S., and Liu, C.C., A new logic-based alarm analyzer for on-line operational environment, *IEEE Trans. Power Syst.*, 11(3), 1600, 1996.

12. Vale, Z.A. and Moura, A., An expert system with temporal reasoning for alarm processing in power system control centers, *IEEE Trans. Power Syst.*, 8(3), 1307, 1993.

13. Vale, Z.A. et al., SPARSE—an expert system for alarm processing and operator assistance in substations control centers, *ACM Appl. Comput. Rev.*, 2(2), 18, 1994.

14. Vale, Z.A. et al., SPARSE: an intelligent alarm processor and operator assistant, *IEEE Expert*, 12(3), 86, 1997.

15. Kirschen, D.S. et al., Controlling power systems during emergencies: the role of expert systems, *IEEE Comput. Appl. Power*, 2(2), 41, 1989.

16. CIGRÉ Task Force 39.03, Practical aspects of intelligent machines in the control room, *Électra*, 159, 102, 1995.

17. Faria, L. et al., An ITS for power systems staff training: matching knowledge representation and learner interaction, Proceedings of the 2001 International Conference on Intelligent Systems Applications to Power Systems (ISAP 2001), Budapest, Hungary, June 18–21, 2001, 335.

18. Silva, A. et al., User modelling concerning control centre operators training, Proceedings of the 2001 IEEE Porto Power Tech Conference (PPT 2001), Porto, Portugal, September 10–13, 2001.

19. Vale, Z.A. et al., Man-machine interaction in power system control centers, in *Proceedings of the 1994 IEEE International Conference on Systems, Man and Cybernetics*, San Antonio, TX, 1994, 1092.

20. Vale, Z.A., Ramos, C., and Faria, L., User interfaces for control center applications, Proceedings of the 1997 International Conference on Intelligent Systems Applications to Power Systems (ISAP '97), Seoul, Korea, July 6–10, 1997, 14.

21. Liu, C.C. et al., *A Tutorial Course on Knowledge-Based System Techniques with Applications to Power Systems*, IEEE, Piscataway, NJ, 1993.

22. Kezunovic, M. et al., Fault analysis using intelligent systems, *IEEE Power Eng. Rev.*, 16(6), 7, 1996.

23. Handschin, E. et al., Experiences with two different fault diagnosis systems applying expert system and device-specific ANN techniques, in Proceedings of the 1997 International Conference on Intelligent Systems Applications to Power Systems (ISAP '97), Seoul, Korea, July 6–10, 1997, 111.

24. Lee, H.J. et al., An advanced fault diagnosis system and SCADA simulator, Proceedings of the 1997 International Conference on Intelligent Systems Applications to Power Systems (ISAP '97), Seoul, Korea, July 6–10, 1997, 122.

25. Saha, M.M. and Kasztenny, B.Z., AI methods in power system protection, *Int. J. Eng. Intelligent Syst. Electrical Eng. Commun.*, 5(4), 183, 1997.

26. Kezunovic, M. and Rikalo, I., Automating the analysis of faults and power quality, *IEEE Comput. Appl. Power*, 12(1), 46, 1999.

27. Tan, J.C. et al., Fault section identification on a transmission network using action factors and expert system technology, Proceedings of the 13th PSCC—Power Systems Computation Conference, Trondheim, Norway, June 28–July 2, 1999, 820.

28. Lee, H.J. and Venkata, S.S., An advanced fault diagnosis system and SCADA simulator, *Int. J. Eng. Intelligent Syst. Electrical Eng. Commun.*, 8(1), 39, 2000.

29. Lee, H.J., Park, D.Y., and Ahn, B.S., Dealing with uncertainties in the fault diagnosis of power systems, *Int. J. Eng. Intelligent Syst. Electrical Eng. Commun.*, 7(4), 169, 1999.

30. Min, S.W. et al., Fuzzy expert system for fault section diagnosis considering the operating sequences, Proceedings of the 2001 International Conference on Intelligent Systems Applications to Power Systems (ISAP 2001), Budapest, Hungary, June 17–21, 2001, 81.

31. CIGRÉ Task Force 38.06.02, Fault diagnosis in electric power systems through AI techniques, *Électra*, 159, 50, 1995.

32. Maizener, A. et al., The use of AI techniques in the design of an alarm specification aid, Proceedings of the Third Symposium on Expert Systems Applications to Power Systems, Tokyo-Kobe, Japan, April 1–5, 1991, 33.

33. Khosla, R. and Dillon, T., Learning knowledge and strategy of a neuro-expert system architecture in alarm processing, *IEEE Trans. Power Syst.*, 12(4), 1610, 1997.

34. Kádár, P., Kovács, A., and Mergl, A.K., A tolerant event recogniser and alarm filter based on sequential pattern matching under introduction into the national dispatching center, Proceedings of the 1997 International Conference on Intelligent Systems Applications to Power Systems (ISAP '97), Seoul, Korea, July 6–10, 1997, 142.

35. Kirschen, D.S. and Volkmann, T.L., Guiding a power system restoration with an expert system, *IEEE Trans. Power Syst.*, 6(2), 558, 1991.

36. CIGRÉ Task Force 38.06.04, A Survey of expert systems for power system restoration, *Électra*, 150, 86, 1993.

37. Frenken, R. et al., Neural networks, fuzzy logic and genetic algorithms in the electricity supply industry, *Int. J. Eng. Intelligent Syst. Electrical Eng. Commun.*, 5(3), 127, 1997.

38. Stenzel, J. and Zimmer, K., Comparison of a rule-based expert system and artificial neural networks for alarm processing in power systems, Proceedings of the International Conference on Electrical Power Systems Operation and Management (EPSOM '98), Zurich, Switzerland, September 23–25, 1998, 44–1.

39. CIGRÉ Task Force 38.06.06, Artificial neural networks for power systems, *Électra*, 159, 76, 1995.

40. Khosla, R. and Dillon, T.S., A neuro-expert system approach to power systems problems, *Int. J. Eng. Intelligent Syst. Electrical Eng. Commun.*, 2(1), 71, 1994.

41. Cho, H.J., Park, J.K., and Lee, H.J., A fuzzy expert system for fault diagnosis of power systems, Proceedings of the 1994 International Conference on Intelligent Systems Applications to Power Systems (ISAP '94), Montpellier, France, September 5–9, 1994, 217.

42. Tang, S.K., Dillon, T.S., and Khosla, R., Application of an integrated fuzzy knowledge-based, connectionistic architecture for fault diagnosis in power systems, Proceedings of the 1996 International Conference on Intelligent Systems Applications to Power Systems (ISAP '96), Orlando, FL, January 28–February 2, 1996, 188.

43. CIGRÉ Task Force 38.06.03, Practical use of expert systems in planning and operation of power systems, *Électra*, 146, 30, 1993.

44. Vale, Z.A. et al., Knowledge-based systems for power system control centers: is knowledge the problem?, Proceedings of the 1997 International Conference on Intelligent Systems Applications to Power Systems (ISAP '97), Seoul, Korea, July 6–10, 1997, 231.

45. Vale, Z.A. et al., TEMPUS: a knowledge automatic extraction tool, Proceedings of the 2001 International Conference on Intelligent Systems Applications To Power Systems (ISAP 2001), Budapest, Hungary, June 18–21, 2001, 329.

46. Santos, J. et al., VERITAS—an application for knowledge verification, Proceedings of the IEEE 11th International Conference on Tools with Artificial Intelligence, Chicago, IL, November 9–11, 1999, 441.

47. CIGRÉ Task Force 39.03, Exploring user requirements of expert systems in power system operation and control, *Électra*, 146, 68, 1993.

48. CIGRÉ Task Force 38.06.03, Testing and maintenance procedures for expert systems in power system operation and planning, *Électra*, 173, 92, 1997.

49. Vale, Z.A. and Ramos, C., Reasoning about time in AI applications for power control centers, *Int. J. Eng. Intelligent Syst. Electrical Eng. Commun.*, 7(2), 91, 1999.

50. Malheiro, N. et al., Enabling client-server explanation facilities in a real-time expert system, Proceedings of the 12th International Conference on Industrial and Engineering Applications of Artificial Intelligence and Expert Systems (IEA/AIE-99), Cairo, Egypt, May 31–June 3, 1999, 331.

51. Bergström, W., The inhumanity of HMI, Seminar on Human Machine Interface in Control Centers—Trends in Electricity Generation, Transmission and Distribution Companies, Copenhagen, Denmark, October 21–22, 1996.

52. Preece, A. and Shinghal, R., Foundation and application of knowledge base verification, *Intelligence Syst.*, 9, 683, 1994.

53. Zlatareva, N. and Preece, A., An effective logical framework for knowledge-based systems verification, *Int. J. Expert Syst.*, 7(3), 239, 1994.

54. Ginsberg, A., A new aproach to checking knowledge bases for inconsistency and redundancy, Proceedings of the IEEE Computer Society 3rd Annual Expert Systems in Government Conference, Washington, D.C., 1987, 102.

55. Preece, A., Towards a methodology for evaluating expert systems, *Expert Syst. (UK)*, 7(4), 215, 1990.

56. Gonzalez, A. and Dankel, D., *The Engineering of Knowledge-Based Systems—Theory and Practice*, Prentice-Hall, Upper Saddle River, New Jersey, 1993.

57. Rodríguez, P., Morales, E., and Vadera, S., A knowledge-based framework for learning, applying and consulting engineering procedures, 3rd International Conference on Intelligent Tutoring Systems, Montreal, Canada, 1996, 392.

58. Clancey, W.J., Qualitative student models, *Ann. Rev. Comput. Sci.*, 1, 391, 1986.

59. Overbye, T.J., Reengineering the electric grid, *Am. Scientist*, 88, 220, 2000.

4

Intelligent Fault Diagnosis in Power Systems

4.1 Introduction VI-113
4.2 Power System Fault Diagnosis Problem VI-114
   Control Center and Data Acquisition in the Italian Power
   System • Power System Fault Diagnosis
4.3 Protection System Description VI-116
   Transmission Line Protection Scheme • Transformer Protection Scheme • Bus Protection Scheme • Breaker Failure
   Device • Examples of Protection Operation
4.4 Generalized Alarm Analysis Module (GAAM) VI-120
   Capability of GAAM • Diagnostic Process of the GAAM
   • Diagnosis Method of the GAAM
4.5 Software Design VI-128
   GAAM Software Structure • Man–Machine Interface of
   the GAAM
4.6 Test Results VI-131
   Case Study of Italian Power System
4.7 Conclusions VI-133
4.8 Acknowledgments VI-133
References VI-133

Marino Sforna
GRTN — Italian Independent System Operator

Chen-Ching Liu
University of Washington

Dejan Sobajic
Electric Power Research Institute

Sung-Kwan Joo
University of Washington

4.1 Introduction

Control centers are responsible for the supervision, monitoring, and control of power systems. They are equipped with energy management systems (EMS) supporting software applications such as load flow, state estimation, and security analysis. In addition, control centers have access to a large amount of information regarding the power system states, e.g., currents, and voltages, that is acquired and transmitted through the SCADA (supervisory control and data acquisition) systems. With that information, operators in a control center monitor the power system's real-time conditions and maintain the system operation at a normal state subject to economic and security constraints.

When a fault occurs in a power system, operators in a control center need to interpret a large amount of information including data, alarms, and status change messages from the protection system and identify the event that triggered the alarm messages. In such an emergency condition, it is difficult for the operators to quickly identify the actual fault location, fault type, and malfunction types of protective devices such as relay and breaker. Consequently, there is a need for decision-making support to help the operator take proper actions following a fault event in the power system.

This chapter provides an overview of power system fault diagnosis and intelligent fault diagnosis systems. The main results described in this chapter are based on a generalized alarm analysis module (GAAM), developed at the University of Washington, as an example. The GAAM will be integrated into GRTN's new EMS environment at the national and regional control centers in Italy. The remainder of this chapter is organized as follows. Section 4.2 illustrates a general outline of power system fault diagnosis. In Section 4.3, various power system protection schemes are described together with examples of the sequence of relay operations against a fault. This is followed by an introduction of the malfunction analysis algorithms associated with various protective devices. In Section 4.5, software design issues of fault diagnosis are discussed along with the software structure of the GAAM. Section 4.6 uses the test results of GAAM with the field data from the Italian network to illustrate the effectiveness of the introduced fault diagnosis system.

4.2   Power System Fault Diagnosis Problem

4.2.1   Control Center and Data Acquisition in the Italian Power System

A power system includes system components such as electric transmission lines, substations, transformers, and the control and protection systems. Transmission lines are essential to transmit electric power over long distances to consumers spread over a wide area. High voltage (HV) $400 \div 220$ kV substations are located at the nodes of the transmission network. The control system is designed to help maintain the power system at its normal operating conditions by adjusting system variables such as voltages, currents, and power flows on the network. Power system components such as lines, busbars, and transformers can be affected by different faults such as short circuits and lightning strikes. The protection system on the transmission lines is designed to detect and isolate faults by opening the breakers. There are various protection schemes and protective devices for protection of the power system components. In actual protection systems, protective devices such as breakers and relays are coordinated to ensure reliable operation.

The power system in Italy is operated by a control structure organized in three hierarchical levels: control point (CP), regional control center (RCC), and national control center (NCC). The lowest level is a control point, which acts as a local switching center. Every CP is responsible for 10–20 HV substations located nearby and for the connected transmission lines. The main task of a CP is to perform actions issued by higher level control centers through telephone calls. The secondary task of a CP is to supervise the substation apparatus. The middle level is regional control centers. Presently, there are 8 RCCs in the entire Italian system. Each of them is in charge of 1 or 2 control points and closely controls the networks at 400, 220, 150, and 132 kV. RCCs are also responsible for identification of the line faults and suggesting the actions to be carried out in order to recover functionality of the transmission network. These actions are sent to the control points through communications and are performed by these CPs. RCCs are not able to directly operate the network, except for remote load shedding in emergency situations. The highest level is the national control center. It is in charge of general supervision of the network, coordination of actions that involve two or more regions, and the short, medium, and long term operation planning. During large disturbances, the NCC also coordinates all the actions in order to restore a normal operating condition.

The current energy management system (EMS) at GRTN in Italy receives from the HV electric network about 5000 telemeasurements (analog measurements) and 7000 telesignals (status signals). Analog measurements and status signals are recorded permanently at the NCC or RCCs every 15 minutes. The SCADA system at each RCC receives data (telesignals and telemeasurements) from the substations under its supervision. Every 4 seconds, RCCs also receive voltage, active, and reactive power measurements from all the substations.

The SERs (i.e., sequence-of-events recorders) are installed at the substations of an Italian power system network and chronically record the operations of the substation relays and breakers in real time. Typically, an operation is recorded by a textual message about the relay or breaker location and relay or breaker operation time, type (e.g., relay tripping), and phase (e.g., phase A, B, or C). The precision of timing in the SER messages is within a few milliseconds. The types of operation recorded include relay starting, relay tripping, relay zone of tripping (primary or backup), relay reclosure, and breaker opening. During a fault event, the SER data (i.e., the SER messages) generated from the different substations are made available in real time to the power system control centers through telecommunication links. Based on the SER data from the different substations, system-level fault diagnosis can be performed at the power system control center. Due to the level of details contained in the SER data, the fault diagnosis can produce accurate results on the fault and malfunctions. Online fault diagnosis using the SER information can provide important guidance for power system operation.

4.2.2 Power System Fault Diagnosis

Power system fault diagnosis acts to reconstruct the actual fault scenario by interpreting the available information at the control centers and substations. This task is typically performed by operators based on EMS data acquired from the substations. The types of information commonly used for fault diagnosis include breaker contact status and relay operation information. Based on the online information of protective device operations, the operators in a control center have to identify

- Fault locations: transmission line, busbar, transformer, or their combinations
- Fault types: single-phase, double-phase, or three phase faults

and decide what remedial actions have to be taken. The task of identifying the actual fault scenario can be complicated by relay malfunctions, breaker failures, and the complexity of a power system. The most common types of malfunctions include the following:

- Breakers: stuck breaker, failure to trip all three phases
- Relays: failure to trip, misoperation, wrong tripping zone, improper operating time

Such protective device malfunctions make it more difficult for operators in a control center to analyze a fault event within a short period of time. Consequently, there is a need for a fault diagnosis system to help the operator identify fault locations/types and take proper actions during a fault event in the power system. Since the beginning of the 1980s, several artificial intelligence techniques have been proposed for power system fault diagnosis. A feasibility study of alarm processing is given by Wollenberg.[1] A survey of fault diagnosis methodologies for power systems is provided in Kirschen and Wollenberg.[2] There have been attempts to use expert systems based on the conventional knowledge representation and inference procedures such as rule-based[3–5] and mode-based approaches.[6] The rule-based approach has some limitations because it cannot handle unexpected complex fault scenarios. A model-based expert system is more robust in general, but the inference process requires a time-consuming search. The artificial neural network (ANN) approach has been used for fault diagnosis in power systems.[7] Fushuang and Chang[8] proposed a probabilistic approach using a genetic algorithm (GA) for identification of the faulted section of a power system.

Much of the fault diagnosis system development has concentrated on the effects of faults in an electric transmission network. The possibility of a protective device failure must be included in the fault diagnosis system. In other words, the fault diagnosis system can provide a general description of the malfunctions of protective devices in addition to fault locations and types. To handle malfunctions of various protective devices such as distance relays, differential relays, overcurrent relays, and breaker failure devices, malfunction analysis of these protection devices has to be included in the fault diagnosis system. In recent years, several authors have proposed logic-based systems for fault diagnosis involving various fault types and protective device malfunctions.[9,10] The fault and malfunctions are analyzed by evaluating the embedded logic formulas which are based on protection logic. A logic-based approach can deal with complex fault/malfunction scenarios that cannot be predicted in advance.

It is also possible that some alarm messages are missing as a result of communication failures at substations. Missing or incomplete information can affect the accuracy of diagnostic results. Fault diagnosis systems should be able to deal with message problems such as missing information, repetitive messages, and inconsistent information.

The generalized alarm analysis module (GAAM), developed by the University of Washington and GRTN, is designed to perform fault and malfunction analysis using the information available at the control centers.[11] The diagnosis method of the GAAM for fault and malfunctioning device analysis relies on logic-based algorithms, which are based on the physical protection devices.[12] The GAAM has the capability to handle missing or incomplete information. The GAAM is integrated with a preprocessor, which is able to identify the approximate time of the missing (or incomplete) messages and reconstruct the missing (or incomplete) messages.[13]

The capability of fault diagnosis is dependent on the availability of data embedded knowledge and the reasoning process. The analysis of fault and malfunctioning devices is performed by evaluating the embedded logic formulas using the SER information and breaker status from the EMS. The logic in the GAAM has to be updated only when the protection logic changes. In case of power system topological changes, there is no need for modification of the logic of the GAAM.

4.3 Protection System Description

A power system is exposed to faults created by natural calamity and equipment or operator failures. Such faults are unavoidable in power system operations. Once a fault occurs in a power system, it is necessary to isolate the faulted components from the rest of the network as soon as possible in order to minimize the system outages. Although there are variations in power system protection mechanisms throughout the world, the protection schemes have to detect faults and disconnect only faulted components from the network by opening the circuit breakers. Each protective device protects one component but may also act as backup for the nearby protective devices. In practice, a modern power system network has the following main protection schemes:

- Distance protection
- Differential protection
- Overcurrent protection
- Breaker failure protection

This section will illustrate how these protection schemes are applied to protect transmission lines, busbars, and transformers.

4.3.1 Transmission Line Protection Scheme

Distance protections are the most widely used for transmission lines even though they are also used for transformers and generators as backup protection. Distance relays estimate the direction and distance of the fault. They are set up to intervene only when the fault is located in its protection direction and within the designated distance.

Each distance relay continuously measures the voltage and current magnitudes and calculates the impedance seen by a relay. Since the impedance of the line is known, the impedance seen by the relay gives information on the distance and direction of the fault. When the calculated impedance is higher than a threshold, a fault condition is detected and the protection device starts. The line distance protection is characterized by four (five in some cases) intervention delays, called *zones* (or steps). Figure VI.4.1 shows the relationship of the zones, fault distance, and relay operation time.

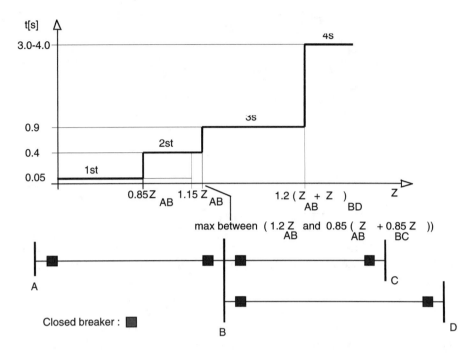

FIGURE VI.4.1 Zones in distance protection.

The closer the fault is, the faster the relay operates. Further technical details of distance relay can be found in Horowitz and Phadke[14] and Blackburn.[15]

In Figure VI.4.1, the small boxes represent the line breakers. Breakers and protective devices are located inside the substations. The Z symbols are the electric impedances of the transmission lines that are used for protection setting.

4.3.2 Transformer Protection Scheme

The most common protection method for transformer internal faults is by differential relays. In general, differential relays are employed to protect large MVA transformers against internal faults. Overcurrent relays can be applied to primary protection for small transformers or for a transformer without differential relays. They may be also used as backup protection for large MVA transformer units that are usually protected by differential relays.

For illustration, consider the following protection scheme for a transformer, depicted in Figure VI.4.2. It consists of a single differential relay coordinated with an overcurrent relay. The basic philosophy of differential protection is to compare the sum of the currents entering and leaving the protected transformer by current transformers (CTs). Under normal conditions, the sum of currents coming into the transformer should be equal to the sum of currents going out of the transformer. On the other hand, if a transformer internal fault occurs within the protected area of the differential relay, this condition will no longer be satisfied. The differential relay can trip breakers on both sides of the transformer once it detects an internal fault in the transformer.

If the differential relay fails to trip, the overcurrent protection provides backup protection for the transformer. When a fault occurs on a power system, the fault current is usually greater than the normal load current. An overcurrent relay is designed to sense the fault current. If the differential relay fails to trip the breakers, the overcurrent relays can trip breakers on both sides of the transformer. In case of a failure of the overcurrent relay, the transformer is protected by distance relays from zone 1 to zone 3.

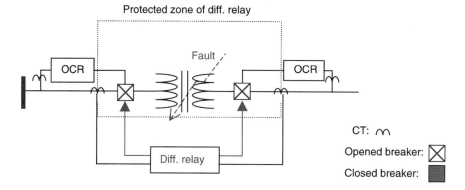

FIGURE VI.4.2   Differential protection for transformer.

In Figure VI.4.2, the protection devices collect the measurements of the currents by means of the apparatus shown as CT. Once a fault is detected, the protection devices trip the breakers within tens of milliseconds.

Another application of the overcurrent relays is the protection of transforms from external faults, and the relays are usually on both sides of the transformer. External faults are defined as faults that occurred in other power system components such as transmission lines and busbars, causing overcurrent through the transformer. In addition, an overcurrent relay on each side of the transformer provides backup protection for downstream devices.

4.3.3   Bus Protection Scheme

A busbar is a critical component to protect because it is the connecting point for lines and transformers. Busbar faults can be caused by animal contacts, broken insulators, wind-driven objects, and contamination. Once a busbar fault occurs, all circuits connected to the bus must be opened to isolate the fault and, consequently, it can result in considerable loss of service. Although busbar faults are not common, they can develop into critical system failures.

As in transformer protection, the most effective and preferred protection method for buses is differential protection. These relays are fast, selective, and sensitive. The basic idea of differential protection for buses is the same as that of transformer differential protection. Figure VI.4.3 illustrates a differential protection of buses.[15] The differential relay can trip all the breakers of the faulted bus once it detects an internal fault in the bus. If the differential relay fails to isolate the internal fault in the bus, the distance relays at 2nd or 3rd zones are supposed to operate.

4.3.4   Breaker Failure Device

When there is a fault on a transmission line, it is possible that distance relays attempt to trip the associated breakers but a certain breaker fails to clear the fault as a result of a breaker internal fault, trip coil failure, or loss of DC trip supply. Therefore, the possibility of breaker failure must be considered in the protection scheme. The function of a breaker failure device (BFD) is to provide a local backup operation in the event of breaker failure.

A breaker failure device monitors breaker operations following a tripping command from a protective relay. If the tripped breaker does not open within a predefined period of time and the fault current continues to flow after relay tripping, the breaker is considered to have failed. If the breaker failure device detects a breaker failure, it sends a tripping command to all other breakers around the substation to isolate the fault from the rest of network.[14,15]

Figure VI.4.4 describes the operation of a breaker failure device. Suppose that there is a single-phase to ground fault between bus A and bus B. The relay tripped the breakers, but the

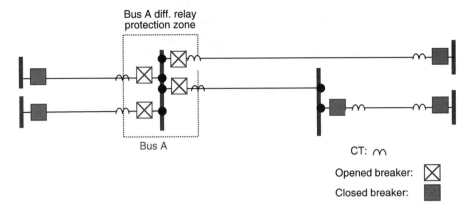

FIGURE VI.4.3 Bus protection.

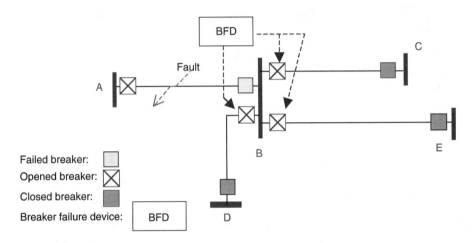

FIGURE VI.4.4 Operation of breaker failure device.

associated breaker around bus B failed to open. As a local back-up protection, the breaker failure device attempts to trip all the other breakers connected to bus B to minimize the outage area.[15] If the back-up distance protection intervened against the breaker failure in the second or higher zone, the power supplied to tapped loads between bus B and bus E would be interrupted. Consequently, there is a need for local backup protection that prevents the intervention of backup distance protection at the second or third levels in case of breaker failure.

For line, bus, and transformer faults, the sequences of operations of protective devices together with breaker operations are shown in Figure VI.4.5.

4.3.5 Examples of Protection Operation

Suppose that there is a permanent fault on the line between bus A and bus B. Distance relay A tripped the breaker correctly, but distance relay B failed to tripped the breaker. Distance relay C intervened in zone 2 as backup protection. As shown in Figure VI.4.6, the fault current goes through the transformer as a result of the distance relay failure. It is necessary to interrupt the fault current through the transformer. In this case, the breaker on the high voltage side of the transformer can be

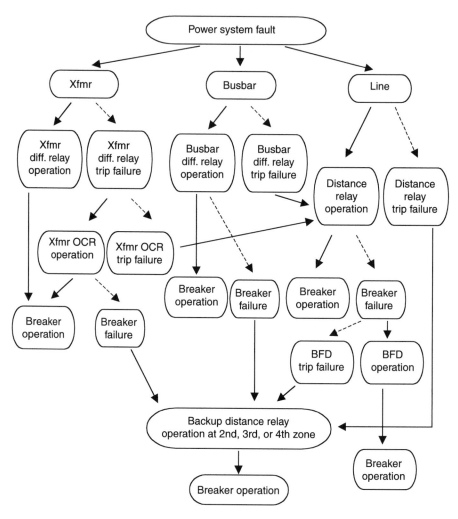

FIGURE VI.4.5   Sequences of protective device operations.

opened by the overcurrent relay so that the fault current through the transformer can be interrupted (see Figure VI.4.7).

4.4   Generalized Alarm Analysis Module (GAAM)

4.4.1   Capability of GAAM

The generalized alarm analysis module (GAAM) was developed for online fault diagnosis. The GAAM is designed and implemented in an object-oriented environment. The GAAM has the capability to handle:

- Fault scenarios involved in line faults, busbar faults, or transformer faults
- Malfunctions of various protective devices such as distance relays, differential relays, overcurrent relays, and breaker failure devices
- Uncertainties due to missing alarm messages

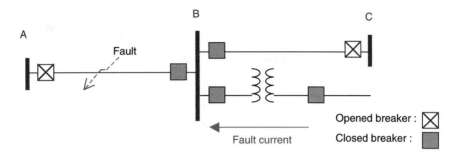

FIGURE VI.4.6 Example of distance relay protection failure.

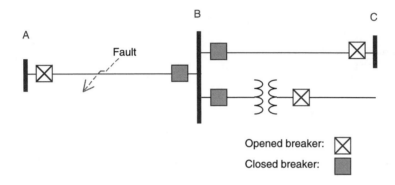

FIGURE VI.4.7 Example of the possible operations of the transformer overcurrent protection.

Based on the sequence-of-events recorder (SER) information* and breaker status from EMS data, the GAAM is able to diagnose faults on power system components such as transmission lines, busbars, and transformers. It also provides the ability to analyze possible malfunction of the protection system. The diagnostic process of the GAAM is based on multiple-hypothesis analysis so that it can deal with complex fault scenarios that involve multiple faults and multiple malfunctioning devices.

If designed well, object-oriented approaches result in software that can be easily modified, extended, and maintained. It is impossible to anticipate all possible maintenance tasks during the future usage of the fault diagnosis system, since it is difficult to foresee all possible modifications in protection systems. The logic embedded in the GAAM has to be updated only when the protection logic changes. In the case of power system topological changes, there is no need for modification of the logic of the GAAM.

4.4.2 Diagnostic Process of the GAAM

System structure of the GAAM is shown in Figure VI.4.8. Its database contains the network topology, the relay settings, and breaker contact status. Due to message problems, the GAAM requires an additional data preprocessor that examines SER messages before they are used for fault diagnosis.

---

*SER information is a set of data generated when a protection device detects a fault. This set provides details of when and how a specific protection operates and when and if the associated breakers open.

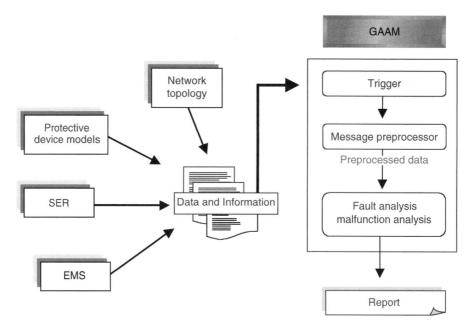

FIGURE VI.4.8   Schematic diagram of the GAAM.

The preprocessor can handle message problems such as timing errors due to substation clocks, repetitive messages, inconsistent information, and missing information. Technical details of how those problems are resolved before fault diagnosis is performed can be found in Jung, J. et al.[13]

The trigger module initiates fault and malfunction analysis when SER messages indicate that a relay tripped and the breaker opened or EMS messages indicate that a breaker opened. A triggering signal, which makes the GAAM start its analysis, could be one of the following:

- Relay tripping message and breaker opening message from SER
- Breaker opening message from EMS*

A salient feature of the GAAM is that it is designed to perform multiple hypothesis analysis based on a given set of SER messages and breaker contact status. The GAAM is triggered by a relay tripping message or breaker opening message. Identification of faulty components is an essential step in the fault diagnosis procedures of the GAAM. To identify faulty components, the GAAM searches the de-energized zone(s) where the possible fault location(s) exists. Each component in the de-energized zone(s) is a fault suspect. Therefore, each system component inside the de-energized area is selected for analysis of fault(s). Then, the GAAM generates respective hypotheses for each component in the de-energized zone(s). Each hypothesis contains a fault candidate component and a number of malfunctioned protective devices. It is necessary to indicate the credibility of each hypothesis based on the evidence. The multiple hypothesis analysis of the GAAM is designed to rank the true hypothesis with a higher value of credibility index. The credibility index of a hypothesis is calculated based on the degree of information relevance to the fault event. See Jung, J. et al.[16] for the method to calculate the credibility index of a hypothesis.

---

*Messages from the EMS regarding the status of the breakers are called telesignals.

Another important function of the GAAM is the malfunction analysis of protective devices. This requires a diagnostic method to analyze the operations of protective devices using relay and breaker information. In previous research,[12] the malfunction analysis algorithms were developed for distance relays, and, therefore, they are not applicable to other protective devices such as differential relays, overcurrent relays, and breaker failure devices. As a result, the earlier version of the GAAM was not able to handle the operation of protection devices such as OCR, differential relay, and BFD. The relay malfunction algorithms are extended in the current version of the GAAM to include malfunctions analysis of additional protective devices. The fault/malfunction analysis algorithms of the GAAM are based on predefined protection schemes and the logic relations between operations of protective devices. The logic-based approach and formulas embedded in the GAAM do not rely on heuristics that are usually derived from the experience of human experts.

The functional procedures of the GAAM can be summarized as follows.

Task 1: Read data files, which contain network configuration and protective device settings.
Task 2: Scan SER messages and telesignal messages for identifying the triggering message.
Task 3: Go to Task 4 if the GAAM is triggered by the triggering message. Otherwise, go to Task 2.
Task 4: Preprocess SER messages.
Task 5: Identify de-energized zone(s).
Task 6: Perform fault/malfunction analysis for each candidate in de-energized zone(s).
Task 7: Rank hypotheses based on credibility index.
Task 8: Report diagnostic results.

A flow chart of the GAAM functional procedures is shown in Figure VI.4.9.

4.4.3 Diagnosis Method of the GAAM

Power system faults can occur on the transmission lines or other power system components such as transformers and busbars. A logic-based reasoning technique is proposed in a previous work[12] for proof of the transmission line fault(s) based on information of distance relay and corresponding breaker In order to handle the fault related to transformers and busbars, it is necessary to involve the (mal)function analysis of OCR, differential relay, and BFD in the diagnosis of transformer and busbar faults. It is possible to derive the logical relationship between operations of protective devices from predefined protection schemes for diagnosis of the transformer and busbar faults.

4.4.3.1 Diagnosis of Transformer Faults

When an internal fault occurs in a transformer, the corresponding differential relay trips breakers on both sides of the transformer. If the differential relay fails to trip the breakers, then the overcurrent relay trips both breakers or only the breaker on the high-voltage side of the transformer. Table VI.4.1 shows the possible relationship between primary protection (differential relay) and backup protection (overcurrent relay for transformer internal faults). It is assumed that the primary relay must operate first for a fault within the protection zone to isolate a fault, and the backup relay is expected to operate only when the primary relay fails.

Suppose that a transformer differential relay (TDR) protects transformer T against internal faults as a primary protection and an overcurrent relay OCR protects transformer T against internal faults as a backup protection. Let each breaker on both sides of the transformer be X. It is assumed that breakers operate correctly according to associated relay operations.

4.4.3.1.1 *Malfunction Analysis of a Differential Relay for Transformer Protection*
Based on Table VI.4.1, a differential relay failed to trip if a transformer internal fault occurred and the differential relay should have tripped but it did not. The English version of the corresponding rule follows.

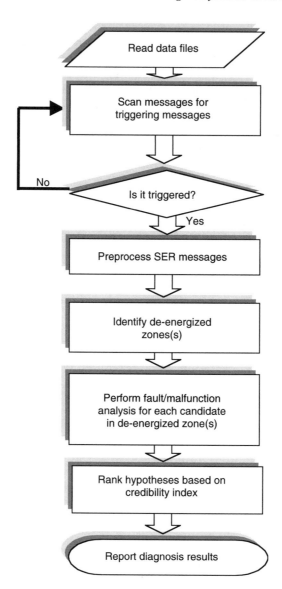

FIGURE VI.4.9   Flowchart of the GAAM functional procedures.

TABLE VI.4.1   Malfunction Analysis of Transformer Protection

| | | | Malfunction Analysis | |
| | Primary Protective Device (Differential Relay) | Secondary Protective Device (OCR) | Differential Relay | OCR |
Fault Type				
	Trip	Not trip	Normal	Normal
Transformer	Trip	Trip	Normal	Misoperation
Internal Fault	Not trip	Trip	Failure to trip	Normal
	Not trip	Not trip	Failure to trip	Failure to trip

Rule 1: Failure to Trip of a Differential Relay If {[*Transformer T is faulted*] *and* [*No SERs show differential relay TDR tripped breaker X*]}, *Then Differential Relay R failed to trip*.

The equivalent logic formula for "failure to trip of a differential relay" is:

$$Malf1(TDR) = \{Fault(T) \wedge [\sim SER_{TDR}(Tripping)]\}$$

Based on Table VI.4.1, a differential relay is considered to have misoperated if the transformer differential relay TDR tripped when it should not have. The English version of the corresponding rule is as follows.

Rule 2: Misoperation of the Differential Relay If {[*Transformer T is not faulted*] *and* [*SER outputs show differential relay TDR tripping*]}, *Then Differential Relay R misoperated*.

The equivalent logic formula for "misoperation of the differential relay" is:

$$Malf2(TDR) = \{\sim Fault(T) \wedge SER_{TDR}(Tripping)\}$$

If at least one of these two malfunction conditions is satisfied, the differential relay is considered to have malfunctioned. That is,

$$Mal(TDR) = Mal1(TDR) \vee Mal2(TDR)$$

4.4.3.1.2 *Malfunction Analysis of OCR for Transformer Protection*

For transformer internal faults, an overcurrent relay OCR is considered to have "misoperated" if OCR should not have tripped but it did. The English version of the corresponding rule is as follows.

Rule 3: Misoperation of an OCR If {[*Transformer T is faulted*] *and* [*SER outputs show transformer differential relay TDR tripping*] *and* [*SER outputs show overcurrent OCR tripping*]}, *Then overcurrent relay OCR misoperated*.

The equivalent logic formula for misoperation of an OCR is:

$$Mal1(OCR) = \{Fault(T) \wedge [SER_{TDR}(Tripping)] \wedge [SER_{OCR}(Tripping)]\}$$

An overcurrent relay is considered to have failed to trip if a transformer internal fault occurred, the differential relay failed to trip as primary protection, and the overcurrent relay should have tripped as backup but it failed to do so. The English version of the corresponding rule is as follows.

Rule 4: Misoperation of an OCR If {[*Transformer T is faulted*] *and* [*No SERs show differential relay TDR tripped breaker X*] *and* [*No SERs show overcurrent relay OCR tripped breaker X*]}, *then overcurrent relay OCR failed to trip*.

The equivalent logic formula for "failure to trip of an OCR" is:

$$Mal2(OCR) = \{Fault(T) \wedge [\sim SER_{TDR}(Tripping)] \wedge [\sim SER_{OCR}(Tripping)]\}$$

If at least one of these two malfunction conditions is satisfied, an overcurrent relay is considered to have malfunctioned. That is,

$$Mal(OCR) = Mal1(OCR) \vee Mal2(OCR)$$

4.4.3.2 **Diagnosis of Busbar Faults**

When a busbar fault occurs in a substation, the corresponding differential relay trips its associated breakers. If the differential relay fails to trip the breakers connected to the busbar, both distance relays and overcurrent relays may operate as backup protection. Overcurrent relays are employed to protect the transformer against an external fault such as a busbar or line fault. Table VI.4.2 shows the possible relationship between primary protection (differential relay operation) and backup protection (overcurrent relay operation) for busbars.

TABLE VI.4.2   Malfunction Analysis for Busbar Protection

Fault Type	Primary Protective Device Differential Relay	Secondary Protective Device Distance Relay or OCR	Malfunction Analysis	
			Differential Relay	OCR (or Distance Relay)
Busbar Fault	Trip	Not trip	Normal	Normal
	Trip	Trip	Normal	Misoperation
	Not trip	Trip	Failure to trip	Normal
	Not trip	Not trip	Failure to trip	Failure to trip

Suppose that a busbar differential relay (BDR) protects busbar B and its associated breaker Y. Let the overcurrent relay be OCR, which protects transformer T attached to the busbar B, and let each breaker on both sides of the transformer be X.

4.4.3.2.1 *Malfunction Analysis of Differential Relay for Busbar Protection*
A differential relay is considered to have failed to trip if a busbar fault occurred and the differential relay should have tripped as primary protection but it did not. The English version of the corresponding rule is as follows.

Rule 5: Failure to Trip of a Busbar Differential Relay   *If {[Busbar B is faulted] and [No SERs show differential relay BDR tripped breaker Y]}, Then Busbar Differential Relay BDR failed to trip.*
The equivalent logic formula for "failure to trip of a differential relay" is:

$$Malfs1(BDR) = \{Fault(B) \wedge [\sim SER_{BDR}(Tripping)]\}$$

A busbar differential relay is considered to have "misoperated" if busbar differential relay BDR should not have tripped but did. The English version of the corresponding rule is as follows.

Rule 6: Misoperation of a Busbar Differential Relay   *If {[Busbar B is not faulted] and [SER outputs show differential relay BDR tripping]}, Then Busbar Differential Relay BDR misoperated.*
The equivalent logic formula for misoperation of a differential relay is:

$$Malf2(BDR) = \{\sim Fault(B) \wedge SER_{BDR}(Tripping)\}$$

If at least one of these two malfunction conditions is satisfied, a busbar differential relay is considered to have malfunctioned. That is,

$$Mal(BDR) = Mal1(BDR) \vee Mal2(BDR)$$

4.4.3.2.2 *Malfunction Analysis of OCR for Busbar Protection*
For a busbar fault, an overcurrent relay OCR is considered to have misoperated if the overcurrent relay OCR should not have tripped but did. The English version of the corresponding rule is as follows.

Rule 7: Misoperation of a Busbar OCR   *If {[Busbar T is faulted] and [SER outputs show busbar differential relay BDR tripping] and [SER outputs show overcurrent OCR tripping]}, Then overcurrent relay OCR misoperated.*
The equivalent logic formula for misoperation of a busbar OCR is:

$$Mal1(OCR) = \{Fault(B) \wedge [SER_{BDR}(Tripping)] \wedge [SER_{OCR}(Tripping)]\}$$

An overcurrent relay is considered to have failed to trip if a busbar fault occurred, the busbar differential relay failed to trip, and the overcurrent relay should have tripped but it did not. The English version of the corresponding rule is as follows.

TABLE VI.4.3   Malfunction Analysis of BFD

Fault Type	Primary Protective Device (Distance Relay)	Breaker B	Local Backup Protective Device (BFD)	Malfunction analysis of BFD
	Trip	Open	Not trip	Normal
	Trip	Open	Trip	Misoperation
Line Fault	Trip	Not open	Not trip	Failure to trip
	Trip	Not open	Trip	Normal
	Not trip	Don't care	Not trip	Normal
	Not trip	Don't care	Trip	Misoperation

Rule 8: Failure to Trip of a Busbar OCR   *If {[Busbar B is faulted] and [No SERs show differential relay BDR tripped breaker Y] and [No SERs show overcurrent relay OCR tripped breaker X]}, Then overcurrent relay OCR failed to trip.*

The equivalent logic formula for "failure to trip of overcurrent relay" is:

$$Mal2(OCR) = \{Fault(B) \wedge [\sim SER_{BDR}(Tripping)] \wedge [\sim SER_{OCR}(Tripping)]\}$$

If at least one of these two malfunction conditions is satisfied, the overcurrent relay is considered to have malfunctioned. That is,

$$Mal(OCR) = Mal1(OCR) \vee Mal2(OCR)$$

4.4.3.3   Diagnosis Method of BFD

A breaker failure device (BFD) is supposed to intervene only when there is a distance relay tripping message but the associated breaker does not open due to breaker malfunction. If a breaker has not received a tripping command from a distance relay, the BFD is not supposed to send a tripping command to all other breakers connected to the busbar. Malfunctions of BFDs can be classified into two categories, i.e., failure to trip and misoperation.

Table VI.4.3 shows the possible operation combinations of a distance relay with the breaker failure device. Suppose that distance relay R protects line L between busbars A and B and their associated breakers, breaker A and breaker B. Let the BFD be the breaker failure device attached to busbar B.

For a line fault, BFD is considered to have misoperated if one of the following conditions is satisfied:

- The distance relay tripped, the associated breaker opened correctly, and BFD should not have intervened but it did.
- The distance relay failed to trip and BFD should not have intervened but it did.

The English version of the corresponding rule is as follows.

Rule 9: Misoperation of a BFD   *If {[Line L is faulted] and [SER outputs show distance relay R tripping] and [EMS status data indicates that breaker B_b is open] and [SER outputs show BFD tripping command]} or {[Line L is faulted] and [No SER outputs show distance relay R tripping] and [SER outputs show BFD tripping command]}, Then BFD misoperated.*

The equivalent logic formula for misoperation of BFD is:

$$Mal1(BFD) = \{Fault(L) \wedge [SER_{BFD}(Tripping)]$$
$$\wedge \{\{[SER_R(Tripping)] \wedge [TS_{Bb}(Open)]\} \vee [\sim SER_R(Tripping)]\}\}$$

BFD is considered to have failed to trip if a line fault occurred, the distance relay tripped, but the associated breaker did not open due to breaker malfunction and BFD should have intervened but did not. BFD is also considered to have failed to trip if a line fault occurred, the distance relay did not trip but BFD intervened. The English version of the corresponding rule is as follows.

Rule 10: Failure to Trip of a BFD   *If {[Line L is faulted] and [SER outputs show distance relay R tripping] and [NO EMS status data indicates that breaker B_b is open] and [No SER outputs show BFD tripping command]}, Then BFD failed to trip.*

The equivalent logic formula for "failure to trip" of BFD is:

$$Mal2(BFD) = \{Fault(L) \wedge [SER_R(Tripping)] \wedge [\sim TS_{Bb}(Open)] \wedge [\sim SER_{BFD}(Tripping)]\}$$

If at least one of these two malfunction conditions is satisfied, BFD is considered to have malfunctioned. That is,

$$Mal(BFD) = Mal1(BFD) \vee Mal2(BFD)$$

4.5   Software Design

4.5.1   GAAM Software Structure

The GAAM software is implemented in an object-oriented environment using $C++$ programming language. Object-oriented programming (OOP) provides a useful structure in developing and modifying software. OOP allows power system components and various protective devices to be represented in class modules. To achieve the goals of OOP, the class module selection and hierarchy implementation have to be considered at the initial stage of system design procedures.[17] The class structure of the GAAM software can be classified in two parts. The first part is the attributes of power system components such as line, busbar, and transformer. The other part is the definition of classes that represent physical behaviors of protection devices such as relays and breakers.

The following classes are created to represent different protection schemes for protected components such as transmission lines, busbars, and transformers:

- LineProtectionNet: object that represents the protection procedure related to lines
- BusbarProtectionNet: object that represents the protection procedure related to busbars
- XfmrProtectionNet: object that represents the protection procedure related to transformers

On the other hand, there are common characteristics in the operating principles and methods of protective devices such as distance relay, differential relay, OCR, and BFD. Therefore, class modules for these protective devices contain common procedures. To increase the degree of inheritance among these protective devices, a desirable class hierarchy can be developed so that code can be re-used. The following classes are created and inherited from other classes to form a good hierarchical structure:

- Breaker (objects representing circuit breakers)
- ImpRelay (objects representing impedance relays with 1st step transition)
- BusbarDiffRelay (objects representing busbar differential relays)
- XfmrDiffRelay (objects representing transformer differential relays)
- OverCurrentRelay (objects representing overcurrent relays)
- BFD (objects representing breaker failure device)

Class hierarchies of the GAAM for protective devices are shown in Figure VI.4.10. The fault and malfunction analysis algorithms of protective devices are implemented as a member function in each class module. Whenever there is a need for modifying the relay operation logic and protection scheme or adding new types of protection devices, a programmer can take an existing object and add features to it without major change.

The class hierarchies of the GAAM for the protection net are shown in Figure VI.4.11.

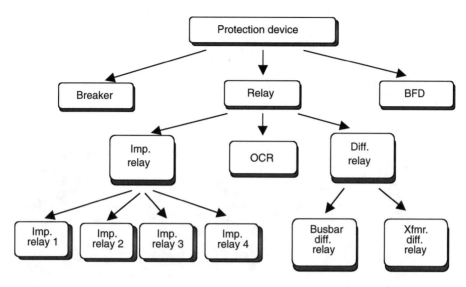

FIGURE VI.4.10   Class hierarchies of the GAAM protective devices.

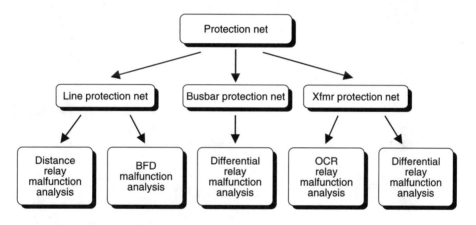

FIGURE VI.4.11   Class hierarchies of the GAAM for protection net.

4.5.2   Man–Machine Interface of the GAAM

As shown in Figure VI.4.12, four windows are available for the graphical user interface of the GAAM. The main window displays the part of the area of the power system where the possible fault location(s) exist. This area is automatically displayed based on the SER messages. Specifically, operators can use zoom commands in menu items to perform functions for zooming in and out of the network. The item "Full Screen" allows the window that displays the network to grow to the size of the main window. The "Find" menu allows the user to search and find a substation entering the name in a dialog box. The upper right window is designed to show fault diagnosis results, i.e., all hypotheses generated by the GAAM. Each hypothesis contains name(s) of fault location(s) or a list of malfunctioning devices. Two other windows are used to show relay and breaker operation messages from SER and EMS. Every message in those windows can be stored in a dedicated file.

The transmission lines in the main window are colored according to the protection operation. The purpose of coloring is to help operators understand where the fault is located and to confirm the textual results listed in the upper right window. To help the operator understand the fault scenario

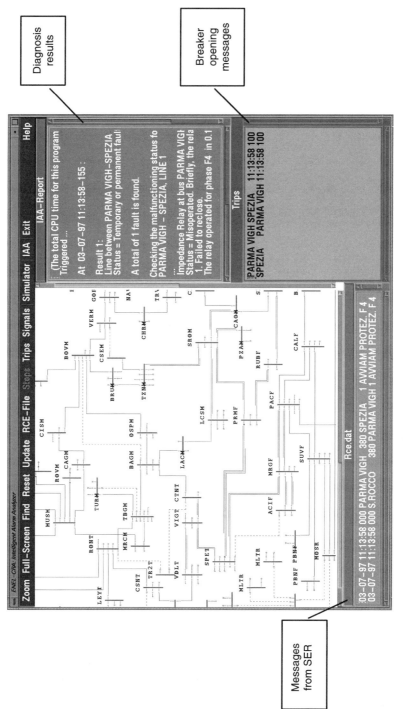

FIGURE VI.4.12   MMI of the GAAM.

and to support the textual results, the MMI provides a drawing of the protection operation on the scheme of the power system. In particular, each protection that operated to clear the fault is shown. For example, a yellow line segment represents the protection starting while a light blue line segment represents the operation of zone 1 of the distance relay. A green line segment represents the operation of zone 2 of the distance relay.

From the colors, the operator can understand which protection intervened and for which zone. According to the distance protection logic, all the protection devices in the faulted area should see the fault. So, the overlapping of colored lines coming from different protections identifies the component(s) where the fault is located. Finally, the coloring technique also allows the visual diagnosis of the protection devices (online and off-line). From the colored scheme, it should be evident to the expert which protection operated incorrectly.

The MMI of the GAAM also provides a simulator that was developed for testing and operator training purposes. With the simulator, operators can specify the fault(s) or malfunctioning device(s) to be simulated. Then, the simulator generates SER and EMS messages for a given fault/malfunction scenario. In the simulator, the power system network is represented by data obtained from an off-line database. The off-line database is updated after a simulation is completed. The graphic display is also updated as the network status changes.

4.6   Test Results

4.6.1   Case Study of Italian Power System

Validation and verification of the proposed fault diagnosis system was required before the GAAM could be installed in a power system control center. The proposed method of GAAM has been validated through testing with various test cases from the simulator as well as 51 actual fault scenarios from the 220–400 kV transmission network of the Italian power system. The test system consists of 96 400-kV substations and 20 220-kV substations. In total, it includes 403 components, i.e., 116 substations, 160 transformers, and 127 generators. A contextual representation method was developed for verification of logic-based fault diagnosis systems.[18,19] The purpose of the contextual representation method is to identify the equivalence relation among all the domain scenarios and derive a minimal set of basic contexts. An exhaustive testing of all possible scenarios can be avoided by testing the scenarios in the minimal set.

For illustration, consider the following test case that includes the operation of an overcurrent relay to protect the transformer against an external fault. As can be seen in Figure VI.4.13, a permanent fault occurred on the 380-kV transmission line between substations Nave and Flero, the NAVE relay sent a tripping signal toward FLERO, and then the associated breaker opened.

Table VI.4.4 shows part of the SER outputs at substations connected to the Flero substation.

In Table VI.4.4, the Signals column shows the SER short messages in Italian as they are sent from the substations to the control center. The Comments column reports their meaning.

According to SER outputs and EMS status data, there is no relay-tripping event from FLERO to NAVE. Due to malfunction of the distance relay at FLERO toward NAVE, breaker FLERO-NAVE remains closed. As backup protection, therefore, the remote distance relay protection intervenes at its second zone and overcurrent relays related to ATR1 and ATR2 also send an opening command toward the breaker on the low side of each transformer. The transformer overcurrent protection intervened to protect the transformer from an external fault. The final scenario of the breaker opening is reported in Figure VI.4.13. It should be noted that the specific sequence of opening is because the Flero substation is not equipped with the breaker failure device.

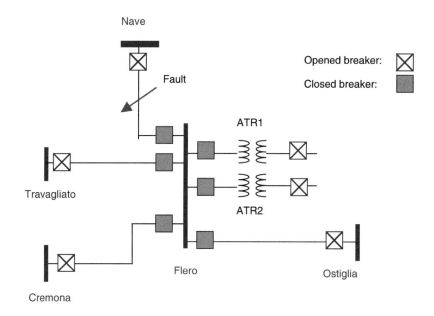

FIGURE VI.4.13   Test case.

TABLE VI.4.4   SER Outputs for Test Case

Date and Time (msec)	Location	Signals	Comments
10-03-95 22:14:30 350	FLERO → NAVE	AVVIAM PROTEZ.OMOP	Start protection
10-03-95 22:14:30 350	NAVE → FLERO	AVVIAM PROTEZ.OMOP	Start protection
10-03-95 22:14:30 405	NAVE → FLERO	SCATTO PROTEZ. F 8	Trip phase B
10-03-95 22:14:30	NAVE → FLERO	APERTO INTERR. P 8	Beaker open for phase B
10-03-95 22:14:31 210	OSTIGLIA → FLERO	APERTO INTERR. P 8	Beaker open for phase B
10-03-95 22:14:31 210	OSTIGLIA → FLERO	APERTO INTERR. P12	Beaker open for phase C
10-03-95 22:14:31 210	OSTIGLIA → FLERO	APERTO INTERR. P 4	Beaker open for phase A
10-03-95 22:14:31 210	CREMONA → FLERO	APERTO INTERR. P 8	Beaker open for phase B
10-03-95 22:14:31 210	CREMONA → FLERO	APERTO INTERR. P12	Beaker open for phase C
10-03-95 22:14:31 700	FLERO → ATR1¼	APERTO INTERR. P 8	Beaker open for phase B
10-03-95 22:14:31 700	FLERO → ATR2¼	APERTO INTERR. P 8	Beaker open for phase B
10-03-95 22:14:32 400	NAVE-FLERO	INTERV RICH. AUTOM.	Auto reclosure intervened

Based on the SER outputs and EMS status data, the GAAM generates the following hypothesis that represents the true scenario of an event.

Fault diagnosis result from the GAAM:

Hypothesis No. 1

At 10-03-95 22 : 14 : 30-350

> *Line No.* 1 *between FLERO-NAVE*
>
> *Status* = *Permanent fault on phase F*8
>
> *Hypothesis Index* = 0.800
>
> *The malfunctioning relays/breakers are:*
>
> *Impedance Relay at bus FLERO (line no. 1 between FLERO-NAVE)*
>
> *Status* = *Misoperated. Briefly, the relay:*
>
> 1. *Failed to trip. (Redundant relay also failed.)*
>
> *A Total of 1 fault is found.*

The previous diagnosis is consistent with the SER messages and data received and it is also correct.

4.7   Conclusions

Due to advances in computer and communication technologies, event messages from recording devices such as SERs and digital fault recorders can be made available in real time to the power system control center for online fault diagnosis. Based on the hierarchical structure of the protection system, SER information, and breaker status from EMS data, the GAAM is able to diagnose faults on power system components such as transmission lines, busbars, and transformers. It also provides the ability to analyze possible malfunctions of the protection system. The proposed method uses logic formulas to perform the analysis of fault and malfunctioning devices such as relays and breakers. The embedded logic formulas in the GAAM can be updated when the GAAM is applied to different power systems that may have different protection schemes. In addition, object-oriented design features of the GAAM software are portable from one power system to another. The current version of GAAM has been completely validated through testing available network scenarios for both 400 kV and 220 kV on the Italian electric network. The test results have shown the effectiveness of the proposed fault diagnosis system.

4.8   Acknowledgments

This research was sponsored by GRTN and EPRI under Contract WO8858-01. A number of graduate students at the University of Washington have contributed to the development of GAAM during the project, in particular, Mingguo Hong and Juhwan Jung. From the GRTN side, Rossella Baffa developed the MMI and engineered the system. We thank Mauro Mocenigo, GRTN, and Dario Lucarella and Massimo Gallanti, CESI, for their strong support of this project.

References

1. Wollenberg, B.F., Feasibility study for an energy management system intelligent alarm processor, *IEEE Trans. Power Syst.*, 1, 241, 1986.
2. Kirschen, D.S. and Wollenberg, B.F., Intelligent alarm processing in power systems, in *Proceedings of the IEEE*, IEEE, New York, 1992, 663.
3. Fukui, C. and Kawakami, J., An expert system for fault section estimation using information from protective relays and circuit breakers, *IEEE Trans. Power Delivery*, 1, 83, 1986.
4. Vale, A.Z. and Machado, M.A., An expert system with temporal reasoning for alarm processing in power system control centers, *IEEE Trans. Power Syst.*, 8, 1307, 1993.

5. Minakawa, T. et al., Development and implementation of a power system fault diagnosis expert system, *IEEE Trans. Power Syst.*, 10, 932, 1995.

6. McArthur, S.D.J. et al., The application of model based reasoning within a decision support system for protection engineers, *IEEE Trans. Power Delivery*, 11, 1748, 1996.

7. Rodriguez, C. et al., A modular neural network approach to fault diagnosis, *IEEE Trans. Neural Networks*, 7, 326, 1996.

8. Fushuan, W. and Chang, C.S., A probabilistic approach to alarm processing in power systems using a refined genetic algorithm, in *Proceedings International Conference on Intelligent Systems Applications to Power Systems*, IEEE, New York, 1996, 14.

9. Park, Y.M. et al., A logic based expert system for fault diagnosis of power system, *IEEE Trans. Power Syst.*, 12, 363, 1997.

10. Sidhu, T.S. et al., An abductive inference technique for fault diagnosis in electrical power transmission networks, *IEEE Trans. Power Delivery*, 12, 515, 1997.

11. Liu, C.C., Sforna, M., and Miao, H., On-line fault diagnosis using sequence-of-events recorder information, in *Proceedings International Conference on Intelligent Systems Applications to Power Systems*, IEEE, New York, 1996, 339.

12. Miao, H., Sforna, M., and Liu, C.C., A new logic-based alarm analyzer for on-line operational environment, *IEEE Trans. Power Syst.*, 11, 1600, 1996.

13. Jung, J. et al., Logic and validation techniques for handling of missing information in fault diagnosis, in *Proceedings International Conference on Intelligent Systems Applications to Power Systems*, IEEE, New York, 1999, 278.

14. Horowitz, S.H. and Phadke, A.G., *Power System Relaying*, 1st ed., RSP, New York, 1992.

15. Blackburn, J.L., *Protective Relaying*, 2nd ed., Marcel Dekker, New York, 1998.

16. Jung, J. et al., Multiple hypotheses and their credibility in on-line fault diagnosis, *IEEE Trans. Power Delivery*, 16, 225, 2001.

17. Pollinger, S.J., Liu, C.C., and Damborg, M.J., Design guidelines for object-oriented software with an EMS man-machine interface application, *Electrical Power Energy Syst.*, 14, 122, 1992.

18. Hong, M. et al., Verification of logic-based fault diagnosis systems using contextual representation of domain instances, in *Proceedings International Conference on Intelligent Systems Applications to Power Systems*, IEEE, New York, 1997, 127.

19. Jung, J., Liu, C.C., and Gallanti, M., Automated fault analysis using intelligent techniques and synchronized sampling, in *Tutorial on Artificial Intelligence Applications in Fault Analysis*, IEEE, New York, 2000, 24.

5

Application of Intelligent Computing for Short Term and Online Electric Load Forecasting

5.1 Introduction VI-136
5.2 Description of the Problem of Load Forecasting ... VI-136
5.3 The Traditional/Standard Algorithms for Load
    Forecasting VI-137
    Time Series • Regression
5.4 The Application of Intelligent (Soft) Computing
    to Load Forecasting VI-138
    The Tools
5.5 Searching for the Electric Load–Weather
    Temperature Function VI-140
    Electric Load as a Function of the Weather Temperature •
    Results
5.6 A System for Online and Short-Term Load
    Forecasting VI-145
    The Artificial Neural Network Architecture • The Sys-
    tem Engineering and Functionality • The Neural Network
    Architecture of the Short-Term Load Forecaster • The
    Online Error Corrector • The Online Error Estimator • Field
    Test of the Online Corrector • The System Man–Machine
    Interface • Results of the Field Test
5.7 A Hybrid System for Short-Term Load Forecasting
    of Anomalous Load Periods VI-160
    The ANN 1 Basic Model • Day-Type Classification
    by Means of Unsupervised Learning • The Unsuper-
    vised/Supervised Approach • Case Studies • Discussion
    of the Results
5.8 The Neural Network Interpolation of the Load
    Demand VI-169
    The Problem of Numeric Interpolation • An Initial
    Approach to Neural Interpolation • The Final Neural
    Network Interpolator • Results
5.9 Conclusions and Recommendations VI-175
Acknowledgments VI-176

Marino Sforna

GRTN — Italian Independent System Operator

References VI-176

Appendix A: The Group Method of Data Handling

    Algorithm VI-178

5.1   Introduction

In the early 1990s, ENEL, the former Italian national power company, addressed R&D in the field of artificial intelligence. One of the branches that proved to be successful was load forecasting leading to some applications that are presently in use at the Italian independent system operator (called GRTN). Artificial intelligence techniques are now internationally recognized as ripe and essential tools to face a set of power system problems whose solutions would not be satisfactory with traditional approaches. Neural networks and fuzzy logic are user-friendly and effective tools. They can be used alone, in combination with one another, or with traditional techniques. This versatility can successfully confront a number of problems in the areas of prediction, regression, classification, optimization, and control, which cover almost every field of the power system, especially load forecasting. The following is a comprehensive description of the activities in this field that in 10 years time have led to several prototypes and two stable working software applications and have also paved the way to several other minor tools.

To reach a wider audience, the following description is intentionally qualitative and academic rigor has sometimes been sacrificed in the interest of stimulating intuition and understanding among readers.

5.2   Description of the Problem of Load Forecasting

The aim of short-term electric load forecasting (STLF), limited to the 24 to 48 hours of the day following the forecast, is to provide information to be adopted in operational scheduling and in security functions of the energy management system such as unit commitment, economic dispatch, hydro-thermal coordination, and load management. Considerable economical and operational factors, in particular during the rapidly increasing load in the early hours of the morning and the daily load peak, are also related to the accuracy of the forecast.

Traditionally, STLF is a statistics problem where a lot of data (the past loads) can be used to predict a limited number of future values. One of the tricks in STLF is to find a way to get the maximum information from the past series of load data. Behind the simple weekly and yearly periodicities in the series, there are spasmodic irregularities that make this task quite difficult. The main irregularities are correlations with weather variables and unusual consumption patterns for public holidays. In addition, for each company, specific disturbances caused, for example, by special tariff days or partial public holidays complicate the forecast. Considering these influences, to build an efficient forecasting model, several conditions must be satisfied:

- The relevant variables which have a strong correlation with the load time series must be clearly identified and recorded, e.g., temperature and cloudiness.
- The accuracy of weather variable forecasts must not have too big an impact on the accuracy of load forecasting.
- The model must be able to extrapolate extremes such as cold weather conditions or heat waves.

Additionally, whatever method is used to model the load series, two difficult steps remain:

1. The identification of the model and the estimation of its parameters from a set of load data (also called the training set). This requires a large amount of data, especially for non-linear models that generally have a high number of degrees of freedom.
2. Assessing the model generalization ability that is usually evaluated using an independent set of data (i.e., those not used for the identification and the estimation of the parameters). This second set (also called the test set), like the training set, must be representative of the phenomenon.

The Italian load is approximately 49,000 MW (year 2000) during the national winter peak. This power is supplied by a prevalence of thermal production and by a strongly regulated energy exchange with the bordering countries. This huge load economically justifies the presence of a team of skilled human forecasters working in shifts to cover a period ranging from early morning to late evening every day.

Previous studies show that the Italian national load is weakly affected by independent weather variables. This characteristic is mainly due to the latitude extension of the country, the limited use of domestic appliances for home temperature correction, and the prevalence of industrial components (75% of the total) which contribute to the load inertia. On most days, the forecast follows the values of the related days located at a fixed time-series periodicity (i.e., seven days before), a condition that helps the forecasters' job. Unfortunately, sometimes a mix of weather conditions and socio-cultural events induces a still unpredictable influence on the forecast with appreciable errors. Despite these difficulties, the average error of the forecasters ranges between 1and 2% of the hourly load for normal days. Note that an estimation of the white noise related to the national Italian load ranges from 0.5 to 1%, values that meet international experiences and leave the field open for improvement by approximately 1%.

In the 1980s, short-term human forecasting activity was also supported by an elaborate adaptive bivariate Box–Jenkins statistical analysis. Considering the forecasters' performances, an error margin of between one and two percentage points was desired from that tool. This objective was partially reached during normal days by using human experts and traditional statistics already in use (ARIMA Box–Jenkins models). The quality of these results was largely due to the repetitiveness of the shape of the daily electric load because of its huge importance and because of the national demand inertia. Less satisfactory results were usually obtained on days characterized by social-cultural events of national interest such as strikes, bank holidays, or sporting events. In any case, a better understanding of the influence of weather factors (temperature, cloud, wind speed, etc.) on the electric load should lead to more precise forecasts. Over the years, the ARIMA model provided results close to human performance for normal days, but since 1995, it has been abandoned due to the complexity of the technique regarding benefits.

5.3 The Traditional/Standard Algorithms for Load Forecasting

Traditionally, short-term load forecasting is achieved by statistical methods such as multiple linear regression and stochastic time series, mainly in the form of Box–Jenkins analysis, but exponential smoothing and Kalman filter are also used. Despite the huge effort in this field, the ambiguous results obtained paved the way to the recent alternatives offered by artificial intelligence and, in particular, neural networks. The standard methods can be split into two approaches: time series methods and regression methods.

5.3.1   Time Series

The idea behind time series methods is that all the necessary information to forecast a series is in this series itself. In this case, the electric load series is described as a signal with periodicities that enable the forecast to be made. The most famous methods are Kalman filtering, the Box–Jenkins method, ARMA (auto-regressive and moving average), and spectral analysis (especially Fourier series decomposition). Generally, time series methods work well unless a rapid change in the environment occurs. This irregularity could be, for instance, a weather change or a public holiday. The problem comes from the fact that the load series is not, in practice, a regular periodic series.

5.3.2   Regression

In contrast, regression models assume that the series has high correlation with exogenous variables. In our situation, the forecast of the load series would be built using weather variables. The procedure is composed of three steps: first, identify the most correlated exogenous variables; second, choose an appropriate function; third, evaluate the parameters. The easiest model is linear regression, but the accuracy of such a model for the load series is insufficient because the relation between electric load and weather conditions is not linear. The hybrid solution, which consists of taking into account both time-lags in the series and exogenous variables, is better.

5.4   The Application of Intelligent (Soft) Computing to Load Forecasting

The state of the art due to the large number of projects in the area of electric load forecasting is the following:

- Using Kohonen networks to somehow cluster raw data and, by this, to decide on the relevance of input data.
- Promising results involve multi-layer perceptrons trained with some back-propagation methods.
- Hybrid systems seem to perform better than the use of single tools.

Specifically, neural networks used as numerical interpolators are slightly outperforming the traditional approaches in normal weekdays. However, it is proven that for special days or special conditions, the potential flexibility introduced by neural networks leads to obvious advantages if they are correctly applied.

5.4.1   The Tools

Soft computing differs from conventional computing because, unlike hard computing, it is tolerant of imprecision, uncertainty, and partial truth. The role model for soft computing is the human mind. In many ways, soft computing represents a significant paradigm shift in the aim of computing, a shift which reflects the fact that the human mind possesses a remarkable ability to store and process information which is pervasively imprecise, uncertain, and categorically lacking. The guiding principle of soft computing is to exploit the tolerance for imprecision, uncertainty, and partial truth in order to achieve tractability, robustness, and low solution cost. Its strength is due to the fact that soft computing (even if it is based on hard computing) is an instrument of higher level in which some work of information/knowledge organization toward solving the problem has already been done.

The principal constituents of soft computing are fuzzy logic, neural networks, and probabilistic reasoning, with the latter subsuming belief networks, genetic algorithms, chaos theory, and parts of learning theory. What is important to note is that soft computing is not a collection of tools, rather, it is

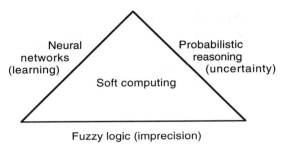

FIGURE VI.5.1

a partnership in which each of the theories contributes a distinct methodology for addressing problems in its domain. In this perspective, the principal contributions of the theories are complementary rather than competitive. The soft computing domain can be represented by Figure VI.5.1.

Being complementary has an important consequence on the methods: in many cases, a problem can be solved more effectively by using the three methodologies in combination rather than individually. A striking example of a particularly effective combination is what has come to be known as a neuro-fuzzy system. In the following subsections, a short introduction of every technique will be provided.

5.4.1.1 Load Forecasting Using Neural Networks

There is a large amount of publications about the use of neural networks for STLF; they mainly regard the use of the multi-layer perceptron in a surprising number of ways. This is due to the relative simplicity of the use of this neural network in comparison with more hostile statistical techniques. Unfortunately, it is only after a number of attempts that the scientific community recognized that the simple use of a neural network could not improve the knowledge of the human forecaster. It is probably in this abused field that the conviction was developed that the features of a neural network had to be linked to other tools in order to compensate for deficiencies. Besides the experience with recurrent neural networks, self-organizing maps, functional-link networks, and not fully connected networks, experts are oriented toward the application of traditional statistics and neural networks, different models of neural networks together, and neural networks with rule-based systems.

5.4.1.2 Load Forecasting Using Fuzzy Logic

There are two reasons for using fuzzy logic for load forecasting, both to improve the capabilities of the existing tools such as neural networks. The first possibility is to embed fuzzy logic inside neural networks to obtain fuzzy neural networks. The second alternative is, more properly, to apply fuzzy logic as a tool to get and use the knowledge that the human forecasters are able to explain, leaving other tools (neural networks) the possibility to gather only the hidden (implicit) information. The first approach is a matter of machine learning (hybrid solutions) that lets the system do all the work of getting the knowledge and providing a suitable architecture. More interesting is the second approach where an effort is made to gather knowledge from the forecasters and to embed it inside fuzzy rules. This approach has the power to work like a forecaster, particularly for special days for which the decision problem is less addressed. In fact, several fuzzy factors act simultaneously to determine the load shape: common weather variables such as heavy rain, very clear day, very warm weather, etc., and peculiar variables such as load shapes quite similar to Christmas, or almost equal to the day before Labor Day, etc.

Fuzzy rules can easily represent the percentage of the load that is influenced by independent variables such as weather conditions. These rules can be extracted by correlation functions, by forecasters' knowledge, or by rule extracting mechanisms, but it must be considered that the rules might not cover all the possible load-independent variable relationships. If this is the case, the gap can be fulfilled by a neural network working in parallel with the fuzzy system.

As a last valuable application, membership values can be associated to each day of the year by using a clustering technique (an unsupervised neural network, for example) which can also screen the pure pattern for each cluster. Then, the forecasting of a particular day can be obtained with an estimation of the memberships of the day for which a prediction is required by composing a fuzzy blend of the predicted pure load shapes.

5.5   Searching for the Electric Load–Weather Temperature Function

It has already been stated that weather factors strongly influence the accuracy of the forecast. Experience suggests that for Italy, weather temperature seems to be the most influential factor. Temperature also has the characteristic of being relatively simple to precisely measure, so detailed data on a national scale are available and reliable; these features are more difficult to obtain regarding other weather factors.

Prior to this study, the link between load and temperature in Italy had not been defined even though human forecasters have always applied some heuristic rules. This uncertainty was due to a number of factors. The most important is the latitudinal extension of Italy, which involves an average difference of 5°C between temperatures in the north and the south. Consequently, the following have been observed:

1. The marginal influence that temperature has on the national load
2. The lack of representative temperature variables associated with areas that are not homogeneous electrically and geographically
3. The necessity to consider the temperature measurement recorded times to avoid a peak load related to an off-peak temperature

An analysis of the particular characteristics of the electric load was not performed, only a statistical study of its past measurements. However, to support observation 1 above, it seems that, in terms of quality, mainly the domestic load is influenced by temperature (in 1990, the year of the analysis, the Italian power company sold approximately 200 TWh of energy, 65 TWh, i.e., 35%, to domestic customers). In addition, the widespread use of fuel heating during the winter and the marginal use of air conditioners during the summer suggests that the domestic electric load correction of weather temperature is rare.

To fully understand this marginal influence, a study was done on a monthly basis throughout 1990 (for which a complete historical series of past variables was available) trying to correlate the maximum and minimum temperatures with the peak load and the total energy of the corresponding day. This study regarded both national and pluri-regional areas (departments). The reason to extend the study to departmental areas comes from the observation that there is a greater link between temperature and electric load when they refer to small geographical areas. In breaking down the territory into subdivisions, a single temperature—that of the departmental headquarters—was correlated with that of the entire departmental load.

To carry out this study, a commercial computer code with mathematical heuristics associated with a regressive analysis of recent development and based on a theory known as the group method of data handling (GMDH) has been adopted. It should be noted that the aims of the study are both:

1. To determine the relation between independent and dependent variables representative of the physical phenomenon in question, i.e., the transfer function between the temperature and the electric load
2. To assess the potentiality, accuracy, and effectiveness of the method adopted

In particular, the innovating feature of the mathematical model used has permitted one person to develop about 450 transfer functions (including data management) in 40 working days. It automatically constructs various functions of one or more independent variables and then chooses the function which is both the simplest and most approximate to the proposed set of dependent variables.

5.5.1   Electric Load as a Function of the Weather Temperature

As stated before, due to the latitudinal extension of Italy (from 37° to 46° north latitude), a marginal influence of only one temperature on the national load was experienced. So, the study focused on regional or homogeneous multi-regional areas in order to increase the connection between the two variables. In particular, Italy has been subdivided into 8 areas with similar socio-cultural characteristics and similar maximum and minimum temperatures of the towns, within the area and with the main load, as a reference. For the national load, the study has been conducted by using only 4 temperature zones chosen among the eight places of measurement as the most representative both for the load and for their geographic position (see Figure VI.5.2).

Before being used for the study, all the variables were subject to preliminary elaboration to eliminate possible anomalous electric load behavior, which almost certainly does not correlate with temperature changes. In particular, these were eliminated:

1. Saturdays, Sundays, and bank holidays
2. Days between weekly holidays
3. Days on which there was a great difference between the load and the expected characteristic values, while temperatures did not show the same behavior
4. Days on which temperature reports were not available or reliable

From the other normal days, the maximum national and departmental load was extracted from the values per hour (for every day), and the total daily energy was calculated (sum of 24 loads).

The load and temperature values, treated as indicated, were considered both globally (the total year) and in monthly subdivisions and then applied to the aforementioned algorithm in order to determine the following relations (or transfer functions):

$$Load_{max} = F(Temp_{max}) \qquad Load_{max} = F(Temp_{min})$$
$$Energy = F(Temp_{max}) \qquad Energy = F(Temp_{min})$$

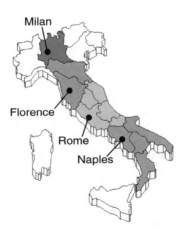

FIGURE VI.5.2   Italian departmental subdivisions. Milan, Florence, Rome, and Naples are chosen as temperature landmarks for the national load. (From Sforna, M., *Electric Power Syst. Res.*, 1, 1, 1995. With permission from Elsevier.)

The software used was able to:

- Choose the variable that has the greatest influence on the dependent variables (load or energy) from the 8 independent variables (4 maximum and 4 minimum temperatures)—this in the case of the national study.
- Choose the variable that has the greatest influence on the dependent variables (load or energy) from the 2 independent variables (maximum and minimum temperatures) locally registered—this in the case of the departmental study.
- Give the transfer function between the dependent and independent variables with the desired approximations.

The result of the elaboration was a polynomial of pth degree with the value of p being dependent on the degree of precision desired. This is the polynomial that, with the chosen precision, fits the series of values given at their best. Alternatively, the polynomial would identify the function of the curve that best approximates the indicated points. The quality of the approximation has been controlled observing the mean squared error (MSE) of both the calculated values and the real values. All values recorded for MSE are between 2 and 5%.

Since one aim of this study was to provide an easy tool for operators in charge of load forecasting, equations of higher orders have been simplified with the aim of eliminating components of a higher frequency from the functions. In fact, high order equations present no physical evidence for the problem studied. Then, the degree of p was limited to the third ($p = 3$). This choice of simplicity has generally led to a negligible increase in MSE.

Another point to be considered is that the reliability of the results with the method used (also with any other tool) depends on the number of examples that are being observed. Preliminary studies have shown that the relationship between electric load and weather variables is essentially a monthly function. This restricted the range of time to the limited number of days of a month, further reduced by the elimination of public holidays, which tend to have different characteristics from working days and which would require separate treatment. These considerations reduced the number of days for the monthly study, for some months (such as January and December) to about fifteen. At best, the number of days is not more than 23. Despite this limitation, it was possible to use results from only 1990 data even for the current situation. This opinion is backed up by observations made in the period prior to the study (April–June 1992) which have not shown any relevant socio-cultural differences from figures for 1990. The validity of the previous hypothesis should remain unchanged for the next years, and if socio-cultural growth maintains the present trend, the correlation obtained should become even stronger. For facility of use by forecasters, the transfer functions obtained from the study have been calculated at a significant temperature interval and the results have been graphically represented. A small group of these will be shown later.

5.5.2   Results

As an initial research activity, an assessment of in how many and in what time periods the year could be subdivided was made in order to make the correlation more effective. The following hypotheses for departmental loads (Milan, Rome, and Palermo) representing the north, the center, and south of Italy have been evaluated:

- Seasonal periods
- Three-month periods within seasons
- Pairs of two-month periods within seasons
- Monthly periods

It has been observed from this study that for the Italian meteorological characteristics, notably variable within the same season, short periods were recommendable. On the contrary, this assumption was in contrast with the need to have data availability as comprehensive as possible. This compromise led to the aforementioned condition to adopt a monthly link that also helped to reduce the influence of increase in load due to the growth in GNP. In addition, the link in proximity to holiday periods, when the load usually decreases for reasons independent of temperature, possibly giving misleading results in case of poor applications of GMDH to data, has been considered. In particular, the reduction of industrial activity during the summer period could produce a negative correlation, i.e., temperature increases and load decreases. Instead, a positive correlation could emerge during the winter period when the coldest days frequently coincide with the Christmas holidays in the northern hemisphere.

It has also been noted that the GMDH is inclined to provide, in some cases and for this study, an over-specialization of the data mainly because they are scarce. Therefore, the functions proposed have almost always been simplified. However, the main aim in the use of this method was to evaluate among several independent variables which was the most influential on the phenomenon examined. This skill has been used for the national study where load and energy have been considered as functions of 8 variables simultaneously in order to select the most significant of them.

The results have also been presented as diagrams for easy consultation by forecasters. Up to 4 curves, showing the double link between load–temperature and energy–temperature, have been drawn for each of the monthly graphs and also for the yearly graph. In order to be able to observe the shape of the transfer function over a wider range, the monthly interval of the actual temperature was raised up to 4 degrees in nearly all cases (2 less than the minimum and 2 more than the maximum).

The curves should be regarded as trends in the load and energy behaviors as functions of temperature and not as a reproduction of their actual shape. The higher the order of polynomials, the more this is true. In fact, a maximum and a minimum of the dependent variables in a temperature function can certainly cause perplexity. In this case, the intermediate interval at both extremes (around the inflection point of the curve) would be considered valid.

The assumption to only consider the straight interval of the curves allows one to extend their validity for incremental variables as well, i.e., the variation of the load due to a specific foreseen variation of temperatures.

Again remembering the aforementioned hypothesis 1 to 4, the x-axis shows the recorded interval ($+4$ degrees) of the local temperatures for the departmental studies and, for the national study, the most significant temperature of the 8 that GMDH has chosen in the year 1990 or in the monthly analysis. The y-axis shows the peak active power (left axis) and the daily energy (right axis). The values of the ordinates give an indication of the range of variation for peak power and daily energy on working days, but they do not represent their exact values.

In Figure VI.5.3, three out of thirteen graphs regarding the national load are reported. Graphs in Figure VI.5.3a show the yearly function between the load and maximum temperatures of Rome and between the energy and maximum temperatures of Naples. Among the eight temperatures of reference, these have the strongest influence.

Trends correspond to the following function, where temperatures of Rome and Naples are considered:

$$L_{\max}[MW] = 33348 - 178 T_{\max RO}$$

$$E[MWh] = 613149 - 1626 T_{\max NA}$$

As other examples, Figure VI.5.3b and c respectively represent the monthly correlation (February, July) for all of Italy. From the diagrams, a general negative link between load (or energy) and temperatures can be noted both for national and departmental areas. It may be explained by domestic temperature correction during the coldest days in the winter. On the contrary, other diagrams showed that during the summer, the city of Rome experiences a positive correlation due to air conditioning. Even if strongly reduced in amount, the same is registered in Sicily, where the presence of tourists is influential as well. Power system operators confirm these trends.

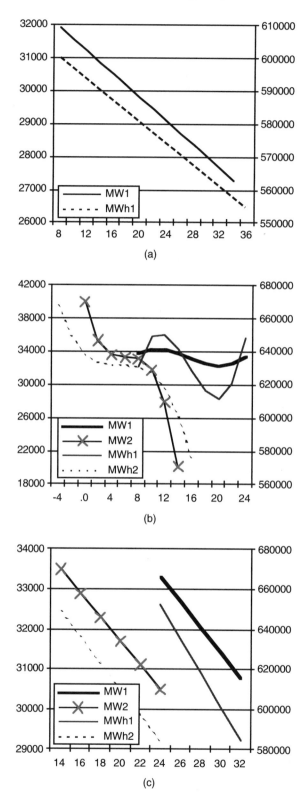

FIGURE VI.5.3 (a) National load (MW1) and daily energy (MWh1) as functions of temperature (°C); (b) Italy–February; (c) Italy–July. (From Sforna, M. and Proverbio, F., *Electric Power Syst. Res.,* 33(2), 139, 1995. With permission from Elsevier.)

By studying the results obtained, it has not been possible to characterize a univocal, general, and reliable law for the Italian electric load–weather temperature function as hoped at the beginning of the study. However, operators confirm that the curves proposed meet their expectations, even if in some cases complex functions may be misleading or lead to uncertain applications.

The validity of the method applied was proven in terms of simplicity of use, sufficient reliability of results, and quickness of reply. For problems such as classification, control, estimation, and diagnosis, the GMDH method is quicker, even if less accurate, than multi-layer neural networks. In addition, the GDMH provides a solution in a numeric form, with the desired degree of approximation, appearing to be an effective method of knowledge representation. Even the characteristic of neural networks to be adaptive can easily be reproduced by the GMDH. For the previous reasons, it is believed that some other power system applications could be successfully fitted to GMDH application as well.

The functions corresponding to the previous diagrams are the following:

$$L_{\max}[MW] = 20176 + 3209T_{\max FL} - 228T^2_{\max FL} + 4.9T^3_{\max FL}$$

$$L_{\max}[MW] = 40011 - 3234T_{\min NA} + 520T^2_{\min NA} - 28T^3_{\min NA}$$

$$E[MWh] = 277415 + 82342T_{\max FL} - 5756T^2_{\max FL} + 124T^3_{\max FL}$$

$$E[MWh] = 638267 - 3796T_{\min FL} + 720T^2_{\min FL} + 46T^3_{\min FL}$$

$$L_{\max}[MW] = 40951 - 318T_{\max RO}$$

$$L_{\max}[MW] = 37667 - 298T_{\min RO}$$

$$E[MWh] = 855676 - 8486T_{\max RO}$$

$$E[MWh] = 739581 - 6473T_{\min RO}$$

5.6 A System for Online and Short-Term Load Forecasting

It is well known that cheap and reliable power system operation definitely is the result of good short-term load forecasting (STLF), i.e., 24 hours ahead, but that is not all. In fact, what has not been predicted a day before must be corrected by the dispatcher in advance during the operation. So, STLF might be subdivided into a cascade of two tasks: the first is off-line (one day before), and the second is online, characterized by a prediction followed by a correction. In spite of what has always been considered theoretically, from the utility's point of view the two steps are strictly linked even if they are performed by two different teams at two different times. This hypothetical link is due to the fact that basically both forecaster and dispatcher have the same knowledge of the physical phenomenon hidden in the load function. The forecaster generates numbers triggered by the typology of the day, the weather variables, and the socio-cultural events. The dispatcher performs a pattern recognition analysis between the predicted load function and those stored in his mind; the error in the forecast primes his next actions. Finally, the supposed link should lead the power system dispatcher to recognize in advance what could be the time period of less credibility in the forecast, essentially due to the increase of unpredictable load components.

The previous considerations forced the testing of an innovative architecture to improve the traditional STLF and to introduce a new system for implementing the online load forecasting (OLLF), e.g., the online STLF correction, from 15 minutes to 2 hours ahead. Artificial neural networks (ANNs) seemed the most suitable tools for both.

This section introduces two NEUral FORecasters embedded in the same environment (called NEUFOR) but able to act separately if needed. The first forecaster is devoted strictly to STLF, while the second one is designed as a tool to support the power system operation and performs online. The reader's attention should be mainly focused on the online architecture, the man–machine interface, and the engineering of the system.

5.6.1   The Artificial Neural Network Architecture

Most ANN applications for STLF regard the use of a multi-layer perceptron as a pattern recognition tool with some generalization skills. For this aim and even in this system, a set of all possible load patterns constitutes the ANN training set for preliminary learning. Moreover, it is subsequently updated as new patterns intervene for further relearning. This approach can present two drawbacks due to the following limitations in the relearning procedure:

- The training set is not comprehensive initially.
- The training set becomes less comprehensive as old patterns are replaced by new ones.

A different approach evaluated and used in this system considers that all the useful load patterns are represented in the final load functions before the forecast, i.e., in the past 1 to 5 weeks. Moreover, the desired load pattern is produced not only as a generalization of the previous patterns, but step by step (hourly or quarterly) as a consequence of considering the load as a continuous function of time.

According to this view, recurrent neural networks seem to cope better with the sequential nature of time-dependent functions, including the power system load. Using this kind of network, the load function can be considered a sequence of frames instead of just a collection of patterns in an order of no influence. In this way, it is believed that the ANN is able to exploit all the possible additional information hidden in the temporal function of the load.

A particular class of recurrent NNs includes networks with a multi-layer architecture containing static and dynamic neurons (with feedback connections). These networks have been shown effective; in fact, they rely on the well known attitude in class discrimination of multi-layer structure, while they also provide temporal features as input. Moreover, these networks present other remarkable temporal features such as the capabilities to remember close sequences of frames and, subsequently, to forget those sequences which no longer introduce useful information in the input signal.

5.6.2   The System Engineering and Functionality

What led to a detailed engineering of the system was the decision to point out the design on an advanced graphic man–machine interface. In fact, when a neural network is under test and when a test regards a forecast, the graphical support is essential in both cases. In addition to these reasons and according to the performance of this system and those of similar international projects, the validation of the forecast by human skilled forecasters still remains the only method to guarantee the local confidence (not the global, e.g., during a month) in the predicted load. To support this human capability, which is mainly expressed as a visual recognition of a known pattern, the system has been provided with a window architecture. A second important aspect of the engineering is the modularity of the system. In fact, it has been conceived in two modules (short-term module and online module), which can be integrated or not according to the users' needs.

From the operational point of view, the system allows the human expert to perform the first learning and then control the following ones (relearning) by using the graphics. In fact, in the learning activity, the curves of the last weekly load function, before the forecast, are displayed with panning facilities, one day at a time together with the feedforward ANN output for the same day (see Figure VI.5.4). The load granularity is chosen by the forecaster hour by hour or 15 minutes by 15 minutes. In this way, the expert is able to control, by comparing the two curves,

1. The accuracy of the global learning for all the days of the past weeks
2. The accuracy of the local learning for the chosen day of the past weeks

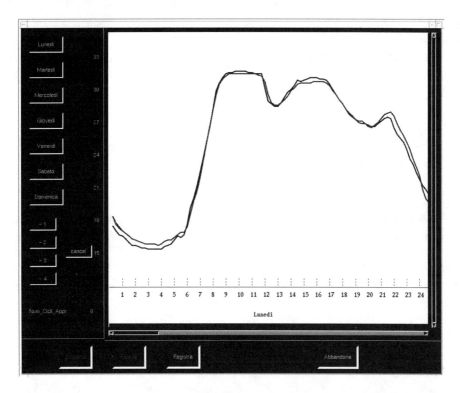

FIGURE VI.5.4   Black and white representation of the window devoted to performing and controlling the ANN learning. One of the curves represents the target load of a past Monday, the second one is the ANN attempt to learn the first curve pattern. From the top left, the first seven buttons allow the display of a particular day included in the training set. Instead, a particular week can be selected by using the next four buttons (from -1 to -4 weeks in the past, the last week is displayed). Other buttons are a set of commands to start and stop the learning and to quit NEUFOR or record the ANN weights. A panning facility to display the entire current week is available at the bottom of the diagram. (From Sforna, M. and Proverbio, F., *Electric Power Syst. Res.,* 33(2), 139, 1995. With permission from Elsevier.)

Moreover, the user has the capability to stop the learning when he believes it is satisfactory for his purposes. During the use of this feature, it has been noted that an accurate learning (where the curves almost perfectly match together) does not always yield an accurate forecasting. This agrees with the conviction that ANN must not copy the training set but generalize it. A second auxiliary window shows the weekly learning error.

When the learning is satisfactory, another window displays the forecasted load both as a shape and as numeric values, with the chosen granularity. Even in this case, the pattern recognition capabilities of the human expert are required. The expert has to decide if the forecast meets his expectations as to shape and values. If it does not, a relearning activity with the previous steps is readily available. On the contrary, the forecast is confirmed. Of course, it is possible to use the incoming real load (up to that moment) during the forecasting activity instead of the previous day's prediction (see the next paragraph for the ANN inputs). This validated load, which represents the link between the past and the future of the load function, substantially improves the accuracy of the forecast. A special starting window allows the definition of the period included in the training set. The longest choice is 5 weeks and the shortest is 1 week, depending upon the human expert's feeling about the influence that the past load has on the future one. This feature is used when rapid changes occur in the load, especially near summer or season holidays. If this is the case, a reduction in the training set period is suggested, while the 5 week period is the default option for a normal period.

The online module, used during the actual power system operation, has a window centered on the current instant, marked with time zero. The entire visualized time period ranges from −16 hours (or 16 quarters) to +16 hours (or 16 quarters) around the current time. Initially, the system displays the following curves:

1. Thirty-two steps of the short-term forecast (produced the last day)
2. Sixteen steps of the actual load as it comes from the field
3. The next 2 hours ahead of online correction with an hour or quarter granularity (refer to Sections 5.5 and 5.6 for a detailed description). This means 2 hours or 8 quarters in advance of correction, respectively. A different box in the same window shows the corrected load values.

In addition, to assess the reliability of the online correction and by request of the power system operator, it is possible to visualize a further curve representing the past predicted correction for a chosen quarter (or chosen hour) in advance. In other words, if the operator is controlling the load behavior, for example, one hour ahead, in order to perform the suitable generation adjustments, he can check the last 15 forecasted load corrections by comparing the "online one hour in advance" past curve with the actual past load curve. From this comparison, it is possible to infer with a degree of confidence the online corrector for that specific future hour (or quarter). As expected, time-closer steps are more reliable than distant ones.

5.6.3 The Neural Network Architecture of the Short-Term Load Forecaster

As usual, previous efforts were addressed to finding the best architecture for network efficiency and quickness in the periodic learning. Early successful attempts suggested the use of a dynamic multilayer network DMLN for each day of the week, according to the idea that each day is related to a specific load function. Further investigation led to dealing with days with similar load functions by using only one DMLN. So the number of ANNs was reduced from 7 to 5, observing that the Italian load shape was quite similar for Tuesday, Wednesday, and Thursday. The process was repeated including Mondays and Fridays and, finally, all the days in one ANN. This simplification stressing the ANN generalization capability introduced a considerable improvement in the system management with minor losses in the forecast accuracy. It has been recognized that essentially only one DMLN works well because there is a common load shape for every day of the year characterized by the double peak of load or *saddle*. This shape is captured in the hidden ANN layers. As a consequence, the characterization of the single day must be assigned to the input synapses.

According to the idea that all the useful information is included in the few weeks approaching the day to be forecasted, even during the feedforward operation the ANN has to work with these data. At least two tasks must be pursued:

1. The reconstruction of the load function, arising from a generalization of previous examples
2. The acknowledgement of recent load trends

Point 1 can be achieved by using as input the hourly load shape of the same day to be forecasted, belonging to the previous week. Point 2 is achieved by using as input the day before the day to be forecasted. Saturdays, Sundays, and Mondays do not follow this procedure because the load functions of these days differ from their antecedents and, as a consequence, they do not actively contribute to the desired load shapes. Following these considerations, the chosen architecture has 11 inputs dedicated to the weekdays (see Figure VI.5.5).

As usual, the number of neurons in the hidden layers comes from the best compromise following a parametric study. A similar approach has been used in choosing the most effective delay for the dynamic layer. To satisfy the effectiveness/time-consuming ratio, the delay has been set at one. A

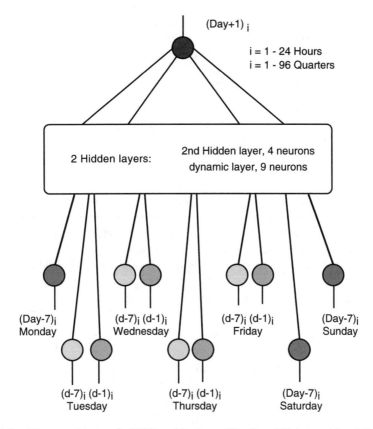

FIGURE VI.5.5 The neural network STLF architecture. "Day" or "d" is considered the day when the forecasting activity is performed. Each of the inputs is marked with the name of the day that will be forecasted. So, d-7 is the day 8 days before the target day of the forecast. Note that only this input was used for the task. (From Sforna, M. and Proverbio, F., *Electric Power Syst. Res.,* 33(2), 139, 1995. With permission from Elsevier.)

longer delay could be harmful for the forecast because of the rapid change in the gradient of the Italian load function, especially in the period around noon. Instead, tests confirm the benefits of a longer delay (not used) during the increasing trend, early each morning, when constant gradients of 7000–8000 MW/hour are common.

From the operational point of view, it must be noticed that on the day during which the forecast is made, e.g., Wednesday if Thursday is the forecasted day, experts have to know the load function of the previous Thursday and the entire load function of the present Wednesday. This information usually comes from Tuesday's forecast. This procedure can be changed by substituting part of the predicted load of Wednesday with the incoming load observed up to that moment, thereby representing the actual link between the past and the future. As expected, there is an improvement in the accuracy of the forecast as a consequence of this opportunity.

5.6.3.1 Performances of the STLF Module

A carpet assessment over a period of six months was performed. In particular, a simulator was set up in order to perform the test automatically in a contracted period of time. In addition, time-saving reasons have been suggested to forecast every whole-week load in only one step (7 steps in common use). Of course, this procedure introduced a cumulative error in the results due to the use of predicted hourly loads as inputs to forecast new hourly loads. In fact, the correct procedure foresees the use of actual incoming hourly loads as inputs. In Table VI.5.1, a list of the weekly errors provided by the use of NEUFOR for the STLF function and for a period ranging from February 15th to July 31st, 1993 (24 weeks) has been reported. In particular, numbers represent:

- First column: the weekly average absolute percentage error in the forecast
- From the 2nd to the 4th column: the average, maximum, and minimum absolute percentage error for the maximum daily load
- From the 5th to the 7th column: the average, maximum, and minimum absolute percentage error for the minimum daily load

Attention was focused on the maximum and minimum loads because both have a strong impact on short-term generation scheduling. Even if the ANN was not set to obtain the best performances for these two extremes, the listed results are comparable with those provided by other neural forecasters proposed in the international bibliography. Not obtaining acceptable results is due to special periods of the year such as during the Easter holidays or during the changes in seasonal weather conditions. These bad results were expected due to the fact that this preliminary study lacks information regarding socio-cultural events or weather conditions. To check whether the ANN could be valuable if it were

TABLE VI.5.1   Absolute Percentage Error for the Short-Term Neural Forecaster

Weeks	Average	Max. Load			Min. Load		
		Average	Max.	Min.	Average	Max.	Min.
1	2.09	2.31	3.40	0.36	1.96	4.11	0.17
2	1.70	0.34	1.54	0.25	2.07	4.67	0.30
3	2.33	1.31	2.44	0.51	3.15	5.93	1.05
4	2.12	1.37	2.98	0.33	1.38	2.21	0.38
5	2.09	1.12	3.22	0.37	4.11	5.75	2.58
6	3.43	3.04	8.19	0.83	1.90	5.71	0.00
7	3.62	2.63	5.64	0.54	3.20	5.45	0.11
8[a]	4.48	4.95	7.47	2.35	4.98	6.47	3.77
9[a]	3.24	4.04	15.5	0.13	4.03	11.6	1.31
10	3.42	1.78	4.56	0.64	3.55	7.44	0.00
11	2.68	1.92	2.83	1.30	4.08	16.5	0.53
12	2.15	1.08	2.94	0.00	2.35	4.11	0.45
13	1.33	1.22	2.61	0.34	1.38	2.91	0.14
14	1.35	1.87	2.64	0.17	1.18	2.66	0.07
15	1.79	1.12	4.11	0.06	2.16	7.84	0.07
16	1.51	2.03	6.02	0.09	1.57	5.55	0.10
17	1.90	1.01	2.96	0.07	3.49	8.24	0.04
18	2.51	1.45	2.41	0.60	4.78	10.1	2.06
19	1.30	1.52	2.45	0.10	1.43	2.42	0.82
20	1.21	1.02	1.99	0.10	1.25	2.56	0.03
21	1.68	1.46	2.24	0.27	1.63	4.06	0.60
22	3.51	3.63	8.50	1.60	4.64	7.69	1.23
23	1.76	2.53	3.54	0.47	1.65	4.85	0.33
24	1.71	1.60	4.70	0.44	1.66	3.40	0.31
Total	2.29	1.93	4.37	0.50	2.65	5.93	0.68

Note: From February 15 to July 31, 24 weeks.
[a] Easter weeks.

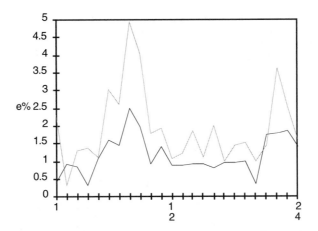

FIGURE VI.5.6   Absolute percentage error for the maximum daily load. Period from February 15 to July 31, 24 weeks. Upper curve: neural forecaster; lower curve: operator. The peak error is for Easter holidays. (From Sforna, M. and Proverbio, F., *Electric Power Syst. Res.*, 33(2), 139, 1995. With permission from Elsevier.)

put into daily operation, a comparison with the operators' performances during the same period of time was made. In Figure VI.5.6, the two curves represent the average absolute percentage error for the maximum load, both for the neural and human forecaster (e.g., the values of the upper curve are listed in the first column of Table VI.5.1).

From Figure VI.5.6 it can be noted that the ANN constantly presents a higher error in comparison with the operators' performances. The total means decline for the operator with an error of 1.18% for the maximum load instead of an error of 1.93% due to the ANN. This result is not satisfactory for Italian needs, but, fortunately, it does not depend on the ANN capabilities to cope with this task but instead stems from a lack of information in the data containing only the time series of loads. Besides these remarks, in the actual use of NEUFOR, it was noted that NEUFOR performs fairly well for the whole daily curve of the load during normal days (the main part of the year), so it can be considered a confirmation of the operator's activity.

5.6.4   The Online Error Corrector

Normally, the operator who controls the power system relies on the knowledge of the last STLF to compare with the current real load for evaluating (even by using optimization software) possible power generation adaptations. Besides, the operator acquires a feeling about the confidence of the forecast and adjusts his attitude to it daily by observing its errors. The aim of the online corrector is similar.

Specifically, it provides a correction of the STLF when compared with the actual load by using an ANN stimulated by the past errors of the forecast. In other words, the ANN first has to capture the past STLF error function; second, it predicts what could be the future STLF error for a limited number of steps. To define this task, it was decided that the knowledge of a possible STLF adjustment influencing the power production, for a maximum of 2 hours (15 minutes by 12 minutes), in advance is in the interest of the GRTN operation. This choice was a compromise between a reasonable confidence of the correction and a useful length of the period of time to set up a generation planning adjustment.

It seemed rational to think about the neural corrector as a regression algorithm working on the time series of STLF error, but with the valuable attribute of building higher level dependencies and the possibility of creating a link in the time series by means of the dynamic layer. Several variables were the object of a parametric study:

1. The architecture of the network and the more suitable time delay for the dynamic layer
2. The range of visibility of the past errors

In particular, the second variable above led to a compromise between a required generalization of the ANN's behavior on a sufficiently large pattern of examples and the observed necessity to adhere to the closest error. This last requirement is derived from the experimental observation that if the STLF makes a mistake, whether the forecaster is human or an algorithm, it persists for some steps with the same sign. A mobile window of the past 4 percentage errors from $e(t)$ to $e(t-3)$, where t is the current instant, proved to be a good compromise. Accordingly, the network architecture proven effective is shown in Figure VI.5.7.

During the operation of the online corrector, it was also noted that the error function could not be easily classified, mainly because it is heavily affected by a white noise component. In addition, the possible predictable component is a local characteristic of the function (if it were not, the STLF could be affected by a strong systematic error which could be easily detected by the first forecaster). This forced the use of only the most recent examples for the training, in particular the last 5 past patterns which alternate, considering that the oldest pattern will be rejected as a newer one arrives. Thus, the network is polarized on a local behavior of the error function as expected. In addition, to avoid the possibility of learning a trend in the error function every pattern is considered single, and the network memories are cancelled, passing from one pattern to another for the 5 available ones.

In the feedforward operation, e.g., for the quarter correction, the first network output $e(t+1)$, which corresponds to the first STLF load adjustment, will be considered the new network input $e(t)$. This involves a one-step shifting for all the input values and the loss of the oldest of them. It should be noted that after the first 4 steps out of the total 8, the network works only with the predicted values for the STLF errors as inputs and, as a consequence, the predicted error (up to the eighth) gradually becomes less accurate. As stated above, during the learning activity no DMLN features were used. On the contrary, during the prediction (and with the maintenance of the memories), the network should link each value with the previous one. This generates a function (curve) for the predicted error according to the conviction that the error should inherit the characteristic of being a time sequence from the load to which it is applied.

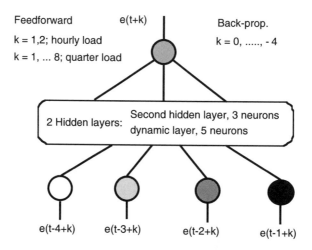

FIGURE VI.5.7   Architecture of the neural network online load corrector. (From Sforna, M. and Proverbio, F., *Electric Power Syst. Res.*, 33(2), 139, 1995. With permission from Elsevier.)

5.6.5 The Online Error Estimator

The need to introduce a further adjustment (after the corrector's action) stems from the observation that even if the Corrector is able to recognize and adjust the STLF systematic errors, the instantaneous errors still remain uncovered. In addition, from previous experience in the use of the system, it has been observed that the presence of a small undetected systematic error in proximity to a change in the gradient of the load function (passing through a maximum or a minimum) causes the corrector to overestimate its adjustments. The corrector usually acts a step after an error is detected. So, at least the first error skips its action. Readers should note that the problem of the correction of a forecast differs from the problem related to the correction of the behavior of a mechanism mainly because the target trajectory is unknown in the former. So, every further improvement must stem only from further external information. Thus, the idea was to check the corrector's errors in the recent past by means of an MLDN, here called an estimator, performing a double action:

1. To estimate the validity of the past corrector's actions, $A(t-j) - C(t-j)$—actual load minus correction—and eventually to propose an adjustment, $E(t+k)$
2. To distribute the adjustment $E(t+k)$ through future instants

The integration between the corrector and the estimator is expressed in Figure VI.5.8. The estimator includes a DMLN, as shown in Figure VI.5.9. In this case, every step ahead from the current instant relies on its own NN. In fact, to produce the $E(t+1)$ addendum, which will be used in the first step, the ANN is trained with examples whose relationship between input and output has a distance of one period of time (one hour, one quarter). Consequently, to obtain the $E(t+k)$ addendum, a training set with a distance of k periods of time between input and output is used.

It was noted that by increasing the number of patterns during the training (from 4 to 6), the estimator's action becomes smoother and equally divided for all the predicted errors. On the contrary, with fewer patterns (form 2 to 4), a resolute action concentrated on the first step ahead has been observed. Power system operation requirements suggest concentrating the actions on the first steps, especially when the maximum and the minimum load are forecasted, because the most important control actions depend on these values. Finally, the job of this network was rendered completely transparent for the operator. In fact, the only signal displayed regards the online load correction $C'(t+k)$, which had already been influenced by the estimator's contribution.

Other approaches have been proved effective for estimator implementation. For example, additional useful information could be located in the short-term forecaster. In fact, it can easily provide,

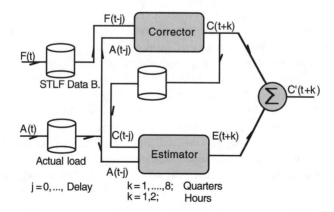

FIGURE VI.5.8 Flow chart of the module for the online correction of the STLF load according to the incoming actual load. (From Sforna, M. and Proverbio, F., *Electric Power Syst. Res.*, 33(2), 139, 1995. With permission from Elsevier.)

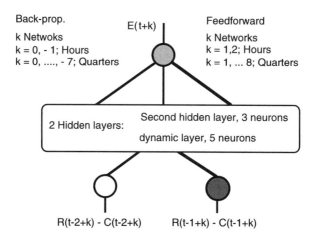

FIGURE VI.5.9 Architecture of the neural network online load estimator. This network represents a sub-module of the online corrector shown in Figure VI.5.8. (From Sforna, M. and Proverbio, F., *Electric Power Syst. Res.*, 33(2), 139, 1995. With permission from Elsevier.)

with sufficient reliability, the instant associated with the change in the load gradient to which the activation of the estimator could be linked.

5.6.5.1 Performances of the Online Corrector and Estimator Modules

In the following, the results regarding the application of both the corrector and estimator are listed. When possible, a comparison has been set with the results obtained by short-term human forecasting, which represents the best in operation up to now. Unfortunately, the online correction performance does not have any alternative procedure to compare to, and, in this case, only the reduction of the STLF error function has been evaluated.

Because the performances of the corrector should be seen globally and not locally, a simple statistical study over the entire test period has been done. From this study, the online corrector shows a considerable capability to adjust the forecast one hour in advance. In fact, there is a reduction both in the absolute and in the relative error made during the human forecasting, as shown in Table VI.5.2. It should be considered that the online adjustment centers the function of the actual load with an average relative error practically null. The best results, however, are noted by comparing the standard deviation of the STLF obtained by the forecasters with its online correction. The latter shows a scarce number of data, mainly centered around the average value. This is a favorable result for the electric operation because it ensures that the online correction of an error made in the forecast is contained with a probability easily measured within a restricted range. For example, it was calculated that 90.2% of all the data of the percentage absolute forecast error are contained within the $+/-\sigma$ extremes ($+/-1.002\%$). So, corrections above these limits, which may be suggested by the neural network, should be considered improbable. In other words, the risk of incorrectly applying very costly large corrections of the production schedule is very low.

From Table VI.5.2, it can be noted that the correction for the instant located 2 hours in advance shows a reduced correctness in comparison with the closest period of time. This was expected and is physically acceptable. Then, the correction 2 hours in advance has meaning only as a trend in the next future load.

5.6.6 Field Test of the Online Corrector

The previous system has been fully proven to be very powerful in the actual operation. However, extensive use and operators' close observations showed that its performance could still be

TABLE VI.5.2   Comparison Between the Human STLF Hourly Error and the Online Adjustment One Hour and Two Hours in Advance

	Percentage Absolute Forecasting Error		
	Online Corrector		Forecaster
	1° hour	2° hour	STLF
Average	0.843	2.851	1.250
Standard Deviation	1.002	1.935	1.268
	Percentage Relative Forecasting Error		
	Online Corrector		Forecaster
	1° hour	2° hour	STLF
Average	0.002	0.300	0.049
Standard Deviation	1.310	3.433	1.780

improved. In particular, dispatchers complained about poor effectiveness during two different situations:

1. When the STLF is extensively wrong
2. When the correction regards steps distant from the present time

Moreover, when the STLF is proven ineffective, dispatchers wish to have a correction that lasts for the rest of the planning (up to the end of the day), not just for 2 hours following.

Generally speaking, the poor performances attributed to the first point above were due to the required fine tuning of the corrector for a limited range of STLF errors $(+/-2\%)$ in order to obtain the best performances for the range of errors the forecasters are likely to obtain. In other words, the MLDN is saturated for errors greater than $+/-3\%$ according to the present project design. The second point, even more intuitively, shows that the correction is as inaccurate as the step is distant from the instant of observation. Even if it is evident that for the first situation a second neural system, tuned for greater errors, could be proposed, the research addressed finding a common solution for both the problems.

Operators suggested that usually, in a critical moment, they mentally rely on the historical series of loads in order to find a past load curve which most closely matches the values showed by the unpredictable actual load function. Recalling the past, the operators hope to have an indication of what the future trends might be. In addition, their search is limited to the close past, ranging from the same week to a few months back. The previous behavior was considered consistent with the assumption that the time series of loads likely reproduces itself even during periods which differ from the common periodicities. So, the corrector was provided with a third module capable of recognizing, throughout the series of loads, the past load shape closer to the actual load function recorded up to the observed instant. In particular, the actual load function could be the load of the entire day or, more likely, part of it, i.e., only the first hours. If the search is successful, the extracted load function is considered as both the correction for the rest of the day (as requested in point 2) and the new STLF, which shows reduced error, by definition (so point 1 is also covered). More precisely, for the first 2 hours ahead, when the action of the corrector is still possible, the extracted past load function will be modified exactly as it was during the old load forecast. The described process is repeated if the same adverse conditions occur again.

Particular attention should be paid to the retrieval of the load function from the load database. The adopted approach stresses the features of a Kohonen self-organizing map (SOM) as an unsupervised classifier and an associative memory all together. It is well known that an SOM has the ability to find clusters of data as well as internal data structures and to perform an ordered map, eventually discovering similarities. Moreover, when clusters are created, a partially noisy or incomplete input pattern can be associated with the representative of the closer class of data (the representative is the set of weights pointing to a particular cluster). The closer concept (or vicinity) deserves some attention because it influences the performance of the map in relation to the problem. The Euclidean distance is frequently used, but, in this case, better results can be obtained adopting one of its variants. Basically, the load assumes different importance during a day depending on its value. Peaks are very important, as are steep slopes, and errors close to the maximum are considered critical for the operation. That simply induces giving up the classical metrics for more sophisticated metrics. Encouraging results are obtained by simply calculating a weighted distance according to the load value, as shown in the following equation:

$$d_i = \sqrt{\sum_k (x_k - m_{ik})^2 \frac{x_k}{\sum_k x_k}}$$

where

d_i = distance of vector i

m_i = weights of vector i

x_k = quarterly load

k = $1, \ldots, 96$

In general, the concept of distance can be used both during the training of the SOM, when only the winner vector (i) with minimum distance is updated, and during the retrieval of the closest pattern (vector) to the partial actual load function. Even the single use during the retrieval produces some advantages.

Figure VI.5.10 shows the architecture of the modified online corrector. The central role of the classifier (the SOM), supplied by the actual load database and optionally also using the errors for a more precise retrieval, is evident. It should be noted that this configuration allows several variations. One of them, for example, provides that after a cluster has been selected, instead of using its representative, a second search for the closest actual past load curve inside the cluster is performed. The supposed advantage of this second strategy is the possibility of using a load curve actually recorded, which includes the white noise. In fact, this is a component that has been filtered by the SOM and should be coherently added after the retrieval in the first approach.

Finally, it should be considered that the job of the ANNs, i.e., their periodical training and the activation of the retrieval functions, is made completely transparent for the operator. The critical point is the correct training of the three ANNs, which is performed offline only initially. However, particular attention has to be paid to the updating of the training set and to periodical training of the SOM. In fact, strategically, two main options are available:

1. The training set includes one year of data, assuming that the pattern searched time by time can also be built by load curves which are distant from the present time.
2. The training set includes a short number of the most recent curves and a set of curves including the same present period of the last year (e.g., one month).

The first option (the one adopted) enlarges the space of the search, but the obtained clusters are greater in number and are more general (e.g., depending on the training, 30–40 clusters are common). Conversely, the second choice is strictly focused on the period of operation. It introduces search boundaries with the possibility of losing important information, such as socio-cultural events (i.e., strikes) that happened in the distant past but are still useful for the present situation. For both choices,

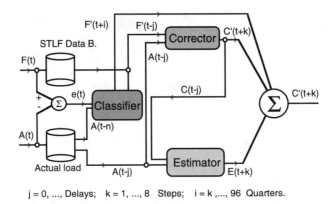

j = 0, ..., Delays;   k = 1, ..., 8 Steps;   i = k ,..., 96 Quarters.

FIGURE VI.5.10 The architecture of the improved online corrector with the introduction of a classification module for the analysis of the historical load series. (From Caciotta, M. et al., *IEEE Trans. Power Syst.*, 11(4), 1749, 1996. With permission.)

the strategy of including the past day in the training set, losing the more distant one, is adopted in order to reduce the number of load curves. The SOM periodical training is usually performed automatically just after midnight as soon as the last load curve has been completed.

5.6.7   The System Man–Machine Interface

What leads to a detailed engineering of the system is the decision to focus the design on an advanced graphic man–machine interface. This is particularly important in the control room environment, where decisions have to be made quickly and the operators usually focus their attention on the decision process, reducing the time for the acquisition of information to a minimum. Therefore, an effective/expressive man–machine interface can make the difference in terms of acceptance.

The online module, now being used during the actual power system operation at GRTN, has a window centered on the current instant, marked with time zero. The entire visualized time period ranges from $+/-16$ quarters around the current time (adaptation to a different time granularity is available as a software option). The system displays the following information (Figure VI.5.11):

1. Thirty-two steps of the short-term forecast (produced the previous day)
2. Sixteen steps of the actual load as it comes from the field
3. The next 2 hours ahead of online correction with a quarter granularity, i.e., 8 quarters (steps) in advance
4. Two sets of boxes show the values of the STLF and the respective corrections

In addition, to assess the reliability of the online correction and, by request, of the power system operator, it is possible to visualize another curve representing the past predicted correction for a chosen quarter in advance. In other words, if the operator is controlling the load behavior, for example, one hour ahead to perform generation adjustments, he can check the last 15 corrections by comparing the "one-hour in advance" past curve with the actual past load curve. From this comparison, it is possible to infer with a certain confidence the online corrector for that specific future step. As expected, closer steps have been found to be more reliable than distant ones.

5.6.8   Results of the Field Test

Due to the fact that there is no analytical way to assess the performances of any ANN forecaster other than comparing with recorded test data, a carpet assessment during the first year of operation

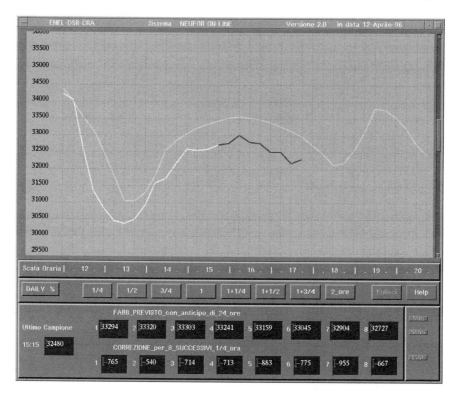

FIGURE VI.5.11 The main NEUFOR window reproducing an example of the online correction. The upper curve represents the STLF of the day before, the first part of the lower curve is the actual load, and the second part of the lower curve is the online correction (curves are colored differently in reality). The two sets of boxes respectively report the STLF values and their correction. It should be noted that the STLF error is approximately 400 MW in this example. (From Caciotta, M. et al., *IEEE Trans. Power Syst.*, 11(4), 1749, 1996. With permission.)

has been performed. Table VI.5.3 reports the results of the monthly average performances, one hour ahead, in comparison with human STLF values.

In terms of the results, it can be noted that there is always a reduction of the relative and absolute errors. Particularly, the yearly average relative error (14.7 MW) is practically null, which shows that the correction is balanced independently from the amount corrected. The same does not apply to the STLF; in fact, the human forecasters seem to support the security of the power system by overestimating their forecast by almost 100 MW. In the mean square deviation columns, the scarce number of errors hovering around zero after the correction has taken place confirms the good performances of the online module. Finally, the mean absolute percentage error (MAPE) column shows a constant monthly reduction with an average value of $\approx -0.7\%$ starting from a value of 1.6% by human forecasters, which is already good. This last comparison leads to the consideration that the corrector can provide advantages for the online operation inversely proportional to the precision of the STLF. Similarly, during the online operation, it has been noted that the greatest advantages are obtained when unexpected variations in the actual load occur. This is the case of sudden socio-cultural events that force a rough modification in the electric generation schedule, so far only estimated by the power system operators.

5.6.8.1 Operators' Acceptance

It is well known that a power system control room is a demanding environment for innovations. Operators constantly require improvements of their software tools for supervision and control. Conversely,

TABLE VI.5.3　Comparison Between the Set of Errors of the Human STLF and the Set of Errors Recorded After the Correction One Hour Ahead During the Period of One Year (May 1995–April 1996)

Months	Relative Error (MW)		Absolute Error (MW)		Mean Square Devation		MAPE	
	Forecast	Correction	Forecast	Correction	Forecast	Correction	Forecast	Correction
May 1995	107.4	−3.56	443.0	235.7	571.0	362.4	1.76	0.94
June	68.8	6.29	345.8	225.7	481.8	330.4	1.34	0.87
July	−8.3	11.00	359.4	238.2	501.7	367.2	1.35	0.91
August	156.5	18.10	389.5	236.4	462.5	356.4	1.79	1.09
September	40.9	−16.70	366.4	274.0	494.6	444.7	1.38	1.04
October	186.1	4.91	501.4	226.5	1103.0	360.7	2.01	0.86
November	−18.0	−4.22	459.3	309.6	695.0	529.6	1.59	1.08
December	90.8	7.38	431.4	263.6	646.8	431.9	1.66	1.03
January 1996	181.7	43.00	476.4	298.5	612.4	396.4	1.79	1.10
February	64.6	34.7	361.4	243.1	515.2	323.9	1.30	0.87
March	171.8	54.3	416.7	236.7	585.4	317.0	1.55	0.88
April	119.2	35.2	414.4	247.7	576.5	324.1	1.67	1.00
Total Year	97.5	14.7	411.8	253.5	640.6	384.3	1.61	0.98

Note: Sequentially, as a monthly average for both the quantities, the relative error, the absolute error, the mean square deviation, and the mean average percentage error (MAPE) are reported. The one-hour ahead step was chosen because operators consider it the most significant for operational planning.

they keep using the applications they are accustomed to. Very briefly, in the case of the corrector, the validity of its validation test strictly depended on the online integration, the results obtained, and the expressiveness of the man–machine interface, which could only be generally assessed in laboratory tests. After the initial expected skepticism of some of the operators, a generalized use of the corrector, which has been totally integrated in the existing energy management system to be available 24 hours a day, was recorded. Operators are inclined to use the corrector only during large STLF error (exactly when they need it the most), even if a glimpse of it every now and then seems to give them greater control and confidence for the current situation.

5.6.8.2　Perspectives

Despite the fact that the performances are beyond expectations, some improvements are still possible. For example, an improvement has to be implemented to face the processing of very poor forecasts and during an occasional temporary lack of information coming from the EMS. In particular, the case of poor forecasts is tricky when the forecasters fail to consider sudden temporary events (e.g., a few-hour strike during the afternoon) which can cause a negative step in the consumption followed by a positive one. In this case, the corrector is obviously unreliable for a period prior to the end of the disturbance. To face that or similar problems, it is necessary for the corrector to receive additional information from the online environment. Basically, information will be imprecise (i.e., "a small reduction of load"), and it can be in the form of textual sentences, typed by the operator according to defined grammar. If this is the case, a fuzzy engine has to be included to interpret this information in order to modify the STLF in advance. However, the fuzzy logic introduction is already under implementation to perform an effective mixture between the extracted load curve and the error corrections calculated by the DMLNs of the corrector. This is desirable considering that, from the tests, it was evident that the effectiveness of the calculus of the error correction fades as the time distance increases. The opposite is true for correction coming from the curve extraction from the past series of loads. Finally, the automatic introduction of additional independent information, such as instantaneous weather conditions not foreseen previously and, therefore, not considered in the STLF, could substantially improve the module performances.

5.7 A Hybrid System for Short-Term Load Forecasting of Anomalous Load Periods

Anomalous load periods are characterized by exceptional circumstances of a national nature such as strikes, holidays, and sports events. In fact, these events represent weak statistical samples for an effective setting of the traditional algorithms used for STLF and for human forecasters as well. The STLF for anomalous periods seems, therefore, an interesting area of research, investigating the potentialities of ANNs as a promising alternative to the traditional methods as well as an effective support to the human forecaster's knowledge.

An extensive application of different ANN architectures to historical load data has revealed basic deficiencies of the models concerning the load forecast of critical periods such as Christmas and Easter holidays, August, as well as some long weekends. So, a new ANN-based procedure (SOM + ANN 1) has been implemented in order to enhance the forecasting accuracy in the above-mentioned periods. The procedure provides a combined (unsupervised/supervised) approach structured in three subsequent stages. The first stage provides some identification criteria of the characteristics of the days through the classification of historical hourly loads, in order to obtain clusters of similar load profiles. The classification is performed by means of a Kohonen SOM. The second stage consists of an actualization process of the information deduced from the previous day's type identification. To do this, human operators give meaning to the load classes. The third stage, performing the proper forecasting task, is done by means of a multi-layer perceptron based on the back-propagation learning algorithm.

In the following, there is the description of two different models aimed at the STLF of normal days (ANN 1) and of anomalous days (SOM + ANN 1).

5.7.1 The ANN 1 Basic Model

The model implemented consists of a totally connected two-layer network (input, hidden, and output layers) having the following structural characteristics. There are 51 inputs, namely:

- 24 hourly load values of the day preceding the forecast day
- $(y_{(i-l,tj)}, j = 1, \ldots, 24)$
- 24 hourly load values of the day preceding the foregoing (two days prior to the forecasted day $(y_{(i-2,tj)}, j = 1, \ldots, 24)$
- 3 binary codes flagging the 2 previous days and the day of the forecast (e.g., 001 Sunday and 111 Saturday)

The number of hidden neurons was parametrically optimized to 31. The number of outputs was set at 24, comprising the simultaneous forecast of the load of the day concerned (see Figure VI.5.12).

The ANN 1 is trained on training sets composed by the hourly daily load patterns of an entire month of the previous year. Therefore, 12 weight sets result from the training stage for the forecasting task of a complete year. No further specialization is required concerning the day type since the binary code included in the patterns permits the network to achieve a sufficient accuracy for all the day types included in periods with normal load conditions. The test sets include the load data of the year to be forecasted. For instance, in order to forecast the daily load for January 4, 1993, the weight set obtained from the ANN training stage relevant to January 1992 must be used. The input used in the forecasting activity will include the daily load data of January 2 and 3, 1993.

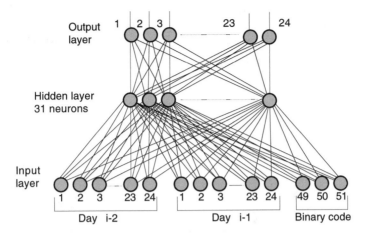

FIGURE VI.5.12   The ANN 1 structure. (From Sforna, M., Fiorino, E., and Melandri, R., *AEI Automazione Energia Informazione*, 86(5), 50, 1999. With permission from IEEE.)

Various tests have been carried out on data concerning the Italian hourly load supplied by the Italian power company in 1992 and 1993. As an example of the results obtained, the hourly load forecasts for the following weeks in 1993 (weeks of normal days) are reported:

- Fourth week of February
- Second week of May
- First week of October

For each of these forecasting periods, Table VI.5.4 reports the following information:

- Maximum value, in the week considered, of the forecasting error of ANN 1 for each hour of the day, calculated by the formula:

$$\varepsilon\%(t_i) = 100 \cdot \max \left| \frac{\hat{y}_j(t_i) - y_j(t_i)}{y_j(t_i)} \right|^{j=1...7}$$

- PRE (percentage relative error) calculated, for each hour of the day, by the formula:

$$PRE\%(t_i) = \frac{1}{7} \sum_{j=1}^{7} (\varepsilon\%(t_i))$$

The results reported in Table VI.5.4 show good performance for most of the hours, but they are less accurate than those obtained by human forecasters. This means that some pieces of information were missing and one reason for that is because ANN 1 did not consider the influence of weather. It is well known that a correct understanding of the influence exerted on electric load by meteorological and climatic factors such as temperature, cloud cover, wind speed, etc. would improve the forecasting results achieved through a neural architecture. However, even though the influence of weather data on the hourly electric load is almost certain, this influence seems less important during the anomalous periods (the goal of this study), when all the social aspects influencing the load profiles become predominant. In these periods, the maximum forecasting error (as shown later) exceeds 50% of the actual value, with a daily average up to 20%. It seems, therefore, unlikely that correctly representing the weather influence could significantly change the forecasting performances. Moreover, the limited occurrence of such anomalous periods makes it difficult to build a training set which correctly reproduces the weather influence on the load profile.

TABLE VI.5.4

$t(h)$	Fourth Week of February		Second Week of May		First Week of October	
	ε (%)	*PRE* (%)	ε (%)	*PRE* (%)	ε (%)	*PRE* (%)
0	1.56	0.86	3.66	1.86	2.76	1.31
1	1.24	0.50	3.84	2.24	2.72	1.61
2	1.21	0.56	4.05	1.80	3.09	1.76
3	1.98	1.02	3.82	1.81	2.44	1.57
4	2.15	0.97	3.78	1.84	2.66	1.78
5	2.18	0.94	2.24	1.47	2.93	1.70
6	4.00	2.57	1.12	0.62	3.00	2.12
7	5.29	3.12	2.41	0.71	3.81	1.61
8	3.03	1.57	1.35	0.77	2.16	1.19
9	1.31	0.54	1.11	0.84	2.93	1.21
10	3.03	1.10	1.28	0.70	2.88	1.35
11	3.72	1.56	1.88	0.84	4.56	1.61
12	3.92	1.60	1.63	0.94	3.94	1.57
13	3.88	1.70	2.30	1.11	3.80	1.50
14	4.29	2.25	3.09	1.03	3.99	1.58
15	5.40	2.74	2.18	1.29	3.86	1.90
16	5.31	2.78	2.02	1.48	4.00	2.39
17	5.56	2.38	3.71	1.45	4.30	1.90
18	2.73	1.24	3.58	1.62	2.70	1.81
19	1.98	1.08	2.25	1.41	3.06	2.17
20	2.32	1.28	1.65	1.11	2.45	1.58
21	2.26	1.28	2.25	0.85	3.74	2.36
22	2.14	0.94	1.96	0.93	3.15	1.67
23	3.30	1.50	2.26	1.23	3.61	1.93

5.7.2   Day-Type Classification by Means of Unsupervised Learning

Vacation periods, holidays, and long weekends are some of the most noticeable factors that can determine anomalous load conditions with respect to typical load patterns. Thus, a specific procedure was developed specifically for anomalous load conditions. This task has required a day-type classification aimed at identifying similarities among patterns of load profiles, and an unsupervised learning approach seems suitable for this task. This is based on the assumption that load patterns already incorporate all the information concerning the external influences affecting the load. This typical pattern recognition task can be adequately achieved by using a Kohonen self-organizing map (SOM).

An extensive classification of load profiles relevant to the years 1991–1993 has been performed through an 8×8 SOM classifier. The input patterns were daily load vectors composed by 24 hourly values. Parametric analyses have been conducted on each parameter of the algorithmic formulation (maximum neighborhood radius, learning rate value, conscience value) as well as on different sizes of the training set (1, 3, and 12 months).

The algorithm tends to cluster the load profiles into the following principal day types: Sundays, Saturdays, Mondays and working days after a Holiday, working days from Tuesdays through Fridays, main holidays such as Christmas and Easter, middle-of-August days, etc. The number of different clusters ranges between 11 and 15 for the years analyzed. The following questionable classifications have been also evidenced:

- The working days of August (generalized vacation period) have been classified together with the Sundays of late fall.
- The days between weekends and bank holidays are classified as both Saturdays and Sundays.

A seasonal classification can be deduced among a spring summer season comprised of April, May, June, July, and September, an intermediate season comprised of March and October, and a winter season comprised of January, February, November, and December.

In Table VI.5.5, the main characteristics of the different daily load profiles within each cluster for the entire year of 1991 have been reported. The table also reports the maximum value of the peak demand, the minimum value of the minimum demand, and the average daily energy value of the various profiles in the cluster. The relevant standard deviation values are also indicated.

The size of the training set of a Kohonen SOM (time horizon) can be considered a convenient means for calibrating the accuracy provided by the classification activity. The detailed results of the classification stage conducted by reducing the training set to a monthly time horizon for the months of December 1991–1993 have been illustrated in Table VI.5.6. The accuracy thus obtained has been considered more satisfactory for a correct identification of anomalous daily load conditions. For instance, December 8 is an Italian holiday, and December 8, 1992 represents a typical case of a long winter weekend. In Table VI.5.6, therefore, Tuesday, December 8 is included in a cluster used for classifying Sundays and Holidays. Monday, December 7 is included in a contiguous cluster, thus indicating that its load shape is closer to a holiday's rather than to a Monday's classified with other working days in a farther cluster. In 1993, Thursday, December 9 is included in a cluster containing two Mondays. The load pattern of this day type is, in fact, similar to a typical working Monday.

5.7.3 The Unsupervised/Supervised Approach

It is expected that a preventive classification of historical load data can enhance forecasting accuracy of the ANN models. Various approaches using this technique share the basic concept that the unsupervised stage tries to map many similar patterns to the same output, thus reducing the burden assigned to the supervised stage which is provided in the forecasting task. In fact, the supervised

TABLE VI.5.5

Cluster	Pmax (MW)	STD (MW)	Pmin (MW)	STD (MW)	Energy (MWh)	STD (MWh)
0	23528	695	12660	857	441021	12856
1	23999	503	13597	493	464061	6390
2	27240	979	14604	533	496391	12092
3	28252	698	16022	856	525884	12284
4	29157	742	13165	1588	539755	12717
5	31656	901	17605	413	586767	16101
6	32154	806	16955	343	601264	11916
7	32492	698	12674	792	584613	15050
8	33729	537	16106	638	625677	8797
9	33797	453	18711	410	644056	9548
10	33872	763	14686	1530	617432	13730
11	36169	580	14567	1323	659632	11841
12	38246	680	15973	834	687094	4987
13	38166	253	19362	207	715047	4599
14	21336	673	11883	672	389865	18471

TABLE VI.5.6

December 1991			December 1992			December 1993		
Day Type	Date	Cluster No.	Day Type	Date	Cluster No.	Day Type	Date	Cluster No.
Sa	07	45	Tu	15	40	Su	05	31
Sa	14	45	We	16	40	We	08	31
Sa	21	45	Th	17	40	Su	12	31
Mo	23	47	Fr	18	40	Su	19	31
Fr	27	50	Tu	01	41	Fr	24	31
Sa	28	50	We	02	41	Sa	25	31
Su	29	50	Th	03	41	Su	26	31
Mo	30	50	Fr	04	41	Mo	27	36
Tu	31	50	We	09	41	Tu	28	36
We	25	51	Th	10	41	We	29	36
Th	26	51	Fr	11	41	Th	30	36
Su	01	52	Mo	14	41	Fr	31	36
Su	08	52	Mo	21	41	Sa	04	37
Su	15	52	Tu	22	41	Sa	11	37
Su	22	52	We	23	43	Sa	18	37
Tu	24	52	Sa	05	45	Th	23	40
Tu	10	55	Sa	12	45	Mo	06	41
We	11	55	Sa	19	45	Th	09	41
Th	12	55	Mo	07	46	Mo	13	41
Fr	13	55	Su	06	47	Mo	20	41
Tu	17	55	Tu	08	47	We	01	43
We	18	55	Su	13	47	Th	02	43
Mo	02	57	Su	20	47	Fr	03	43
Tu	03	57	Th	24	47	Tu	07	43
We	04	57	Fr	25	47	Fr	10	43
Th	05	57	Sa	26	47	Tu	21	43
Fr	06	57	Su	27	47	We	22	43
Mo	09	57	Mo	28	47	Tu	14	44
Mo	16	57	Tu	29	47	We	15	44
Th	19	57	We	30	47	Th	16	44
Fr	20	57	Th	31	47	Fr	17	44

stage can perform its learning procedure on the clustered data rather than on the entire data set. Thus, the information provided to the supervised stage is alternatively composed:

- By all the load patterns grouped in the same cluster
- By a representative feature of the various load patterns of the same cluster (cluster center, average load pattern)

The implementation of a methodology well tailored to the capabilities of the ANN 1 model has suggested an unconventional use of the information deriving from the classification stage. Figure VI.5.13 shows the implemented structure. The Kohonen SOM provides a cluster identification code for each input pattern. The daily load profiles and the binary codification of their cluster constitutes the new training set of the ANN 1 previously illustrated. The cluster codification substitutes the original day-type coding used for the ANN 1 model. The application of the procedure illustrated requires a labeling procedure for each future day based only on its *a priori* characteristics such as

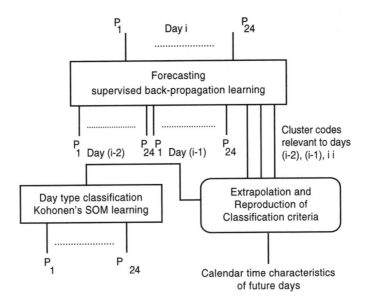

FIGURE VI.5.13 Unsupervised/supervised procedure adopted (SOM + ANN 1). (From Sforna, M., Fiorino, E., and Melandri, R., *AEI Automazione Energia Informazione*, 86(5), 50, 1999. With permission from IEEE.)

calendar variables. This task that aims to replicate the classification obtained by the unsupervised ANN is assigned to human operators. The calibration activity of the Kohonen SOM parameters for anomalous periods is performed through a feedback process driven by an expert user who calibrates the parameters with the results of the forecasting stage activity. If the cluster codes obtained permit achieving adequate forecasts for all these periods, the accuracy obtained by the classification stage can be considered suitable.

5.7.4 Case Studies

Extensive tests have been conducted on load data of the years 1991–1993. In order to test the potentiality of the proposed procedure, some anomalous periods of the above-mentioned years have been selected for comparing the results provided by the basic method (ANN 1) and the combined (SOM + ANN 1) method. December 1992 presents a typical anomalous load situation in regards to December 7, which is part of a typical long winter weekend. Furthermore, December 31 represents a typically critical day for load forecasting. Figures VI.5.14 and VI.5.15 illustrate comparatively the forecast results for these days as obtained with both the illustrated procedures. Figures VI.5.16 and VI.5.17 comparatively illustrate the performances of the two procedures for the week December 5–11 as well as for the Christmas week (December 25–31). For these weeks, an analytical comparison among the results, based on the calculation of the previous coefficients, is reported in Table VI.5.7.

The combined procedure (SOM + ANN 1) determines a noticeable increase in forecast accuracy in all the anomalous load conditions. The forecast results of the combined procedure have been obtained by using a day-type classification such as that shown in Table VI.5.5 (monthly basis).

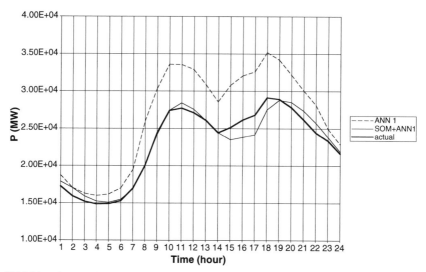

FIGURE VI.5.14   Curves show the comparison among actual and forecasted data for December 7. (From Sforna, M., Fiorino, E., and Melandri, R., *AEI Automazione Energia Informazione*, 86(5), 50, 1999. With permission from IEEE.)

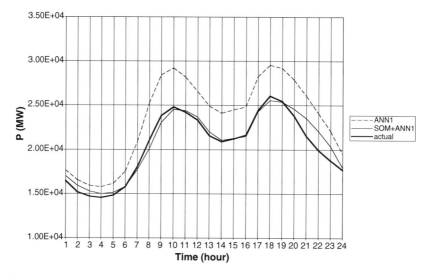

FIGURE VI.5.15   Curves show the comparison among actual and forecasted data for December 31. (From Sforna, M., Fiorino, E., and Melandri, R., *AEI Automazione Energia Informazione*, 86(5), 50, 1999. With permission from IEEE.)

5.7.5   Discussion of the Results

The previous figures and tables give clear evidence that the combined procedure (SOM + ANN 1) provides much better forecasts for all the anomalous load periods of the year that was analyzed. Therefore, the following basic considerations can be inferred:

- The ANN 1 model shows better performances for all load patterns that reproduce the typical day-type characteristics (normal days).
- For each anomalous load situation examined, the combined approach generally gives better forecasts than those obtained through the ANN 1.

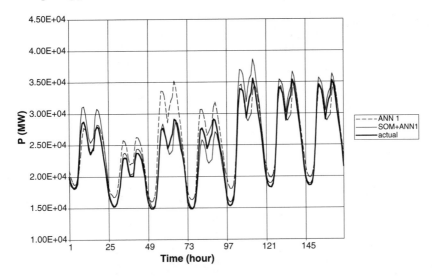

FIGURE VI.5.16 Curves show the comparison among actual and forecasted data for the week December 5–11. (From Sforna, M., Fiorino, E., and Melandri, R., *AEI Automazione Energia Informazione*, 86(5), 50, 1999. With permission from IEEE.)

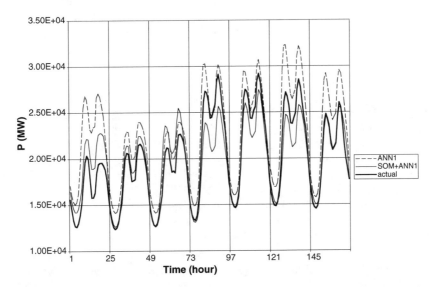

FIGURE VI.5.17 Curves show the comparison among actual and forecasted data for the week December 25–31. (From Sforna, M., Fiorino, E., and Melandri, R., *AEI Automazione Energia Informazione*, 86(5), 50, 1999. With permission from IEEE.)

- A previous identification of anomalous periods should include in this category, in addition to all the holidays, the long weekend days, each day perturbed by special events, and all the days coincident typically with the days before and after these periods. In fact, the holiday periods can generate an inertia effect on the social activities that can result in load conditions that are difficult to predict.
- An input data set of the day-type classification activity as limited to a monthly time horizon seems to be the most effective both for the typical Italian load patterns and for the characteristics of the implemented Kohonen's SOM.

TABLE VI.5.7

| | December 5–11 | | | | December 25–31 | | | |
| | ANN 1 | | SOM + ANN 1 | | ANN 1 | | SOM + ANN 1 | |
$t(h)$	ε (%)	PRE (%)	ε (%)	PRE (%)	ε (%)	PRE (%)	ε (%)	PRE (%)
0	8.84	6.18	6.47	3.48	20.36	8.78	5.45	2.54
1	9.98	6.11	6.28	3.29	21.90	9.79	5.95	2.78
2	11.09	6.70	6.49	2.74	25.24	10.18	8.69	2.92
3	11.97	7.07	6.38	2.59	27.42	12.33	10.44	2.90
4	13.96	7.87	3.25	2.20	29.03	13.19	11.19	3.06
5	20.66	10.27	4.53	1.31	30.37	15.54	10.55	4.14
6	34.67	18.37	10.99	4.80	41.27	22.09	9.76	6.09
7	44.37	24.45	17.23	6.07	52.58	26.6	17.26	9.57
8	41.94	21.45	17.83	6.28	49.98	24.02	16.18	9.78
9	37.25	17.95	13.69	5.97	44.34	20.90	14.39	8.15
10	35.12	17.83	10.73	5.23	41.20	18.79	14.37	7.47
11	32.13	16.53	6.71	4.38	37.47	16.96	14.62	7.59
12	32.64	15.84	7.98	4.90	37.82	17.72	14.35	7.72
13	38.67	20.95	9.20	6.11	45.14	20.99	18.08	10.20
14	40.61	23.82	8.44	6.35	46.60	21.23	20.29	10.98
15	41.88	24.53	8.03	6.49	48.62	21.44	20.64	11.60
16	39.29	21.20	12.94	6.24	45.97	20.91	17.12	10.06
17	35.51	17.23	14.88	6.00	41.09	18.86	13.64	8.96
18	32.13	14.23	12.58	5.99	37.22	17.94	14.25	7.45
19	29.25	12.15	8.60	6.29	33.51	17.06	14.00	6.32
20	26.50	11.18	9.78	6.26	29.49	16.92	14.32	6.42
21	23.64	10.01	10.24	3.68	25.39	14.19	12.24	6.89
22	22.79	9.41	8.79	3.68	26.06	14.35	10.10	5.24
23	23.13	9.04	8.18	3.80	25.98	13.88	7.52	2.99

The results obtained suggest the adoption of different specific procedures for STLF of anomalous and normal days. An ANN-based forecasting tool can, therefore, be implemented by integrating the two different procedures. This tool provides that the ANN 1 be applied to forecast the normal load conditions periods, while the combined procedure should be applied to each anomalous period that must have been identified through an accurate preventive classification activity conducted on various years of historical load data. The preventive classification permits achieving the detailed knowledge relevant to the typical anomalous load conditions that is required from the end user of the integrated tool. For the Italian case, the months of January, February, May, June, July, September, October, and November can be considered almost entirely composed by normal periods, with only a few anomalous load conditions. The application of the ANN 1 model to these months results in the following accuracy statistic:

- Maximum hourly error = 7.1%
- Maximum average error = 2.6%
- Maximum standard deviation value = 2.8%

The application of the proposed integrated procedure (ANN 1 for normal periods and SOM + ANN 1 for anomalous periods) has given the following results for the months of December and April, typically the months that include the largest occurrence of anomalous load conditions:

December:

- The maximum hourly error decreases from 52.5% to 14.9%.
- The maximum average error decreases from 12.9% to 7.7%.
- The maximum standard deviation decreases from 14% to 4.5%.

April:

- The maximum hourly error decreases from 33.2% to 9.8%.
- The maximum average error decreases from 6.9% to 2.7%.
- The maximum standard deviation decreases from 6.1% to 2.1%.

The benefits of the architecture (SOM + ANN 1) with respect to a basic multi-layer perceptron neural network (ANN 1) are clear for some highly anomalous conditions of Italian load patterns. The results obtained show that architecture can provide a considerable improvement of the forecast accuracy, which can be assumed to range up to a factor of three.

5.8 The Neural Network Interpolation of the Load Demand

The load demand is a continuous time function that could present substantial changes within each hour. Figures VI.5.18 and VI.5.19 compare the national demand curves recorded hourly and quarterly. The difference between the 2 curves in terms of energy is 3.4 GWh per day with a maximum absolute error of 850 MW (at 8.15 a.m. during the morning ramp of load).

A detailed analysis of this problem shows that during the year, the demand variations within the hour are comparable to the forecast error. Moreover, for a small number of days (5%) the afternoon quarterly peak is 780 MW above the linearization of the hourly curve with an absolute maximum error of more than 1000 MW.

The common activity of unit commitment considers that the production plan possibly remains constant within each hour. Actually, the demand being highly variable, as shown previously, there could be differences that must be corrected during the online operation with possible problems for the energy market and/or the network security. So, there is a need to calculate a unit commitment more precisely, relying on a quarterly load forecast with 96 values for each day.

All the efforts and the attempts to adopt algorithms for load forecasting showed that, at least for the Italian load, the results obtained are comparable with the performances of human forecasters. The average error for both is, at best, 1.5% for the peak load of normal days. However, human beings have limits in dealing with large amounts of numbers, so, practically, the manual forecast is effective only for 24 hourly values. This fact suggests adopting the manual forecast of 24 values as the basis of the load forecasting activity. Studies were performed to automatically build a forecast of the missing 72 values that, together with the 24 (manually forecasted), are the daily demand quarter by quarter. Of course, the goal of the analysis was that the algorithm had to perform better than the linear interpolation between the hourly values.

5.8.1 The Problem of Numeric Interpolation

Statistically, the interpolation problem is described as the mathematical procedure to substitute a series of numbers with a theoretical function such as $y = f(x)$ crossing the given data or passing among them. In the first case, the solution can possibly be an equation of the maximum degree allowed

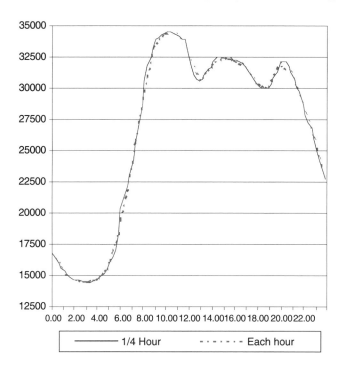

FIGURE VI.5.18   Comparison between the curves of the national load demand recorded quarterly and hourly. Differences are within 1–2% and they appear small, but they are relevant because they are sited close to the changes of the load trends exactly when the quickest reaction of the generation pool is requested (see Figure VI.5.19 for details).

by the problem, i.e., the number of data minus 1, with coefficients determined as the solution of the system obtained substituting the couples of data y, x in the equation. Conversely, the interpolation among the data is to find a function $y = f(x) + \varepsilon$ among those that better fit the data of the families of straight lines, parabolas, etc. with the aim of reducing the error ε by using an optimization algorithm. The case of load forecast is an interpolation passing data, because the resulting curve still has to keep the input data, i.e., the hourly values.

Then, to calculate the 3 quarters within each couple of hourly forecasted values, the following auto-regressive model was initially adopted:

$$F(k + n) = A(n, k) \cdot F(k) + B(n, k) \cdot F(k + 1)$$
$$\text{being: } k = 0, \ldots 24; \quad n = 1/4, 1/2, 3/4$$

The $A(n, k)$, $B(n, k)$ coefficients are different for each quarter and are calculated with the minimum squared method for:

- Each of the 24 couples of hourly values
- Each kind of 12 daily courses representing typical days of the 4 sets: weekdays, days before holidays, holidays, and days between holidays
- The 3 periods of summer, winter, and fall/spring together

This method gave good results and proved to be robust; however, the static classification of curves in the 12 classes proved unable to represent all the possible load demand characteristics. A possible improvement goes through a larger number of classes to be defined automatically. The uncertainty of how to find the most effective number of classes led to an alternative solution which translates

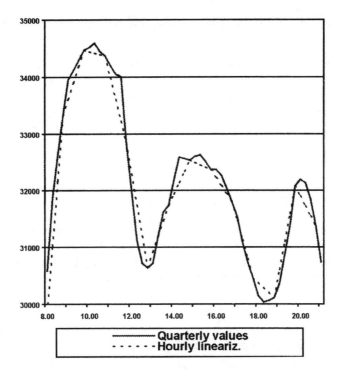

FIGURE VI.5.19 Details of the period from 8.00 a.m. to 21.00 p.m. of the load demand reported in Figure VI.5.18. Differences between the curves are more evident. Physically, this means that a certain amount of energy, represented by the area included between the two curves, is not taken into account during the unit commitment stage.

the problem of interpolation to that of recognition among past curves of load profiles. These curves, having the same hourly values, should also have the most suitable characteristics to be the searched quarterly load forecast. The hypothesis behind this method is the supposed periodicity of the load demand for itself and especially for those curves having the same hourly values.

Then, the solution of the interpolation problem follows two logical stages:

- Recognition of the past hourly load curves that are closer to the load forecast to be interpolated
- Generation of the $96 - 24 = 72$ missing values by using for them the load values actually recorded and extracted at the previous stage

The load recognition problem is naturally located in the set of the pattern recognition analysis for which statistic techniques and neural networks are specifically devoted.

5.8.2 An Initial Approach to Neural Interpolation

The first approach to the interpolation problem was done with a multi-layer perceptron with the following characteristics:

- Inputs of the ANN were the 24 hourly forecasted values for a given day.
- Outputs were the $96 - 24 = 72$ missing values to be directly used as the quarterly load values, scaled to the input values if needed.
- There were 25–30 perceptrons for the hidden layer of the ANN, which, theoretically, means splitting the load demand into an equivalent number of classes. That number seemed to be a

good compromise between crowded classes and the need to represent as many load patterns as possible.

Two modules constitute the neural system. They automatically perform both the learning and the association functions. A training set was built for each day for which the interpolation was requested. It was constructed with the load values of the 4 weeks immediately before the given day and the 4 weeks after the same day of the previous year. The values were updated to the present days with the yearly trend. To save time, the learning procedure was split into two steps:

1. A first learning of a preliminary training set with $300 \div 500$ cycles. This is for the set-up of the neural network.
2. A periodic relearning with 5–10 cycles to be performed each day after the update of the training set (two patterns in and two patterns out).

This method performed better than the statistic interpolation. The reasons for that stem from a better classification (due to the number of perceptrons) and, mainly, because the interpolated curve is the result of values taken from actual load curves instead of being calculated. On the contrary, the ANN does not have a deterministic behavior. In fact, for anomalous days or errors in the hourly values used as input, wrong interpolations were recorded. In this case, it should be noted that a wrong pattern is matched with actual load patterns and so an uncertain result is obtained.

5.8.3   The Final Neural Network Interpolator

After a parametric analysis of different ANN architectures and several configurations of input and output data, the best performances were obtained by an SOM with the following characteristics:

- Instead of the 24 hourly values of the previous prototype, the input vector collects the information related to the demand around a specific hour and about the hour itself. It seemed effective to use:

  1. The average demand at a hour k, $\overline{F}(k)$, calculated as the average of the hourly values at time k and $k + 1$
  2. The derivative of the demand at hour k, $\overline{F}'(k)$, calculated as the difference between the hourly load at time $k + 1$ and k
  3. The second derivative demand at hour k, $\overline{F}''(k)$, calculated as:

  $$\overline{F}''(k) = F(k + 2) + F(k - 1) - F(k + 1) - F(k)$$

  To recognize the day zone of interest for the interpolation, the day was split into 5 fuzzy sets with membership functions regarding the zones (see Figure VI.5.20): night, morning, early-afternoon, late-afternoon, evening. The membership degree of the singular zone to the previous sets was included among the input values of the ANN. So, the input vector has 8 numbers. It should be noted that the choice to focus on a subset of data related to a specific zone, instead of the total daily curve, makes the job of the ANN simpler and more effective.

- The output vector, instead of the 72 values of the total day, gives only the 3 quarterly values searched for the period between the 2 specific hours. Specifically, instead of the exact quarterly load, it was found more effective to use as output the 3 differences between the searched quarterly load and the linearized hourly load:

  $$\delta F(k + n) = F(k + n) - F_{linearized}(k + n)$$

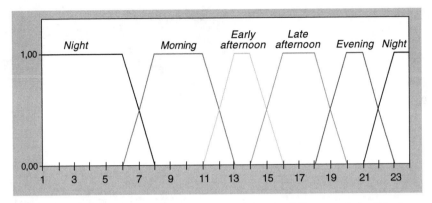

FIGURE VI.5.20 Membership functions defining the different periods of the day.

with $n = 1/4, 1/2$, and $3/4$. The use of that difference (referring to the linearized pattern) allows obtaining the information included in the hourly values more easily, improving the performance of the interpolation system.

- To simplify the managing of the training procedure, the chosen method was to develop four different neural networks, one for each season, in order to train them yearly. This request defines the number of possible acceptable answers of the system. Considering the compromise between the needed accuracy and the fact that each output neuron should represent a substantial number of patterns, it was chosen to adopt 400 neurons for each of the seasonal SOMs.

The 4 neural networks were trained with the actual quarterly data recorded from the EMS system. Less than 100 training cycles of the entire training set (3 months of data) were sufficient for this task.

The developed system was completely integrated within the day-ahead planning procedure at the national control center of GRTN. Then, starting from the 24 hourly values as they are inferred by the human forecaster, the neural network gives the 72 values $\delta F(k+n)$ that are finally summed to the linearization of the hourly forecast to get the 96 values of the total daily searched pattern. It is important to remember that the architecture adopted does not modify the hourly values provided by the human forecaster.

5.8.4 Results

Before the actual use, the system was tested with the data of a period of 6 months, from October 1996 to March 1997. The ANN was trained with the recorded data and supplied with the forecasted data of the same period to reproduce the real operation. Figure VI.5.21 compares the recorded load, the linearized forecast, and the interpolation of the national load. In that example, (see details in Figure VI.5.22) at 17.30 the recorded demand reaches its peak, which is 750 MW larger than the linearized forecast. Definitively, this is a big error from an operational viewpoint. Instead, the interpolation is only 100 MW less and it also generally follows the pattern of the actual load. This performance is particularly appreciated during the winter period when the daily peak load appears in the afternoon and the power system reaches its maximum stress. Again, from Figure VI.5.21, the good system results should be noted during the plateau between 10.00 a.m. and 11.30 a.m. and during 11.45 a.m. and noon, when the demand suddenly drops. These periods are important during the summer when there is peak load due to air conditioning.

Table VI.5.8 reports some statistical analyses of the differences between the $\delta F(k+n)$ values, as the neural interpolator calculated them, and those calculated by using the recorded load. Errors are

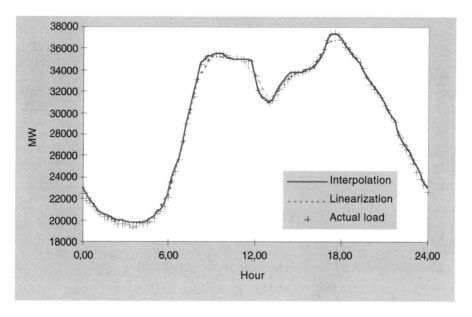

FIGURE VI.5.21   Comparison between the national demand, recorded quarterly, the linearized hourly forecast, and the interpolated curve given by the neural system (see details in Figure VI.5.22).

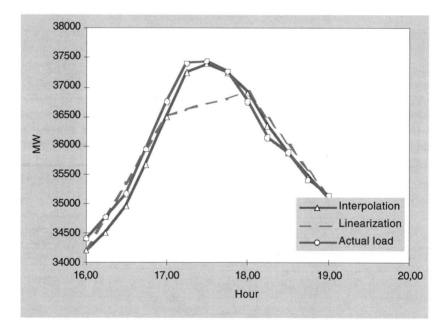

FIGURE VI.5.22   Details for the period between 16.00 and 19.00 for the curves of Figure VI.5.21.

limited to fall under the acceptable values of $200 \div 300$ MW. For the same patterns, the linearization gives errors of $600 \div 800$ MW.

These good results encouraged the adoption of the described neural system for the actual use, which is still in use at the present time.

TABLE VI.5.8   Errors Between the Differences of Neural Interpolation and the Linearized Load, and the Recorded Load and the Same Linearized One

	Average Error (MW)	Standard Deviation (MW)	D (5%) (MW)
On the 72 quarterly values	0	106	174
Max. positive daily deviation	282	142	515
Max. negative daily deviation	−284	128	−495
Max. peak load deviation	41	118	235
Max. morning peak deviation	43	95	199
Max. afternoon peak deviation	46	138	274

Note: The results refer to the period from October 1, 1996 to March 31, 1997. D (5%) represents the values which are overcome in 5% of all the cases.

5.9   Conclusions and Recommendations

Several reasons can be given for the suitability of adaptive algorithms for load forecasting problems:

1. ANNs are a mature technique to be used in load forecasting. Several utilities in the world have already adopted ANN models in daily use, with reported advantages in prediction accuracy. The most adopted practice is the multi-layer feedforward scheme.
2. Conventional regression approaches assume a linear relationship between descriptive variables and load data, which has proven unsatisfactory. ANNs easily implement non-linear relationships among input data.
3. These functional relationships are not stationary, but are affected by temporal dynamics. It is possible to introduce this temporal behavior in neural models.
4. Anomalous situations, such as holidays, special days, or sharp temperature changes cannot be efficiently treated by classical methods as they are not correlated with preceding days. The use of load patterns and day categorization may overcome this drawback.
5. One of the drawbacks of ANNs is overfitting, which comes from a too-high degree of freedom of the neural model (too many connections in the architecture). The use of an automatic search tool such as a genetic algorithm to choose an optimal architecture from the set of all possible ones is very useful for improving generalization capacity.

Figure VI.5.23 shows a display of latest version of the NEUFOR application. It incorporates software modules obtained from those previously described. This version was developed according to the end-users' requirements, to give them the choice to use or not use all the available functions together with access to the database of load curves. Then, human forecasters have the possibility to show, in the same display,

- The ANN forecast
- The load curve of the day with similar characteristics of the day to be forecasted chosen by means of an SOM neural network
- The load curve selected from the load database
- A load curve corrected with the influence of the weather variables, calculated by means of a fuzzy logic approach.

From the comparison of these load curves, forecasters have all the available information to develop their own opinion of the future load pattern. Then, they can directly accept one of the curves or they can modify one either graphically (with the mouse) or numerically (with boxes for data entry), which is quite common.

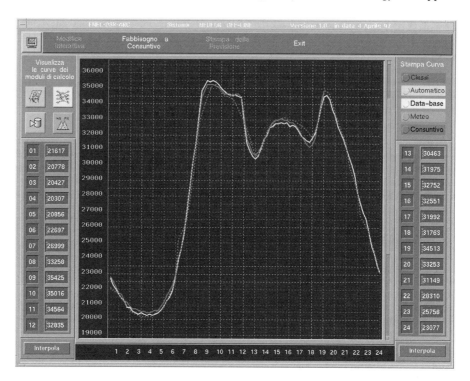

FIGURE VI.5.23   Main display of the latest version of NEUFOR.

Acknowledgments

The described systems are the consequences of extensive R&D activity that lasted from 1990 to 1997 under the author's supervision. During that period, several experts participated in the effort, providing various contributions in system design, software development, and load forecasting expertise. Listing all of them here is not possible, but I sincerely thank them for the passion they put into the job. However, some deserve to be mentioned for their decisive support, knowledge, and practical contributions. They are M. Mocenigo (GRTN), D. Lucarella (CESI), A. Prudenzi (University of Rome), V.O. Cencelli (University of Rome), and E. Fiorino (GRTN). I congratulate them for their skills and sincerely thank them for what they contributed.

References

References listed herein were the theoretical basis and provided some inspirations for the described applications. Additionally, they also provided the most important seminal contributions on intelligence applications to load forecasting by the most prominent researchers, utility experts, and university professors in the field. References are listed according to their date of publication. Generally, titles of papers and books give clear links with the applications previously reported.

1. Fann, K.T., *Pierce's Theory of Abduction*, Martinus Nijhoff, The Hague, Holland, 1970.
2. Makridakis, S., Whellwrigth, S.C., and McGee, V.E., *Forecasting: Methods and Applications*, John Wiley & Sons, New York, 1983.
3. Farlow, S.T., *Self-Organizing Methods in Modeling*, Marcel Dekker, New York and Basel, 1984.

4. Bose, N.K., Jury, E.I., and Zeheb, E., On robust Hurwitz and Shur polynomials, Proceeding of the 25th Conference on Decision and Control, Athens, 1986.

5. Rumelhart, D.E. and McClelland, J.L., *Parallel Distributed Processing, Exploration in the Microstructure of Cognition*, MIT Press, Cambridge, MA, 1986, 547.

6. Gross, G. and Galiana, F.D., Short-term load forecasting, *Proc. IEEE*, 75(12), 1558, 1987.

7. DeSieno, D., Adding a conscience to competitive learning, *Proceedings of the 2nd Annual IEEE International Conference of Neural Networks*, Vol. 1, 1988.

8. Gori, M., Bengio, Y., and De Mori, R., BPS: a learning algorithm for capturing the dynamic nature of speech, *IEEE–IJCNN*, 1989.

9. Pao, Y.H., *Adaptive Pattern Recognition and Neural Network*, Addison-Wesley, Reading, MA, 1989.

10. Damborg, M. et al., Potential applications of artificial neural networks to power system operation, IEEE International Symposium on Circuit and Systems, New Orleans, Lousiana, May 1–3, 1990, 2933.

11. Poggio, T., A theory of how the brain might work, in *Proceedings of the LV Cold Spring Harbor Symposium on Quantitative Biology: The Brain*, Vol. 55, Cold Spring Harbor Laboratory Press, 1990, 899.

12. Conner, J.T., Atlas, L.E., and Martin, D., Recurrent neural networks and load forecasting, Report 91TH0374-9/91, IEEE.

13. Lee, K.Y., Cha, Y.T., and Park, J.H., Short term load forecasting using an artificial neural network, IEEE/PES 1991 Winter Meeting, New York, NY, 1991.

14. Park, D.C. et al., Electric load forecasting using an artificial neural network, IEEE Trans. *Power Syst.*, 6(2), 442, 1991.

15. El-Sharkawi, M.A. et al., Short term electric load forecasting using an adaptively trained layered perceptron, First International Forum on Applications of Neural Networks to Power Systems, Seattle, WA, July 23–26, 1991, 22.

16. Conner, J.T., Atlas, L.E., and Martin, D., Recurrent neural network and load forecasting, First International Forum on Applications of Neural Networks to Power Systems, Seattle, WA, July 23–26, 1991, 26.

17. Brace, M.C., Schmidt, J., and Hadlin, M., Comparison of the forecasting accuracy of neural networks with other established techniques, First International Forum on Applications of Neural Networks to Power Systems, Seattle, WA, July 23–26, 1991, 31.

18. Montgomery, G.J. and Drake, K.C., Abductive reasoning networks, *Neurocomputing*, 2(3), 1991.

19. Pao, Y.H. and Sobajic, D.J., Current status of artificial neural network applications to power systems in the United States, *TIEE Japan*, 111-B(7), 690, 1991.

20. Hsu, Y.Y. and Yang, C.C., Design of artificial neural networks for short-term load forecasting. Part I: self-organising feature maps for day type identification, *IEE Proceedings-C*, 138(5), 407, 1991.

21. Ritter, H., Martinetz, T., and Schulten, K., *Neural Computation and Self-organizing Maps*, Addison-Wesley, Reading, MA, 1992.

22. Lee, K.Y., Cha, Y.T., and Park, J.H., Short-term load forecasting using an artificial neural network, *IEEE Trans. Power Syst.*, 7(1), 124, 1992.

23. Ho, K.L., Hsu, Y.Y., and Yang, C.C., Short-term load forecasting using a multi-layer neural network with an adaptive learning algorithm, *IEEE Trans. Power Syst.*, 7(1), 141, 1992.

24. Chen, S.T., Yu, D.C., and Moghaddamjo, A.R., Weather sensitive short-term load forecasting using non-fully connected artificial neural network, *IEEE Trans. Power Syst.*, 7(3), 1098, 1992.

25. Yu, X.H., Can backpropagation error surface not have local minima?, *IEEE Trans. Neural Networks*, 3(6), 1019, 1992.

26. Lu, C.N., Wu, H.T., and Venuri, S., Neural network based short term load forecasting, *IEEE Trans. Power Syst.*, 8(1), 336, 1993.

27. Djukanovic, M. et al., Unsupervised/supervised learning concept for 24-hour load forecasting, *IEE Proceedings-C*, 140(4), 311, 1993.

28. Peng, T.M., Hubele, N.F., and Karady, G.G., An adaptive neural network approach to one-week ahead load forecasting, *IEEE Trans. Power Syst.*, 8(3), 1195, 1993.

29. Papalexopoulos, A.D., Hao, S., and Peng, T.M., Application of neural network technology to short term system load forecasting, IEEE/NTUA Athens Power Technology Conference on Planning Operation and Control of Today's Electric Power Systems, Athens, Greece, Sept. 5–8, 1993.

30. Pao, Y.H. and Sobajic, D.J., Unsupervised/supervised learning concept for 24-hour load forecasting, *IEE Proceedings-C*, 4, 318, 1993.

31. Sagar, V., Vankayala, S., and Rao, N.D., Artificial neural networks and their applications to power systems—a bibliography survey, *Electric Power Syst. Res.*, 28, 67, 1993.

32. Tejedor, A.G. et al., A neural system for short-term load forecasting based on day-type classification, Proceedings ISAP '94, Montpellier, France, 1994, 353.

33. Bishop, C.M., *Neural Networks for Pattern Recognition*, Clarendon Press, Oxford, U.K., 1995.

34. Niebur, D. (CIGRE TF 38.06.06), Artificial neural networks for power systems—a literature survey, *Electra*, 159, 77, 1995.

35. Sforna, M., Searching for the electric load-weather temperature function by using the group method of data handling, *Electric Power Syst. Res.*, 1, 1, 1995.

36. Sforna, M. and Proverbio, F., A neural network operator oriented short-term online load forecasting environment, *Electric Power Syst. Res.*, 33(2), 139, 1995.

37. Mocenigo, M. and Sforna, M., La previsione del fabbisogno elettrico con le reti neurali, *AEI Automazione Energia Informazione*, 83(12), 50, 1996 (in Italian).

38. Caciotta, M. et al., A neural network based technique for short-term forecasting of anomalous load periods. *IEEE Trans. Power Syst.*, 11(4), 1749, 1996.

39. Sforna, M., Fiorino, E., and Melandri, R., Interpolazione del fabbisogno con metodi neurali, *AEI Automazione Energia Informazione*, 86(5), 50, 1999 (in Italian).

Appendix A: The Group Method of Data Handling Algorithm

The Group Method of Data Handling (GMDH) was developed in the 1960s by the Ukrainian cybernetic mathematician, A.G. Ivaknenko. This algorithm is able to construct a regressive mathematical model of high order which can accept a great number of variables and automatically organize a gradually more complex model until an optimal model is attained. To illustrate the potential of the GMDH algorithm, an evolutionist similarity can be considered. Suppose you are to analyze a complex physical phenomenon whose dynamic is known through a set of measures carried out during its evolution. Also, suppose you are to write a first set of simple equations which try to represent the examined phenomenon. From these, with a certain logic (to be defined), choose those which are best suited to the phenomenon (which survive), and discard the unsuitable ones (which are considered extinguished).

This new generation of more sturdy equations can be combined (reproduction) to give more complex functions. By continuing the process described for many generations, a progeny of mathematical models can be bred that even more faithfully represents the real phenomenon. It is also possible to stop this evolution when the model is complex enough (optimum complexity) for the purposes of study in order to avoid an over-specialization of the phenomenon.

The fundamental principles of the GMDH method are the following: suppose that the phenomenon under observation is regulated by m independent variables $x_1, x_2, \ldots, x_m$ (some may not even be influential, but there is no way of confirming that initially). Thus, record a great number n of observations of variables m and of the dependent variable y for several cases of the phenomenon studied. This set of observations is then divided into two subsets called the training set, e.g., the first

k number of observations, and the checking set, e.g., for observations from $k+1$ to n, with k usually being 80–90% of n. The simplest equations that can be written are those taking all the variables in pairs, as if only one pair was meaningful enough to describe the phenomenon. For example, to observe the generic pair of variables x_i, x_j,

$$y = a + bx_i + cx_j + dx_ix_j + ex_i^2 + fx_j^2 \ldots \qquad \text{(VI.5.A1)}$$

Considering all the pairs of variables for only one observation, a system of $m(m-1)/2$ linear equations in 6 unknown quantities ($a, b, c, d, e,$ and f) is obtained. For example, for $m = 3$:

$$y = a + bx_1 + cx_2 + dx_1x_2 + ex_1^2 + fx_2^2$$
$$y = a + bx_2 + cx_3 + dx_2x_3 + ex_2^2 + fx_3^2$$
$$y = a + bx_1 + cx_3 + dx_1x_3 + ex_1^2 + fx_3^2$$

Considering all training set observations by substituting x with the recorded values, the complete system will have $m(m-1)/2k = NE$ equations in 6 unknown quantities. This system is evidently over-dimensioned; the solution may be obtained by using the method of least squares. The six unknown quantities will have the values of $A, B, C, D, E,$ and F, which should satisfy all the NE equations with the minimum error. Then consider the $m(m-1)/2$ equations with checking set values: substitute these latter in the x variables of Equation VI.5.A1. The dependent variables will be:

$$z_h = A + Bx_{ih} + Cx_{jh} + Dx_{ih}^2 + Ex_{jh}^2 \qquad \text{(VI.5.A2)}$$

where $h = k+1, k+2, \ldots, n$.

The values for z_h, obtained from each of the $m(m-1)/2$ equations of pairs of variables, can be compared with those recorded for y_h to determine a valuation error of r_{hs}:

$$r_{hs} = \frac{\sum_{k+1}^{n}(y_h - z_{hs})^2}{\sum_{k+1}^{n} y_h^2} \qquad \text{(VI.5.A3)}$$

where $s = m(m-1)/2$

Only evaluations z (of y) which are under a predefined error threshold R can be kept; what remains is rejected. The smallest value of r_{hs} is registered and named R_{min1}. After these actions, the first iteration can be considered concluded. Now, the process must be repeated, this time substituting the values for x_i for the new independent variable values z_h. The dimension of the new vector Z may be greater than the vector X. Equations similar to Equation VI.5.A1 should then be built. The new unknown quantities ($a, b, c, d, e,$ and f) will be re-evaluated with training set observations by the method of least squares. From the checking set observations, recalculate the approximate values, $z'h$, of the dependent variable and then the errors, $r'_{hs'}$ of this second iteration, with Equation VI.5.A3.

Equations with errors inferior to the error threshold R that was previously defined should be kept, and the others rejected. The new minimum error R_{min2} will also be recorded.

The process continues until a value R_{minj}, greater than the penultimate one (R_{minj-1}), is obtained; from experience, this usually happens after 3–4 iterations. Among the survivors, the polynomial which best represents the physical phenomenon in question will be that with the minimum amount of error. After a few iterations in the surviving polynomial, only a certain number of initial variables, x_i, with very high orders and some of their combinations are virtually present. Presumably, only the most significant independent variables for the physical phenomenon being studied have been preserved.

6

Intelligent Multiagent Control for Power Plants

6.1 Introduction VI-181
6.2 Control-Software Engineering VI-183
     Need for a Software Engineering Approach • Control-
     Software Problems and Challenges
6.3 Intelligent Agent-Oriented Design VI-186
     Agent-Based Design • Agent-Oriented Technologies and
     Methodologies
6.4 Multiagent Control System for a Power Plant VI-190
     Design Methodology of the ICCS Architecture • Overall
     Power Plant Operation and Control • Context of the ICCS •
     Industrial Hierarchical Structure • Fundamental Functional
     Structure • Intelligence Functional Structure • Architecture
     • Functional Specification • Physical Scope of the ICCS
6.5 Minimum Prototype of an Intelligent Control
     System VI-200
     Definition of the ICCS-MP • Physical Scope of the ICCS-
     MP • Control Structure of the ICCS-MP • Realization of
     the ICCS-MP • Extended Coordinated Control
6.6 Summary and Conclusions VI-207
References VI-207

Raul Garduno-Ramirez
Electrical Research Institute

Kwang Y. Lee
Pennsylvania State University

6.1 Introduction

Commonly, electricity for everyday use is produced at power plants, then carried over long distances through transmission systems, and finally delivered to the end users through intricate distribution systems. Power plants are large industrial processes that must provide the right amount of power to match the current electric load at all times, that is, electric power must be produced at the moment it is needed since it cannot be stored in bulk quantities to be released when required. Therefore, the operation of power plants becomes, in many instances, a critical issue subject to all sorts of requirements (e.g., technical, environmental, political, and economical) that may change frequently, rapidly, and unpredictably. This situation extremely complicates the operation of power plants. Furthermore, achievement of the operation goals is usually far from being optimal because operation is human-based and current control systems lack versatility. In this regard, although the operation of a power plant is carried out through the control system by the operators, the achievement of the operation objectives is still mostly based on their ability and experience, which may vary widely from operator to operator and from shift to shift. This makes it difficult to satisfy the required objectives in a consistently optimal way. Additionally, most current automation and control systems do not provide the means for the process to exhibit the adaptation capabilities required to deal with the changing conditions of their operation environment and at the same time achieve the highest possible performance throughout the process operating range and duty life period.

In response to the above challenges, researches proposed the development of a large-scale intelligent control system for power plants that, in principle, will execute a set of functions (e.g., learning, value judgment, decision making, and knowledge processing) for the controlled system to exhibit the autonomy and self-governing behavior of an intelligent system, with the capacity to consistently satisfy the tightest operation needs. Nevertheless, the realization of such a control system is not a trivial task, mainly due to the inherent complexity of the system and the inadequacy of any artificial intelligence (AI) technique to implement all the required functions in an efficient way. Current intelligent control systems solely based on fuzzy logic, neural networks, and evolutionary optimization have only been succesfully used for controlling small-scale complex plants, and they do not exhibit autonomous behavior. It is the underlaying thesis of this chapter that the development of trully intelligent control systems demand the integration of those and other techniques in a suitable framework to make use of their best characteristics to attain a coherent overall system behavior toward process autonomy. Unfortunately, there are no straightforward methods to engineer this kind of system. To deal with these problems, it has also been proposed to develop a large-scale intelligent control system for power plants in the form of a multiagent system (MAS), where the intelligent agent (IA) paradigm is the building block for its realization.

The MAS technology has created a lot of excitement in recent years in the Internet arena for the development of large complex software systems. It is interesting to investigate the feasibility of the MAS and IA paradigms for the development of large-scale, open, and distributed intelligent control systems. It is the thesis of this work that the MAS approach may allow the creation of a consistent symbiosis toward solving the total automation problem of a large-scale process from an open community of intelligent agents, where each agent solves a partial problem by performing a partial intelligence function. A key issue is that each agent may use the technique that is best suited to solve its corresponding subproblem and may communicate the results to other interested agents to solve the larger automation problem. In this way, the intelligent behavior of the control systems may be achieved through specialized agents acting coherently and grouped along diverse intelligence dimensions, regardless of the techniques used to develop them. The main MAS development problems may be grouped into two categories:

- Synthesis of the multiagent control system, including the formulation and description of the automation problem, its decomposition in subproblems, the definition of intelligent agents and their allocation to subproblems, and structuring of the multiagent system
- Development and integration of the intelligent agents into the multiagent system, including the means to achieve coherency in decision making and action taking with respect to the objectives of the control system

This work focuses on providing solutions to the first category of problems for the development of a large-scale, intelligent, multiagent control system for a power plant. Points of major interest include showing the relationship between the control engineering and software engineering design strategies as a design methodology for MAS-based control systems and the use of intelligence theory and intelligent machines together with control theory concepts to define the structure of the intelligent control system. In what follows, the conceptual model of a large-scale intelligent system for power plant control is presented as a case study. Basically, the proposed intelligent control system greatly extends the scope of current state-of-the-art control systems at fossil-fuel power plants. From an application perspective, the purpose of the proposed system is twofold. First, this kind of system will diminish the drawbacks of human-based operation through the systematic addition of an extensive set of supervisory intelligent functions along various intelligence dimensions, improving the consistency of the control actions under complex operating environments. Second, the multiagent approach will facilitate dealing with the complexity of the control system. The design of the control system is performed through modularization at the highest level of abstraction keeping its complexity within manageable bounds. Therefore, the resultant control system structure, seen as an open organization of intelligent agents, may be regarded as a general framework for the development of large-scale intelligent control systems, where the required operation versatility may be achieved with diverse degrees of intelligence, in accordance with the number and capacity of the intelligent agents being enabled.

The treatment of the above issues in this chapter is as follows. Section 6.2 presents the argument that, since nowadays the development of a large-scale control system mainly consists of the development of a large software system, it is very convenient to apply software engineering techniques from the initial specification and design stages of the control system. Also, due to the potentially very large number of interactions among components to be managed, achieving data consistency is one of the major difficulties in designing large systems. So, there is a need for system design methods to prevent complexity. In Section 6.3, the IA and MAS paradigms are introduced as promising tools to develop large complex systems preventing complexity. A few concepts on the use of the intelligent agent paradigm for the realization of an intelligent multiagent system are presented. Also, a review of the currently available techniques shows a lack of techniques for MAS design. The situation described in Sections 6.2 and 6.3 justifies all efforts aimed at the design of control system structures that simplify the implementation of the desired intelligent control systems. Then, Section 6.4 describes the design of a large-scale intelligent control system for a fossil-fuel power plant in the form of a multiagent system, called the intelligent coordinated control system (ICCS). Details are provided on the amalgamation of concepts from software engineering, control engineering, process engineering, intelligent machines, artificial intelligence theory, and industrial control experience into a general framework for the development of large-scale intelligent control systems. This procedure constitutes a pragmatic design methodology for MAS-based control systems, which is the main contribution of this work. In Section 6.5, the realization of a proof-of-concept minimum prototype of the ICCS, which demonstrates the feasibility of the ICCS paradigm to achieve consistently optimal control and operation versatility under operating environments characterized by multiple operation objectives, is described. Finally, this chapter is summarized and conclusions are drawn in Section 6.6.

6.2   Control-Software Engineering

The development of a large-scale control system primarily involves the development of a large, real-time software system.[1] The interdependence between software engineering and control engineering methodologies is seldom treated in the technical literature, but it is important for the realization stage of simple digital control systems,[2] when it could be very difficult to locate and fix any mistake. As will be shown, it is much better to assist the development of a large control system, considered as a software system, with the software engineering techniques from the very beginning (i.e., specification and design stages) to avoid costly and time-consuming misfits at later development stages. Also, achieving real-time timing and data consistency are among the most difficult problems to be solved for a given design, greatly increasing the system complexity, for which adequate methods need to be devised.

6.2.1   Need for a Software Engineering Approach

Computer-based systems have become a major driving force in modern society. All sorts of information are manipulated through software programs. Coined in 1967, the term software engineering refers to the discipline that provides the means (i.e., process models, methods, and tools) to produce high quality, fault-free software on time and within budget.[3] Software quality may be quantitatively assessed in terms of software metrics (e.g., size, cost, duration, effort, accuracy, reliability, efficiency, integrity, etc.). A software life cycle typically includes a series of general phases: requirements, specification, design, implementation, integration, maintenance, and retirement. The way in which these stages are arranged constitutes a software process or life-cycle model (e.g., waterfall, rapid prototyping, incremental, synchronize-and-stabilize, spiral, etc.).

Software engineering concepts have been successfully used for the development of many software systems. However, after more than thirty years since the inception of software engineering, the flexibility offered by software is still very frequently abused. Much software is developed using the so-called build-and-fix approach, without specifications or any attempt at design. It is very common to find software systems that are extremely customized, which are very difficult to validate and maintain, and, most importantly, for which it is very problematic to guarantee error-free operation. As incredible as this seems, there is still a shocking lack of systematization in the design of software systems. Among many others, one reason for this situation is the prolonged existence of the "early programming culture" that does not give software development the importance it deserves. The misconceptions that anybody can develop software, that software systems are easy to design, generate, and test, and that software can be updated at low cost and in short time are surprisingly still fairly popular in both academic and industrial circles.

The impact of the application of software engineering concepts can be grasped from some quantitative facts regarding the cost of software. The most relevant figures include the cost of software with respect to the total cost of the computer-based system, the relative cost of software per phase of the life cycle, and the relative cost of error detection and correction at every phase of the software process. These figures provide general guidelines for the application of software engineering concepts.

First, the cost of software (application and basic) may easily supersede the cost of equipment by a large amount, ranging from 50–50% (in some telecommunication and military systems) to 90–10% (in some very large systems). The cost of application software (modeling, algorithms, and heuristics) development is usually larger than that of the basic software (scientific libraries, operating systems, compilers, communication, and user interface) in figures that range from 60–40% to 80–20%, respectively. Within the application software, the cost and development effort of heuristics dominates the mathematical part (algorithms and modeling), even in the case of critical mathematical parts, by one or two orders of magnitude.[4]

Second, the average relative cost of software per phase of the software process shows that the requirements phase accounts for 2% of total software cost, specification for 5%, design for 6%, development accounts for 12% (5% in module coding and 7% in module testing), integration for 8%, and maintenance for 67% of total software cost.[3] As shown, maintenance is an extremely time-consuming and expensive phase of the software life cycle. A major concern of software engineering is the application of techniques and tools that lead to a reduction in maintenance costs.

Third, studies have found that between 60 and 70% of all faults detected in large-scale projects are due to specification and design errors.[5] If an error is made in an earlier phase, the resulting fault will propagate into the subsequent phases of the software process. The relative cost of detecting and fixing a fault increases as the software process progresses. Typical figures show a proportion of 1:3:4:10:30:200 in the relative cost of detecting and fixing a fault during the requirements, specification, design, implementation, integration, and maintenance phases, respectively.[6] These facts show the convenience of detecting and fixing faults as early as possible, as well as the relevance of the specification and design phases.

All issues mentioned above directly apply to the development of software for computer-based monitoring and control systems (control software systems), where their relevance is increased to a new dimension when involved in critical applications such as military defense, flight control, health care, power supply, etc. The development of real-time high-risk control software systems poses more challenges than the data processing systems for scientific and office applications.[7]

Real-time requirements may include continuous operation, execution depending on external events or at fixed frequency rates, direct and immediate data entering from outside the system, immediate data processing for process control, direct effect on the controlled process, low rate of printed information, and distributed processing. Consequently, control software systems may easily become very complex systems. Furthermore, the integration of new advanced applications to achieve more versatile and efficient operation of the system under control, and to exploit more fully the digital hardware capabilities, will again notably increase the complexity of any control software system. Every addition or change may become a potential risk and should be thoroughly tested at all the stages of the software process.

Although it has been recognized internationally that the development of high-quality control software systems requires dedicated methods and highly trained personnel,[8] it should always be kept in mind that adherence to standards is not intended to achieve 100% software quality per se, but to reduce the risk of poor quality software. Also, software engineering procedures do not provide the creativity needed for the invention of the essential philosophy and heuristics in the initial design of a large control system. Clearly, the initial design of large control software systems still remains an art that heavily relies on the aptitude and experience of the designers. The application of software engineering techniques may systematically assist the art of control system design to achieve completeness and to prevent conceptual errors through successive refinement. Software engineering may help tackle the complexity of the control system design in a systematic way.

The crux of the matter is that since a control system design will eventually be translated into a software system, it is better to assist its design with software engineering concepts in mind from the very beginning. In turn, this approach will ease the transition from the requirements, specification, and design phases to the development phases, to reduce the time and cost of development and to minimize software errors at later stages in the software process. It is firmly believed that when properly applied, the discipline of software engineering will certainly produce clean-cut overall control system designs, where the benefits of the advanced operation applications can be more easily recognized and successively improved.

6.2.2 Control-Software Problems and Challenges

Around 30 years ago, the early control software systems were developed as direct translations of their analog counterparts. All system functions were performed by a single program, with all system data available to all functions. From the beginning, it was expected to obtain systems that were easier to duplicate and upgrade with more sophisticated control strategies. Nevertheless, the high cost and tight constraints on processing power and memory capacity, together with a lack of software development tools, yielded highly customized software structures with tight couplings between the timing and control functions, which were exceptionally difficult to modify.

Through the years, and with the availability of more powerful hardware and software, the application of computer-based control systems spread out. Control software systems substantially increased in size and complexity. It was then required to design the systems at higher levels of abstraction and to build them using larger development teams. System structures became modular, so different modules could be developed by different groups. All sorts of modules can be created. On the one hand, dependencies between any two modules and a large number of modules yield intricate patterns of communication within the system. Consequently, the integration of modules in a large control system can take as long as their development. On the other hand, although the availability of high level programming languages, specially tailored to the implementation of control systems, has certainly benefited the control engineering community, it has not achieved its expected success. Reasons for this are that the use of high-level languages made control over timing far more difficult than without them, and the dependencies among modules are a far greater problem than the actual programming of the required algorithms. In fact, the problem has never been the coding of the algorithms. Another concern arises with the need to use multiple processors to meet the processing demands of the control system. Concurrent processing raises new problems because parallel processing has not been available before, and the development of procedures for parallel execution requires the opportunity to assess its behavior and correctness. Unfortunately, the presence of multiple processors, even as redundant units for increased reliability, leads to a sharp increase in system complexity.

The traditional approach to control system design presents two major problems: data consistency and timing between the various components of the system. Normally, the relation between two software modules is extremely complicated since not only parameters and functions are shared, but also global variables. In order for the various modules to correctly interpret the data they operate on, it is necessary that the access to the data be well controlled. In many cases, this control is ensured through a procedure-call interface (parameter passing). In other cases, explicit control mechanisms need to be introduced to ensure consistency. This is particularly true in systems where several modules are active at the same time, either through interleaved execution in a single processor or on truly concurrent machines. In complex systems, a large number of interactions between modules need to be managed in terms of both the data that are transferred and control over the transfer itself. The potentially very large number of interfaces to be managed is one of the major reasons for the difficulties experienced in designing large complex systems. Besides making the design process difficult, the built-in connections between modules also cause the resulting system to be static. For flexibility at run-time, which is needed to reconfigure a system, it is vital that the communication between modules is not fixed during the design but can be established as the need arises. As another fundamental problem, the major issue with timing is the instability that can occur if the computational delay between measuring and controlling the process is too large. This physical constraint has to be met by the control system. Other less critical timing constraints arise when human operators are part of the control mechanism. These timing constraints cause problems for system development only if the processing power is a critical issue. If insufficient computing power is available, the only solution is to split the computation into parts that can be executed in parallel, and then add processors in a dedicated configuration. Determining the ability of a system to actually meet its deadlines under all circumstances is very difficult and is currently beyond the state-of-the-art in multiprocessor systems. The effect of the development of complex control systems is to simplify the problem and design for the worst-case situation, in which all processes will have to meet their deadlines concurrently.

Much effort has been spent to solve the problems mentioned above. Nearly all the approaches have concentrated on providing more powerful tools for implementation of the specifications in a function-oriented approach. Typical examples of these efforts are high level programming languages and real-time operating systems. In parallel to these efforts for developing better tools for implementing systems, extensive work has been done in the area of providing support for the development process itself. The so-called software development environments aim at providing mechanisms through which the complexity of the design can be managed. These systems are generally very complex themselves and have not been proven to reduce the design time or achieve higher system quality. From recent publications, it is clear that researchers are increasingly looking for ways to achieve the prevention of complexity, although the majority of effort is still directed to developing better tools. Alternative efforts are aiming at architectures that simplify the implementation of the desired systems. To prevent unnecessary complexity, neither the design method nor the selected architecture should increase the system complexity beyond the intrinsic complexity of the specification to be realized by the system.

6.3 Intelligent Agent-Oriented Design

A major concern with the design of a large-scale control system is the complexity of the system, mainly due to the number of components and their interaction patterns, as explained in the previous section. This complexity is raised to a completely new dimension when the control system is required to become intelligent. A completely new variety of knowledge-processing functions needs to be implemented, and the interaction patterns get more involved. In this regard, the intelligent agent and multiagent system paradigms, as state-of-the-art artificial intelligence software engineering concepts, may provide a comprehensive and unifying framework to handle the design and realization complexity of large-scale intelligent control systems. Nevertheless, a review of the technical literature reveals a lack of design methodologies for multiagent systems, despite the abundance of methods and tools for other development tasks. Thus, this situation indicates the need for an MAS design methodology for large-scale intelligent control systems.

6.3.1 Agent-Based Design

The development of control software systems, considered as real-time high-risk systems, requires the utilization of proper methods, certainly different from those required by scientific and business applications. The development of real-time control software systems was initially carried out with methods that originated in the non-real-time applications with adequate changes. Some of these methods include real-time structured analysis and design[9,10] and real-time object-oriented design.[11]

Methodologies based on the intelligent agent paradigm have been proposed recently to significantly enhance our ability to conceptualize, design, and implement increasingly complex software systems to attack more complex, realistic, and large-scale problems.[12] Even if there is no general consensus on the definition of an agent, a useful working definition states that an agent is an encapsulated software system situated in some environment, capable of flexible autonomous action in that environment in order to meet its design objectives.[13] More simply, an agent is software that can carry out information-related tasks without ongoing human supervision.[14] Furthermore, a multiple agent, or multiagent, system (MAS) can be defined as a loosely coupled network (organization) of problem solvers (agents) that interact to solve problems that are beyond the individual capabilities or knowledge of each problem solver (agent).[15] Multiagent systems tackle complexity through modularity and high-level abstraction by developing a number of functionally specific components (agents) that are specialized at solving a particular problem aspect. Each agent uses the most appropriate techniques for solving its particular problem and can be coordinated with other agents to properly manage their interdependencies. The problem solvers are autonomous and heterogeneous in nature. The motivation for the use of intelligent agents in multiagent systems is due to their expected ability to: (1) solve problems that are too large for a single centralized agent, (2) allow the interoperation of multiple existing systems to keep pace with changing needs, (3) provide solutions that efficiently use information sources that are spatially distributed, (4) provide solutions where expertise is distributed, and (5) enhance performance along the dimensions of computational efficiency, reliability, extensibility, robustness, maintainability, responsiveness, flexibility, and reusability.[16]

Multiagent systems as organizations of intelligent agents provide a framework for agent interactions through the definition of roles, behavior expectations, and authority relations. The organizations are, in general, conceptualized in terms of their structure, that is, the pattern of information and control relations that exist among the agents and the distribution of problem-solving capabilities among them. The issue of organizational adaptivity is crucial. Organizations that can adapt to changing circumstances by altering the pattern of interactions among the different constituent agents have the potential to succeed. An open organization is one in which the structure of the organization is capable of changing dynamically. The information sources, communication links, and components can appear and disappear arbitrarily and unexpectedly. The components may not be known in advance, may change over time, and may be highly heterogeneous. In open organizations, agents may dynamically find their collaborators based on the needs of the task at hand and on which agents are part of the organization at any given time, thus adaptively forming teams on demand pursuing common goals to achieve global system coherence.[16–18]

The MAS paradigm has already been used to implement large software structures for monitoring and control systems. MAS applications first appeared in the mid 1980s.[19] In Durfee and Lesser[15] and Durfee,[20] a set of geographically distributed agents monitor vehicles to track vehicle movements in a global area. In Parunak,[21] an MAS is used to effectively manage the production process in a manufacturing enterprise where each factory is represented as an agent. Ljunberg and Lucas[22] describe a sophisticated agent-based air-traffic control system that represents aircrafts and air-traffic control systems as agents. Here, the agent paradigm provides a useful and natural way of modeling real-world autonomous components. In Schwuttke and Quan,[23] an MAS is used to monitor and diagnose faults in spacecraft control. Boasson[24] explains the MAS paradigm for a radar tracking system. The best-known MAS applications in process control are those described by Jennings, Corera, and Laresgoiti,[25] where MASs were developed for power transmission management and particle accelerator control using the ARCHON software development platform. Velasco et al.[26] designed an MAS for distributed control of industrial processes and applied it to a fossil fuel power plant to improve the heat rate.

Although MASs provide many potential advantages, there are also many design and implementation challenges.[16] The first essential problem consists of the formulation, description, decomposition, and allocation of the overall problem, followed by the synthesis of a group of intelligent agents. Another major challenge consists of engineering practical systems that enable the agents to achieve global problem-solving coherence by providing the means to communicate and interact, to make decisions and take actions, and to recognize and reconcile disparate viewpoints and conflicting intentions. This work reports the results obtained for the later problem. As will be shown, the total automation problem for a fossil fuel power plant is formulated, and an intelligent multiagent control system is proposed. Other research activities are currently dealing with the former problem, where the realization of intelligent agents is being investigated to achieve global problem-solving coherence.

6.3.2   Agent-Oriented Technologies and Methodologies

Driven mainly for Internet applications, agent technology has received a great deal of attention in the last few years; there are different developed agent theories, languages, architectures, and successful applications. As with other paradigms for software development, agent-oriented methodologies should be applied in all the life-cycle phases of an agent-based application. Nevertheless, very little work has been done on the development of methodologies to specify and design applications using the intelligent agents technology. Researchers on agent-oriented methodologies have followed the approach of extending existing methodologies to include the relevant aspects of the agents. Extensions have been mainly proposed to object-oriented (OO) methodologies and knowledge-engineering (KE) methodologies.

Similarities with the object-oriented paradigm consider agents as active objects, that is, objects with a mental state. Both paradigms use message passing for communicating and can use inheritance and aggregation in defining the system architecture. The main difference is the constrained type of messages in the agent-oriented paradigm and the definition of a state in the agent based on its beliefs, desires, intentions, commitments, etc. Agents are not simple objects. Aspects not addressed by object-oriented methodologies include the types of messages used by agents. Agents model messages as speech acts and use complex protocols to negotiate. Also, agents analyze these messages and decide whether to execute the requested action. Another difference is that agents can be characterized by their mental state, and object-oriented methodologies do not define techniques for modeling how the agents carry out their inferences, planning processes, etc. Finally, agents are characterized by their social dimension. Procedures for modeling these social relationships between agents have to be defined. Agent oriented methodologies that extend object-oriented methodologies include:

- Agent-oriented analysis and design — three models for analyzing an agent system are proposed: the agent model, which contains the agents and their internal structure; the organization model, which describes the relationships between agents; and the cooperation model, which describes the interactions between agents.

- Agent modeling techniques for systems of BDI (belief, desire, and intention) agents — this method considers two main levels for modeling BDI agents: the external viewpoint that decomposes the system into agents and their interactions using the agent model and the interaction model, and the internal viewpoint that carries out the modeling of each BDI agent using three models: the belief model, the goal model, and the plan model.
- Multiagent scenario-based (MASB) method — intended for MAS design in the field of cooperative work, this method considers two phases: analysis and design. Activities included in the analysis phase are scenario description, role functional description, data and world conceptual modeling, and system–user interaction modeling. Activities included in the design phase are MAS architecture and scenario description, object modeling, agent modeling, conversation modeling, and system design overall validation.

Knowledge engineering (KE) methodologies provide the techniques for modeling agent knowledge. The definition of the knowledge of an agent can be considered a knowledge acquisition process, and only this aspect is considered in knowledge-engineering based agent methodologies. Designing an MAS shows most of the same problems faced by KE methodologies: knowledge acquisition, modeling, and reuse. KE conceives centralized knowledge-based systems without addressing the distributed or social aspects of agents or their reflective and goal-oriented attitudes. MAS modeling solutions proposed in this direction include extensions to CommonKADS, which can be seen as a European standard for knowledge modeling:[27]

- The CoMoMAS methodology models an MAS through the following models: agent model, expertise model, task model, cooperation model, system model, and design model.
- The MAS–CommonKADS methodology extends the models defined in CommonKADS using techniques from object-oriented methodologies and from protocol engineering to describe the agent protocols. This methodology starts with an informal conceptualization phase to collect user requirements and to obtain a first description of the system from the user point of view. This methodology defines the following models: agent model, task model, expertise model, coordination model, organization model, communication model, and design model. This methodology has been applied in several research projects in different fields.

Some agent developers have successfully developed and applied MAS to practical problems. Although not declared an MAS methodology, these constitute an important approach for the design of MAS that should also be considered since they have provided general guidelines for MAS development. Such is the case of the ARCHON development environment for MAS, which proposes a methodology for analyzing and designing an MAS. The analysis combines a top-down approach that identifies the system goals, the main tasks, and their decomposition, with a bottom-up approach that allows the reuse of pre-existing systems constraining the top-down path. The design is subdivided into agent community design and agent design. The agent community design defines the agent granularity and the role of each agent. The agent design encodes the skills for each agent.

Even if there are no standard definitions for an agent, agent architecture, or an agent language, the methodologies reviewed here show that there exists a conceptual level for analyzing agent-based systems, no matter what agent theory, architecture, or language supports them. This conceptual level should describe:

- Agent models: the characteristics of each agent should be described, including skills (sensors and effectors), reasoning capabilities, and tasks
- Group/society models: the relationships and interactions between the agents

The lack of standard agent architectures and agent programming languages is currently the main problem for MAS model building and implementation. Also, since there is no standard architecture, the design of the agents needs to be customized to each agent architecture. In addition, some methodologies select the agent architecture during the analysis, while others consider that it is a design issue and the agent architecture should be selected depending on the requirements of the analysis.

In this work, a pragmatic approach is followed for the design of a large-scale control system as an MAS. The key underlying concept is that of creating the required control system for the controlled process to behave as an intelligent system, i.e., a large-scale intelligent system. From this, the structure of the control system or MAS is mainly determined by merging the two most prominent theories, to be explained later, for the development of intelligent systems. The definition of this basic structure is complemented by considering some concepts from the field of industrial automation and some previous experience in the development of practical large-scale control systems. This approach is presented in detail in the next section.

6.4   Multiagent Control System for a Power Plant

A successful large-scale intelligent control system design requires the convergence of multiple disciplines. In this work, prominence has been given to control engineering and software engineering because of the focus on the realization of a software system. Process engineering concepts are also required to define the operation objectives of the controlled system, which will be used to evaluate the efficacy of the resultant control system. This section shows how to amalgamate all these concepts, together with some others from intelligence theory and intelligent machines, to produce the design of a full-scope, large-scale intelligent control system for a fossil-fuel power plant. The design approach effectively handles the control system complexity through an intuitive and practical control system organization of intelligent agents. The system is called the intelligent coordinated control system (ICCS).

6.4.1   Design Methodology of the ICCS Architecture

A brief historical note is provided to introduce the approach undertaken in the design of the ICCS architecture. The conceptual model of the ICCS in this work constitutes the most recent in an evolving series of control systems structures that the author has been involved with for power plant automation. This effort started several years ago, motivated by the development of a structured control-software interpreter for batch process automation of the water demineralization plant in a fossil-fuel power unit (FFPU). This interpreter implemented a generalized functional structure, comprising modules for logic signals handling, analog signals handling, protections, regulatory control, logic sequences, recipe handling, and operation interface.[28] In a parallel project, a general control structure for FFPUs under automatic generation control through reconfigurable schemes was developed. This work pointed out the need for more emphasis on the implementation of supervisory functions to attain higher levels of automatic operation. After these two projects were completed, the functional structure implemented by the interpreter was modified to automate a whole power unit. The recipe handling module, necessary for batch-process control, was replaced by a more general supervision module that incorporated set-point generation functions and non-linear characterizations, necessary for operation of an actual power plant. This new structure was used to develop the control software system for small size (1 MW) and medium size (60 MW) turbogas units.[29] Soon after, it was realized that even this modified structure should be additionally enhanced to achieve greater operation versatility. Artificial intelligence techniques were suggested to achieve self-governing characteristics.[30] Thus, a general control structure for FFPUs that incorporates the agent software paradigm as the fundamental design premise for implementing large-scale intelligent control systems is proposed in this work.

The design of the ICCS architecture as a multiagent system organization incorporates several interwoven concepts. To appreciate that, in what follows, the definition of the multiagent system organization is described as a sequential process that incorporates several paradigms step-by-step. This methodology starts with a conceptualization phase to locate the ICCS in the more general context of a power system or power market enterprise and to obtain a first description of the system from the user's point of view that collects both user and operation requirements and objectives. Then, this methodology describes the ICCS architecture from various perspectives, including physical context situation, industrial hierarchical structuring, extended process automation, and intelligence functional grouping. The resultant control system structure is an MAS in the form of an open organization of intelligent agents, which can be regarded as a general framework for the development of large-scale intelligent control systems.

6.4.2 Overall Power Plant Operation and Control

Currently, most electric power is produced at power plants, which are usually located close to the primary source of energy—most of them far away from cities or consumption centers. Then, the electric energy is carried over long distances through transmission systems based on thick cables and huge towers or poles. Finally, the electric energy is delivered to end users through the distribution systems at the specified conditions for domestic or industrial use. Power plants may be of different types: nuclear, fossil-fuel, geothermal, hydroelectric, etc. Units burning some sort of fossil-fuel (oil, diesel, carbon, etc.) produce about 60% of all electric power. A fossil-fuel power unit (FFPU) produces electric power from fossil-fuel through several energy conversion processes, using water as a working fluid. The chemical energy of the fossil-fuel is transformed into steam thermal energy by the boiler, is then transformed into rotational mechanical energy by the turbine, and is finally transformed into electric energy by the generator. Concurrently, the working fluid is alternately vaporized and condensed in a closed circuit following a thermodynamic cycle.

A power plant is intended to supply the electric power demanded by the consumers in a reliable form with high quality characteristics. Normally, the load is not under direct control and follows daily, weekly, and seasonally cyclic variation patterns. Additionally, the connection and disconnection of individual loads cause random fluctuations about these patterns. Since there are no practical means to store large quantities of electric energy, it should be produced as needed by the consumers. Consequently, a power plant participating in frequency regulation never really operates in steady state; it is always trying to match power generation with the load. Thus, FFPUs participating in load-frequency control are always subject to changing load demands and load disturbances as part of their normal operation regime.[31]

At a glance, the operation of an FFPU is as follows. First, the power needed by the load at a given frequency determines the counteracting electromagnetic torque that is to be equalized by the actuating mechanical torque, produced by the steam turbine as power at a given speed. So, frequency is used to determine the generation–load balance (mechanical–electrical torque balance). Frequency above the nominal value (e.g., 60 Hz) indicates that generation is larger than the load, frequency below the nominal value indicates that generation is below the load, and frequency at the nominal value indicates that the generation matches the load. Second, the prime mover torque is a function of the steam flow energy into the turbine, which, in turn, is a function of the steam pressure and temperature. Hence, the steam pressure before the throttle valves is used to determine the boiler–turbine energy balance. If the throttle pressure increases, the boiler is producing more steam than the turbine requires. If the throttle pressure decreases, then the boiler is producing less steam than required. When the throttle pressure is constant, the boiler is matching the steam needed by the turbine. Third, the rate at which steam is generated is determined by the rate at which fuel is burned. Adjusting the firing rate requires adjusting both the fuel flow and the air flow to maintain safe and complete combustion. The turbine valve position and the throttle pressure determine the rate at which steam is consumed. At all moments, feedwater should be supplied in adequate quantities to sustain the steam flow.

In summary, operators in an FFPU must satisfy two global operating requirements: generate the MW needed to satisfy the load demand, and maintain the energy balance within the unit. Operators satisfy these requirements by supervising, through the BTG (boiler-turbine-generator) board instrumentation, the adequate regulation of generated power, main steam pressure, and drum water level. To do so, they primarily adjust the main steam flow, fuel and air flows, and feedwater flow. There are many other variables that need to be regulated in accordance with the generated power and operation practices; nevertheless, the variables just mentioned account for the dominant behavior of an FFPU.

Therefore, an intelligent control system is required to basically emulate the operator actions described above. These constitute the basic requirements for the control system. In addition, the control system must also be able to exhibit the intelligent behavior of skillful operators and process engineers in terms of a series of knowledge-processing capabilities to be described later.

6.4.3 Context of the ICCS

Being part of a power system, an FFPU demands from its control system the capability to be incorporated into a larger control system. Thus, it is very important for the ICCS to provide the means to be coordinated with all other power system control elements. Pyramidal automation models for power systems have been around for several years.[32,33] These models promote the implementation of large-scale, vertical hierarchical automation structures consisting of several levels, with the power units at the bottom level. Nevertheless, the current trend toward market liberalization of the electric sector predicts substantial changes in the management and arrangement of the upper levels in these structures.[34]

Since power plant physics are not subject to these trends, it is expected that power units used for load-frequency control in deregulated power systems will be commanded as self-supported energy production cells through the sole specification of unit load commands. That is, the power system restructuring trend will put even more emphasis on the role of power units as power system actuators, more in the way that control valves are used to regulate process flow variables, to control power generation.

Thus, for the purposes of this chapter, it will be assumed that a power unit is commanded through a single unit load demand profile, regardless of being located either at the bottom level of a larger hierarchical power system or as part of a generation-only utility in a deregulated power system. Also, optimal power unit operation is considered a local affair, procured at the sole discretion of the unit operators and motivated by survival under competition and profit making.

6.4.4 Industrial Hierarchical Structure

In essence, the architecture of the ICCS is formulated as a hierarchical system, taking into account the concepts from industrial process automation engineering and basic principles for the implementation of intelligent systems. In brief, the architecture of ICCS comprehends the bottom levels of a general industrial automation hierarchical model, where each level performs the basic functions of an intelligent system to different degrees. The hierarchical levels considered by ICCS are the direct control level and the supervisory control level (Figure VI.6.1), which correspond to levels 1 and 2, respectively, in the reference model for computer integrated manufacturing (CIM) for a continuous industrial process.[35] According to this reference model, the direct and supervisory levels concentrate

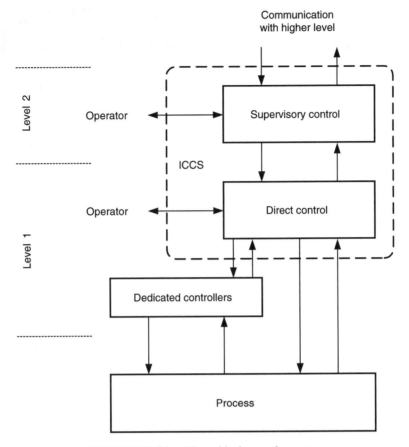

FIGURE VI.6.1 Hierarchical control structure.

on performing control computation and control enforcement tasks, while the upper levels concentrate on production scheduling and management information processing tasks.

6.4.5 Fundamental Functional Structure

The fundamental functional structure of the ICCS is based on a rather general structure model for industrial batch-process automation. This structure model has been applied successfully for batch process automation[36] and for building highly flexible batch process control software.[28] In this model, all functions performed by a control system are grouped into seven basic functional groups, as shown in Figure VI.6.2, where the recipe management functional group has been replaced by what was named the supervision functional group, since, from the batch process control point of view, a power unit produces only one product: electric power.

- Logic/analog input/output signal handling functions constitute the interface between the control functions and the process. They are responsible for entering and sending contact signals, as well as continuous signals, to and from the process instrumentation at regular intervals or under demand.
- Interlocks and protection functions monitor critical variables to avoid the process entering unsafe operating regions. They depend heavily on the physical characteristics of the plant and on the safety requirements of the process.

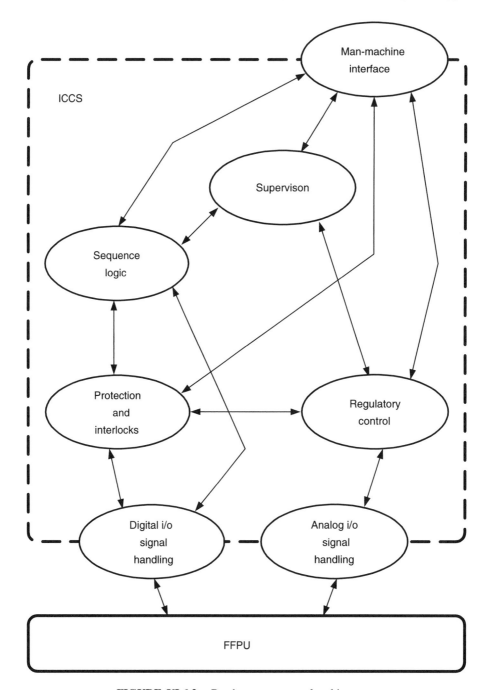

FIGURE VI.6.2   Batch process control architecture.

- Sequence logic functions specify how and in what order the plant operations should be executed. They allow the transition between the various operating states. They enable and disable the continuous control functions as required.
- Regulatory control functions evaluate control algorithms for driving the continuously varying signals in the process.

- Supervision functions provide the set-point values and non-linear characterizations needed at different operating states and to deal with non-linear process dynamics.
- Operator interface functions allow for interaction between operator and controller. They must be able to provide the operator with relevant information about the plant behavior. They must also provide the means to modify the control parameters and strategies.

At least implicitly, all functional groups are currently performed in many industrial control systems. The ICCS embraces all functional groups except the signal handling functions due to their high dependence on the implementation hardware.

6.4.6   Intelligence Functional Structure

The fundamental functional structure in the previous section has demonstrated its value to implement highly structured control software systems.[29] Extension of the capabilities of this structure requires the use of artificial intelligence techniques. The two most prominent principles to implement intelligent systems are used as guidelines to expand the scope of the fundamental control structure in a systematic way.

First, four basic intelligence functions are identified as primary components of an intelligent system: sensory processing, world modeling, value judgment, and behavior generation.[37] Sensory processing implements perception functions and compares observations with expectations to compute physical and dynamic attributes of the process and its environment. World modeling obtains an estimate of the state of the world (process, environment, and control system) based on information from sensory processing to generate expectations and predictions to be used by the other functions. Value judgment evaluates and grades the system behavior, computing costs, risks, and benefits of observed situations and of planned activities. Behavior generation selects goals and plans, executes tasks, and monitors execution of plans and modifies them when required.

Second, in the ICCS, the two-level hierarchical structure is also considered, but different emphasis is placed on the intelligence functions at each control level, in accordance with the principle of increasing precision with decreasing intelligence[38] and with the generic functions of the CIM model. At the direct control level, which is more precise but less intelligent and closer to the power unit, emphasis is placed on rapid and accurate computing over data information for plant reactive control. This level comprises the functions performed by current direct-digital control systems: logic sequences, control algorithms, and protection routines. At the supervisory control level, which is less precise but more intelligent, emphasis is placed on computations over knowledge information for plant proactive control. This level comprises some of the functions currently performed by the operator and process control engineer such as performance monitoring, controller tuning, production optimization, and set-point generation. Figure VI.6.3 shows the resulting intelligence functions as considered in the ICCS.

6.4.7   Architecture

The ICCS is proposed to realize a multiagent system for which a multidisciplinary approach, that amalgamates control, process, and software engineering concepts, has been followed in its design. The ICCS goals are identified using power plant process engineering concepts, and intelligent control systems engineering concepts are used to identify main tasks and to functionally decompose the system. A software engineering agency concept is used to identify and group agents according to their knowledge and purpose interactions.

The proposed ICCS organization is an open superset of functionally grouped agent clusters in a two-level hierarchical system (Figure VI.6.4). The upper level, which is mainly characterized for knowledge-driven processes, performs the supervisory functions needed to provide self-governing operation characteristics, while the lower level, which is mainly characterized for data-driven

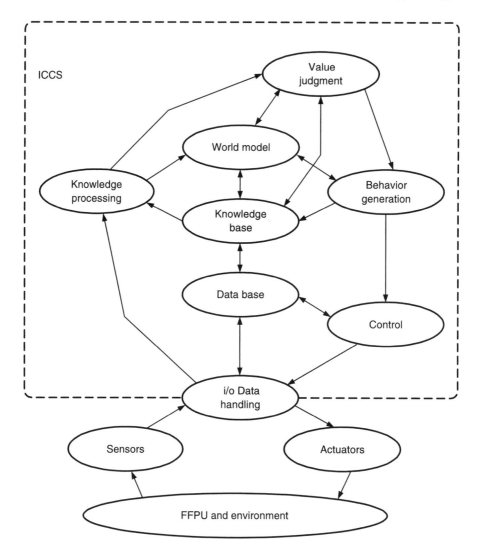

FIGURE VI.6.3   Generic intelligence functions of the ICCS.

processes, performs the fast reactive behavior functions necessary for hybrid (discrete and analog) real-time control and protection. The term organization is preferred in order to emphasize the soft nature of the system structure over a rigid inflexible architecture. Agents are loosely clustered, using the intelligence functions as guidelines. A generic control cluster takes account of the sequence control, regulatory control, protection, and input-output handling agents. A self-awareness cluster is introduced to group the system operating state determination, fault diagnosis, and test assistance agents. The world-modeling cluster comprehends the learning, model building, and adaptation agents. The value judgment cluster comprehends the online performance monitoring, control tuning, and reconfiguration agents. The memory cluster is introduced to include the data (sensory) and knowledge processing agents, as well as the system knowledge and database agents. The behavior generation cluster groups the process optimization, sequence generation, and set-point generation functions.

Agent clustering is introduced to simplify the representation and to indicate that the agents in a cluster use closely related system knowledge or data and have mutual commitments and beliefs.

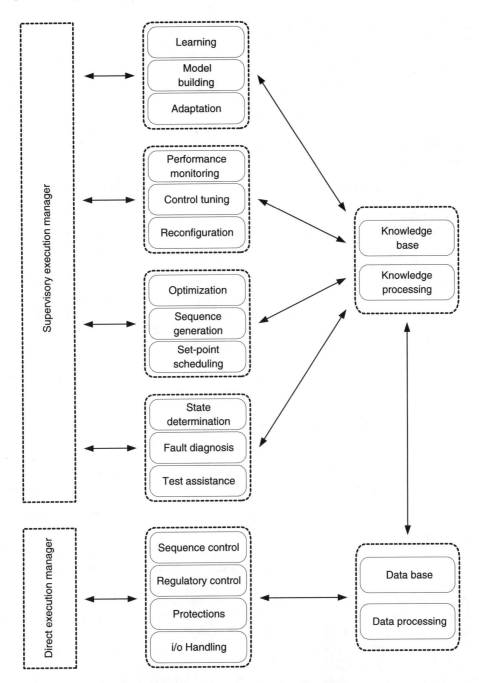

FIGURE VI.6.4 ICCS multiagent organization.

In reality, all agents may coexist as parallel processes with random access to system information. As required for an open system, the ICCS exhibits organizational adaptability mediated by the supervisory execution manager agent and the direct execution manager agent. In principle, the ICCS organization can adapt to changing circumstances by activating or deactivating agents, incorporating new agents or dismissing old agents, or modifying the pattern of interactions among the current agents in the organization. The ICCS agents should dynamically find their collaborators based on the system requirements at hand and which agents are present in the organization at any given time. Clusters should be formed adaptively as required.

6.4.8 Functional Specification

The system functional decomposition into agents in Figure VI.6.4 is not exhaustive in any way; it only shows what is considered a basic set of tasks that should be taken into account to achieve a more general design toward truly intelligent control systems and how they should be organized. In the spirit of an open system, this set of tasks may be augmented or decreased as required by the application at hand.

- Input/output signal handling constitutes the interface between the control system and the plant instrumentation. It is responsible for entering and sending contact signals, as well as continuous signals, to and from the process instrumentation at regular intervals or under demand. In a more advanced application, it should also implement the dialogs with new intelligent instrumentation. Also, it can take care of simulating a virtual environment for the entire system by halting actual inputs and imposing arbitrary input values.
- Protection and interlocking monitor critical variables to prevent the process entering unsafe operating regions or shutdown of the total or partial process when already in unsafe conditions. These functions depend heavily on the physical characteristics, equipment configuration, and protection requirements at different operation stages of the process.
- Continuous regulatory control evaluates control algorithms for driving the continuously varying signals in the process according to predefined references.
- Sequence control allows the transition between the various operating states of the process. It enables/disables the continuous control functions as required.
- Operating state determination evaluates key signals to declare the operating state of the FFPU. This information is to be used by other functions in decision-making and evaluation of permissive conditions.
- Fault diagnosis identifies features of faults before they occur, and determines their causes when they have already occurred. This generates information for fault accommodation.
- Test assistance sets and verifies all necessary conditions to perform operation tests from a given catalog of tests.
- Process optimization determines the optimal operating conditions by solving optimization problems based on physical principles.
- Operation sequence scheduling performs time-sequenced decisions for automatic plant operation, for instance, unit scheduling of power generation based on AGC demands and physical conditions of equipment at the FFPU.
- Set-point scheduling generates set-points for the continuous control functions and limits and threshold values for the sequence control and protection functions are generated according to the different operative stages and optimization routines.
- Performance monitoring evaluates behavior of the process under control to generate meaningful performance indications for adaptation and optimization.
- Control tuning decides, based on performance value, whether or not the current control configuration needs to be tuned. Performs tuning by updating parameters and knowledge of the direct control scheme if enabled by the human supervisor.
- Control reconfiguration decides, based on performance values, whether or not a change in the current control configuration should be made. Suggests a different control strategy from a given catalog, and conditions for switching are set if enabled by human supervisor.
- Learning allows the supervisor to build and modify the knowledge and databases for inferences and decision-making required at both the supervisory and direct control levels

based on observations of the input-output behavior of both the process and the control system itself.

- Model building constructs and updates internal representations to detect changes in the plant and its environment and to predict plant behavior.
- Adaptation in a broad sense provides the mechanisms to deal with changes in the plant and its environment at the supervisory level, such as updating the operating sequences and nonlinear characterizations required for wide range operation.

What is relevant in the ICCS structure is the decomposition of the supervision function shown in Figure VI.6.2 into several other tasks, in the form of agents, to provide the control system with the capability to satisfy increasing performance demands, keeping the complexity of the system within manageable terms. Also, the software agency concept provides the necessary mindset to integrate very dissimilar technologies in a systematic and harmonious way to achieve a practical and effective system by making use of the best characteristics each technology has to offer.

6.4.9 Physical Scope of the ICCS

As pointed out earlier, the ICCS is intended to extend the capabilities of current coordinated control schemes at FFPUs to attain wide range cyclic operation with optimization of multiple operation objectives (e.g., load tracking, duty life, heat rate, and pollutant emissions) during normal operating state. To this end, the ICCS may be seen as an open system in which direct control functions can be arbitrarily incorporated into an overall unit control strategy, that is, the ICCS can modify its control scope over the unit by absorbing into the coordination strategy as many direct control functions as required to satisfy the operating objectives under consideration. The control functions can be either

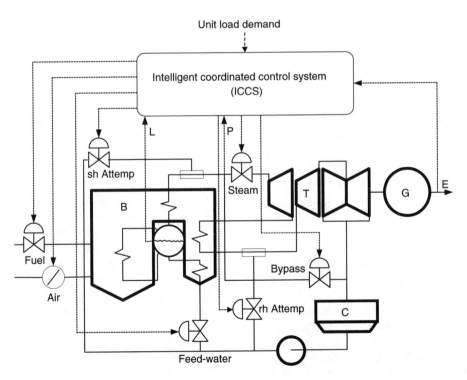

FIGURE VI.6.5   Physical scope of the ICCS. B = steam generator (boiler), T = steam turbine set, G = electric generator, C = condenser.

some of the generic control loops already existing in all power plants or new ones proposed to fit particular requirements of any given plant. What matters here is the capacity (openness) of the control system to modify its scope in a systematic way. This openness characteristic is also carried out into the supervisory level, where new knowledge-processing functions, currently not available at FFPUs, can also be incorporated arbitrarily. These high-level functions are needed to exhibit effective and versatile self-governing operation characteristics such as those performed by dexterous operators and process engineers. Because of its relevance, a more detailed discussion of this issue is separately provided.

Figure VI.6.5 illustrates the openness characteristic of the ICCS. In this case, the control functions comprehended by the ICCS would be the current coordinated control strategy, the steam flow and turbine speed/load control, the fuel/air flow and combustion rate control, the feedwater flow and drum level control, and the water spray flow and steam temperature control. Obviously, many other control loops and supervisory functions can still be included in the ICCS.

6.5 Minimum Prototype of an Intelligent Control System

The development of a comprehensive ICCS for an actual power plant is a formidable task; its magnitude and complexity are well beyond the scope of this chapter. Instead, a prototype with a minimum set of functions that preserves the essentials of the ICCS is developed to demonstrate the feasibility of the proposed ICCS paradigm. The prototype is called the ICCS minimum prototype (ICCS-MP). The ICCS-MP is an organization of a small number of intelligent agents in which advanced and conventional control and supervision techniques are harmoniously integrated to achieve coherent process autonomy regarding wide-range multiobjective optimal operation.

6.5.1 Definition of the ICCS-MP

The ICCS-MP synthesizes an answer to all the main power plant operation issues presented in the previous sections. First, current electric utilities face operation scenarios characterized by a multiplicity of everyday tighter and changing requirements, e.g., load following cyclic operation, plant life extension, heat-rate improvement, and reduction of pollutant emissions. Second, an integral automation approach for operation and control to keep plants running profitably is of paramount importance for the survival of any electric utility. Third, the operation philosophy of the plant is defined and better reflected by the continuous control functions, that is, the overall control strategy comprehends the core continuous control functions needed to drive the unit components as a single entity in its intended main productive state. Finally, since most power units are part of large and complex power systems, an integral automation approach must take into account the global power system requirements to achieve high levels of coordination and maneuverability for large-scale optimal and robust operation. Therefore, the ICCS-MP should provide the means to achieve optimized wide-range cyclic operation by being able to follow any given unit load demand profile, which is issued by upper level economic dispatch and unit commitment systems, and to optimally accommodate an arbitrary number of generally conflicting operating objectives, which are determined by the current operating scenario and specified and prioritized by the plant operators, during normal load operating state. Therefore, the ICCS-MP finally implements an intelligent coordinated control system in the form of a two-level hierarchical multiagent system (Figure VI.6.6). The supervisory agents include optimization and set-point scheduling, learning and control tuning, and performance and state monitoring. The direct level consists of the regulatory control agents. All these agents are as defined in Section 6.4.8.

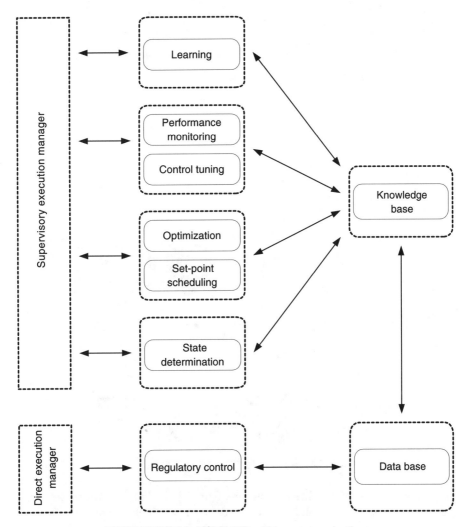

FIGURE VI.6.6   ICCS-MP multiagent organization.

6.5.2 Physical Scope of the ICCS-MP

As described before, in a drum-type FFPU, the essential overall dynamics may be described in terms
of the major inputs (fuel and air flows into the furnace, steam flow into the turbine, feedwater flow
into the boiler, and spray flows at the superheater and reheater) and outputs (electric power, steam
throttle pressure, drum water level, superheater outlet temperature, and reheater outlet temperature).
Electric power and steam pressure are tightly coupled and are heavily affected by the fuel and air
flows and the steam flow. Feedwater flow slightly affects power and pressure but greatly affects the
drum level, which, in turn, is considerably affected by the fuel and steam flows. Similarly, the spray
flows have a minor effect on power and pressure, but greatly affect the heaters outlet temperatures,
which are heavily influenced by the fuel flow. Consequently, fuel and steam flow may be used to
drive the unit to the desired values of power and pressure; this will disturb the drum water level
and heater outlet temperatures, which may then be manipulated with the feedwater and spray flows,
respectively. The interactions among fuel, steam, and feedwater flows as inputs, and power, pressure,
and water level as outputs, greatly affect wide-range operation. Spray flows and temperatures can be
considered for further improvement. The ICCS-MP concentrates on the former situation.

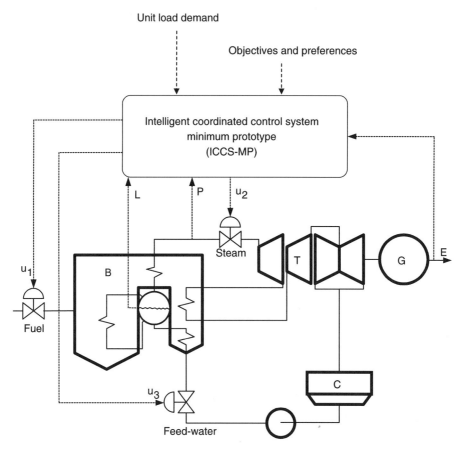

FIGURE VI.6.7   Physical scope of the ICCS-MP.

The physical scope of the ICCS-MP is shown in Figure VI.6.7. From a power system perspective, the power unit is commanded through a unit load demand pattern issued by upper level economic dispatch and unit commitment systems and by an arbitrary number of operating objectives specified and prioritized by the plant operators. The ICCS-MP provides as control signals the demands on position to the valve actuators that control the mass flow rates of fuel (u_1 in pu), steam to the turbine (u_2 in pu), and feedwater to the drum (u_3 in pu). The ICCS-MP receives from the process the electric power (E in MW), drum steam pressure (P in kg/cm^2), and drum water level deviation (L in m).

6.5.3   Control Structure of the ICCS-MP

The realization of the ICCS-MP requires the definition of a control structure to provide the means for wide-range optimal operation. A rather general structure can be formulated for an FFPU by extending the simple feedback control loop through reference-feedforward and disturbance-feedforward control, as shown in Figure VI.6.8, with transfer function:

$$Y = [1 + GG_c]^{-1}[(GG_c + GG_r)R + (G_i + GG_d)W] \qquad (VI.6.1)$$

where, omitting the dependency on the Laplace variable for brevity, Y, R, and W are the Laplace transforms of the output, set-point, and disturbance signals, respectively; G and G_c are the process and feedback controller transfer functions, respectively; G_r is the reference feedforward control

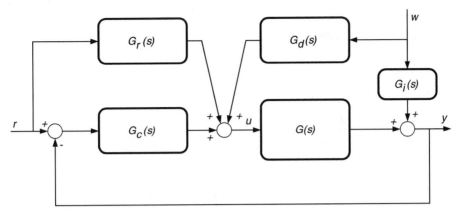

FIGURE VI.6.8 Basic feedforward–feedback control loop.

transfer function; and G_d is the disturbance feedforward control transfer function. Ideally, G_r may be designed as:

$$G_r = G^{-1} \tag{VI.6.2}$$

to take care of changes in reference values to achieve perfect tracking, and G_d may be designed as:

$$G_d = -G^{-1}G_i \tag{VI.6.3}$$

to take care the interaction effects.

Feedforward and feedback control complement each other. Feedforward actions are meant to perform fast corrections due to changes in the reference value or in the disturbance. Feedback gives corrective action on a slower time scale to compensate for inaccuracies in the process model, measurement errors, and unmeasured disturbances. For application in a multivariable process, the single feedforward–feedback control loop can be easily extended to compensate for any number of interaction effects due to the feedback control actions and set-point changes from other control loops. Then, the transfer function Equation VI.6.1 becomes:

$$Y_k = [1 + G_k G_{ck}]^{-1} \left[\sum_j (G_k G_{ck} + G_k G_{rjk}) R_j + \sum_{j \neq k} (G_{ijk} + G_k G_{djk}) U_j \right] \tag{VI.6.4}$$

where $k = 1, \ldots, p$ is the control loop number at hand; $j = 1, \ldots, m$ is the control loop number; Y, R, and U are the Laplace transforms of the output, set-point, and control signals, respectively; G_k, G_{ck}, and G_{rk} are the process, feedback control, and reference feedforward control transfer functions in loop k, respectively; while G_{ijk} and G_{djk} are the interaction and disturbance feedforward transfer functions from loop k to loop j, respectively.

Through adequate manipulations, all the control loops of the form of Equation VI.6.4 can be arranged in the multivariable feedforward–feedback control scheme of the ICCS-MP shown in Figure VI.6.9, where a block for supervisory optimization has been added. Thus, the control structure of the ICCS-MP is determined by the optimization and regulatory control functions. The set-points for electric power, steam pressure, and drum water level, $y_d = [E_d P_d L_d]$, are calculated at the reference governor module. The feedforward and feedback modules provide the feedforward control signal, $u_{ff} = [u_{1ff} u_{2ff} u_{3ff}]$, and the feedback control signal, $u_{fb} = [u_{1fb}\ u_{2fb}\ u_{3fb}]$, respectively. Both control signal vectors are added to provide the demands, $u = [u_1\ u_2\ u_3]$, to the fuel, steam, and feedwater control valves, respectively. Measured variables, $y = [E P L]$, include the electric power, main steam pressure, and drum water level deviation, respectively.

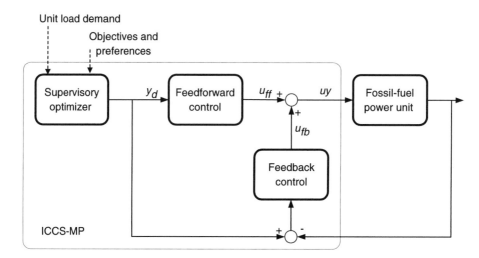

FIGURE VI.6.9   ICCS-MP core control structure.

From a global perspective, the reference governor specifies and coordinates the desired response of the power unit through the set-point trajectories. The feedforward and feedback paths implement a two-degrees-of-freedom nonlinear multivariable controller. The feedforward control path should provide the main contribution to the control valve demands to achieve wide-range operation. The role of the feedback control path is now complementary; it supplies the control signal component necessary for regulation and disturbance rejection in small neighborhoods around the commanded trajectories.

6.5.4   Realization of the ICCS-MP

Following the control structure in the previous section, the implementation of the ICCS-MP is carried out in three modules: reference governor, feedforward control processor, and feedback control processor. The performance and state monitoring and learning and control tuning functions can be executed either under demand on an off-line basis or are implicitly included in the main modules.

The reference governor generates set-point trajectories for the lower level control loops by solving a multiobjective optimization problem for which the objective functions and their priorities can be set arbitrarily, in number and form. This approach allows for process optimization and provides a way to specify the operating policy to accommodate a great diversity of operating scenarios.[39] The feedforward control processor is implemented using a set of multi-input single-output fuzzy inference systems designed from plant input-output data using a neural network paradigm. This approach provides the control system with off-line learning capabilities to attain process optimization under changing operating conditions.[40] The feedback control path is implemented as a PID-based decentralized (multiloop) control scheme with a loop interaction compensator. The compensator is equivalent to a disturbance feedforward compensator and is designed using the relative gain array technique.[41] Both the control algorithms and the compensator are first-order Sugeno-type fuzzy inference systems that achieve satisfactory disturbance rejection and uncertainty compensation during wide-range operation. The process operating window is partitioned to take into account the process non-linear characteristics, and tuning is carried out by a genetic algorithm at the points of interest in the partitions. In Figure VI.6.10, the main modules are expanded up to a second level of detail to show the location of the software agents with respect to the control structure as well as the techniques involved in their development.

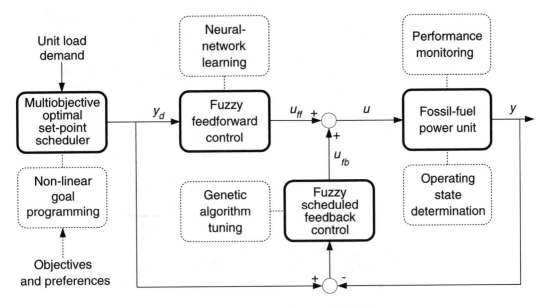

FIGURE VI.6.10 ICCS-MP control block diagram.

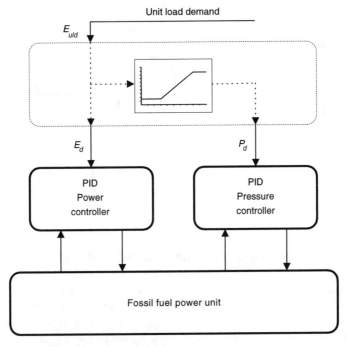

FIGURE VI.6.11 Conventional coordinated control.

6.5.5 Extended Coordinated Control

At fossil fuel power units, the coordinated control (CC) scheme is the uppermost layer of the control system. The CC is responsible for driving the boiler–turbine-generator set as a single entity, harmonizing the slow response of the boiler with the faster response of the turbine-generator, to

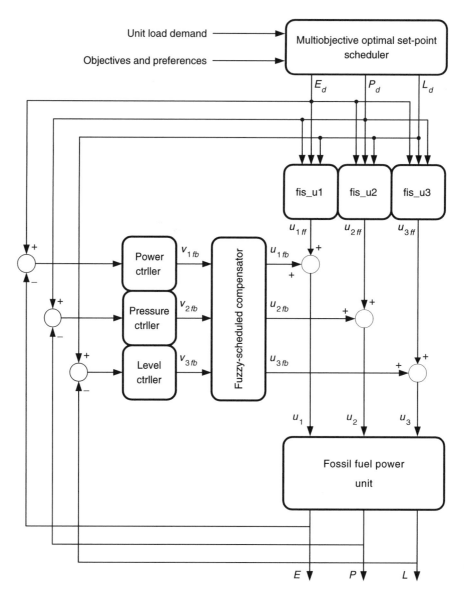

FIGURE VI.6.12 ICCS-MP as extended coordinated control.

achieve fast and stable unit response during load changes and load disturbances. Typically, the CC embraces the power and pressure control loops (Figure VI.6.11). Given a unit load demand, E_{uld}, the CC provides demands to the control loops. Ordinarily, the set-point for the power control loop, E_d, is equal to the unit load demand, and the set-point for the pressure control loop, P_d, is obtained through a non-linear mapping that implements a variable pressure operating policy.[42]

Despite the variations in the implementation of actual CC schemes by different manufacturers, a typical CC scheme basically consists of a decentralized multiloop configuration of SISO PI-based feedback control loops with constant parameters. Also, practical CC implementations show complementary feedforward compensation schemes designed to enhance the system response and to minimize the effect of interaction between the control loops due to the coupled dynamics of the process.[43] Most realizations have evolved through decades of research and practical experience, and most of them are the confidential industrial property of their developers. Current CC schemes are seriously challenged during wide-range load-following operation. Unit performance may decrease due to large non-linear variations and coupling effects of the process dynamics, for which there is no general and systematic way to implement control loop compensations. Also, with current CC schemes, it is not possible to optimally satisfy multiple conflicting operation objectives, as is currently required for power units. Interestingly, there are no mechanisms available to specify the requirements of an operating scenario and incorporate them into the process optimization strategy.

Primarily, the ICCS-MP is intended to extend the scope of current coordinated control schemes in fossil fuel power plants (Figure VI.6.12). The ICCS-MP addresses the above-mentioned needs and conforms itself to a general structure that will allow the control system to satisfy even more complex automation requirements. The ICCS-MP control strategy builds on current coordinated control schemes, which typically account for simultaneous control of electric power output and main steam pressure. The scope of coordination over the internal processes of the power unit is extended to include the control of the drum water level to achieve balanced overall plant operation at all loads. Results on the application of the ICCS-MP control scheme for wide-range multiobjective optimal control of an FFPU demonstrate the feasibility of the ICCS paradigm to achieve consistently optimal control and operation versatility under operating environments characterized by multiple operation objectives.[44]

6.6 Summary and Conclusions

This chapter presented the design of a large-scale intelligent control system for a fossil-fuel power plant in the form of a multiagent system. Motivation for going intelligent is to provide solutions to achieve consistency on the control actions and control system versatility for operation under more demanding conditions. The increased complexity of designing and building an intelligent system is tackled by the multiagent approach through modularization at the highest levels of abstraction based on the intelligent agent paradigm. Major interest was focused on showing that a methodology for MAS design can be formulated by merging concepts from the fields of software engineering, control engineering, and concepts from intelligence theory and intelligent machines. The resultant control system structure, seen as an open organization of intelligent agents, constitutes a general framework for the development of large-scale intelligent control systems, where the necessary operation versatility may be achieved with diverse degrees of intelligence according to the number and capacity of the intelligent agents being enabled.

The details were presented for the design of the ICCS, where the software engineering agency concept is proposed as the fundamental design premise for implementing an intelligent hybrid overall control system for FFPUs. The ICCS organization is an open superset of functionally grouped agent clusters, where intelligence functions are used as guidelines for clustering. Complementarily, the ICCS-MP constitutes a proof-of-concept minimum prototype intended to demonstrate the feasibility of the ICCS paradigm through rapid prototyping. The ICCS-MP realized the main functions of the ICCS and satisfied the main overall operation requirements of an FFPU within a reasonable scope. The ICCS-MP provided the means to consistently achieve multiobjective optimal control actions and versatility to operate in changing environments characterized by multiple operation objectives.

References

1. Garduno, R. and Sanchez, M., Control system modernization: turbogas unit case study, *Proc. IFAC Symp. Control Power Plants Power Syst.*, 2, 245, 1995.
2. Bennett, S., *Real-Time Computer Control: An Introduction*, 2nd ed., Prentice Hall, Englewood Cliffs, 1994.

3. Schach, S.R., *Classical and Object Oriented Software Engineering with UML and Java*, McGraw-Hill, New York, 1999.

4. Benveniste, A., Meetings held with Thomson-CSF, GEC-Alsthom, and Siemens AG, *IEEE Control Syst. Mag.*, 14(4), 86, 1991.

5. Boehm, B.W., Software engineering, R&D trends and defense needs, in *Research Directions in Software Techology*, Wegner, P., Ed., MIT Press, Cambridge, MA, 1979.

6. Boehm, B.W., Developing small-scale application software products: some experimental results, Proceedings of the 8th IFIP World Computer Congress, October 1980, 321.

7. Leveson, N.G., The challenge of building process-control software, *IEEE Software*, 55, 1990.

8. Guidelines for the application of ISO 9001 to the development, supply, and maintenance of software, International Organization for Standardization, 1991, Ch. 2.

9. Ward, P.T. and Mellor, S.J., *Structured Development for Real-Time Systems*, Vols. 1, 2, and 3, Yourdon Press, 1985.

10. Gomma, A., A software design method for real-time systems, *Commun. ACM*, 27(9), 938, 1984.

11. Neidert, R.P., A new paradigm for industrial control systems design, *ISA Trans.*, 225, 1993.

12. Jennings, N., Controlling cooperative problem solving in industrial multiagent systems using joint intention, *Artif. Intelligence*, 75(2), 195, 1995.

13. Wooldridge, M., Agent-based software engineering, *IEE Proc. Software Eng.*, 144, 26, 1997.

14. Knapik, M. and Johnson, J., *Developing Intelligent Agents for Distributed Systems: Exploring Architecture, Technologies, and Applications*, McGraw-Hill, New York, 1998.

15. Durfee, E.H. and Lesser, V., Negotiating task decomposition and allocation using partial global planning, in *Distributed Artificial Intelligence*, Morgan Kaufmann, Palo Alto, CA, 1989, 229.

16. Sycara, K.P., Multiagent systems, *AI Magazine*, Summer, 79, 1998.

17. Lesser, V.R., A retrospective view of FA/C distributed problem solving, *IEEE Trans. Syst. Man Cybernetics*, 21(6), 1347, 1991.

18. Decker, K.S. and Sycara, K., Intelligent adaptive information agents, *J. Intelligent Inf. Syst.*, 9(3), 239, 1997.

19. Jennings, N., Sycara, K., and Wooldridge, M., A roadmap for agent research and development, *Auton. Agents Multiagent Syst.*, 1(1), 1998.

20. Durfee, E.H., A unified approach to dynamic coordination: planning actions and interactions in a distributed problem solving network, Ph.D. thesis, University of Massachusetts, 1987.

21. Parunak, V., Manufacturing experience with the contract net, *Distributed Artif. Intelligence*, 1, 285, 1987.

22. Ljunberg, M. and Lucas, A., The OASIS air-traffic management system, Proceedings of the 2nd Pacific Rim International Conference on AI, September, 1992.

23. Schwuttke, U.M. and Quan, A.G., Enhancing performance of cooperating agents in real-time diagnostics systems, Proceedings of the 13th International Joint Conference on Artificial Intelligence, 1993, 332.

24. Boasson, M., Control systems software, *IEEE Trans. Autom. Control*, 38(7), 1094, 1993.

25. Jennings, N., Corera, J.M., and Laresgoiti, I., Developing industrial multiagent systems, Proceedings of the First International Conference on Multiagent Systems, 1995, 423.

26. Velasco, J.R. et al., Multiagent-based control systems: a hybrid approach to distributed process control, *Control Eng. Pract.*, 4(6), 839, 1996.

27. Iglesias, C.A., Garijo, M., and Gonzalez, J.C., A survey of agent-oriented methodologies, Proceedings of the 5th International Workshop on Intelligent Agents, Agent Theories, Architectures, and Languages, July 1998, 317.

28. Garduno-Ramirez, R. and Mundo-Molina, J.A., Structure of a batch controller, Proc. X Colloquium Autom. Control, 117, 1991.

29. Garcia-Beltran, C.D. and Garduno-Ramirez, R., Control software system of a turbogas power unit, Prep. 5th IFAC Workshop on Algorithms and Architectures for Real-Time Control, 1998, 131.

30. Garcia-Beltran, C.D. and Garduno-Ramirez, R., Gas turbine fuzzy speed control, Proceedings of the 2nd Joint Mexico–USA International Conference on Neural Networks and Neurocontrol, 1997.
31. Dunlop, R.D. and Ewart, D.N., System requirements for dynamic performance and response of generating units, *IEEE Trans. Power Appar. Syst.*, PAS-94(3), 838, 1975.
32. Schweppe, F.C. and Mitter, S.K., Hierarchical system theory and electric power systems, Proceedings of the Symposium on Real Time Control of Electric Power Systems, 1972, 259.
33. Malik, O.P., Hope, G.S., and Fahmy, G., Hierarchical decomposition in power system operation and control, Proceedings of the IEEE Canadian Communication and Power Conference, October 1980, 173.
34. Ilic, M.D. and Liu, S., *Hierarchical Power Systems Control*, Springer-Verlag, New York, 1996.
35. Williams, T.J., *A Reference Model for Computer Integrated Manufacturing (CIM) A Description from the Viewpoint of Industrial Automation*, Instrument Society of America, 1989.
36. Rosenof, H.P. and Ghosh, A., *Batch Process Automation: Theory and Practice*, Van Nostrand Reinhold, New York, 1987.
37. Albus, J.S., Outline for a theory of intelligence, *IEEE Trans. Syst. Man Cybernetics*, 21(3), 473, 1991.
38. Saridis, G.N. and Valavanis, K.P., Analytical design of intelligent machines, *Automatica*, 24(2), 123, 1988.
39. Garduno-Ramirez, R. and Lee, K.Y., Supervisory multiobjective optimization of a class of unit processes: power unit case study, Proceedings of the American Control Conference, 2001, 1497.
40. Garduno-Ramirez, R. and Lee, K.Y., Wide-range operation of a power unit via feedforward fuzzy control, *IEEE Trans. Energy Convers.*, 15(4), 421, 2000.
41. Garduno-Ramirez, R. and Lee, K.Y., Feedforward compensated multiloop control of a power plant, Proceedings of the IEEE PES Summer Meeting, 2000, 206.
42. Garduno-Ramirez, R. and Lee, K.Y., Multiobjective optimal power plant operation through coordinated control with pressure set-point scheduling, *IEEE Trans. Energy Convers.*, 16(2), 115, 2001.
43. Taft, C.W., Performance evaluation of boiler control system strategies, Proceedings of the Conference on Control Systems for Fossil Fuel Power Plants, EPRI CS-6049, 1987, 4–1.
44. Garduno-Ramirez, R. and Lee, K.Y., A multiobjective-optimal neuro-fuzzy extension to power plant coordinated control, *Trans. Inst. Meas. Control*, to be published.

7

Applying Architectural Patterns to the Design of Supervisory Control Systems

Pan-Wei Ng
Rational Software

Chai Quek
Nanyang Technological University

Michel Pasquier
Nanyang Technological University

7.1 Introduction VI-212
   Design Patterns • Architecture Description Languages •
   Objective and Overview
7.2 Supervisory Architectural Description VI-217
   Components • Configuration Formulae • Configuration
   Constraints • Analyzing System Architectures
7.3 Supervisory Schemes VI-230
   Applying Patterns • Relating Architectural Patterns
7.4 Supervisory Frameworks......................... VI-245
   Applying a Framework of Supervisory Schemes • Combin-
   ing and Comparing Supervisory Frameworks
7.5 Conclusion VI-249
   Benefits • Future Work
References VI-250

Abstract. Extending the operating range of a control system often requires employing a number of algorithms together with a supervisory mechanism that selects and activates them appropriately. These algorithms differ in their roles, underlying representations, computational methods, and activating sequences, making the design of the supervisory mechanism a complex task. Although several supervisory schemes have been proposed in the current research literature, discussions have been mostly *ad hoc* and informal. In this chapter, we show that it is possible to apply the concept of design patterns to formally devise a structured approach to the design of supervisory control systems. Design patterns describe typical design problems and their possible solutions. Using patterns transforms the design process into a matching procedure that compares a given control system against a catalog of design patterns and incrementally applies the associated solutions until they are adequately supervised. This chapter, therefore, develops a supervisory architectural description language known as SPADE, which provides an abstract representation of control systems and the catalog of design patterns in a manner that is independent of the underlying formalism. SPADE captures the essential interactions between algorithms and supervisory components through the concept of composition operations and configuration formulae. By operating at a higher level of abstraction, this framework permits the mathematical manipulation, comparison, combination, and analysis of system and pattern descriptions and thereby provides a systematic application of the pattern-based design approach.

7.1   Introduction

An important problem facing the designers of control systems is their ever-increasing complexity, which stems from demands for better process yield, system performance, and robustness to changes within the operating environment. To meet these demands, significant efforts have been undertaken in control engineering research to devise algorithms that deliver adequate performance over an extended period, even when the controlled process is subject to varying operating conditions, parametric drifts, and structural changes. These algorithms are units of computation that collaborate with one another to perform functions that contribute to some system objectives. They originate from different branches of control system research and apply techniques within the realm of robust control,[1–3] adaptive control,[4–7] fault diagnosis,[8–10] and, more recently, intelligent control[11–16] and automatic reconfiguration.[17] However, none of these algorithms is capable of single-handedly coping with the wide variety of situations that may occur in practice. Consequently, extending the operating range of a control system requires employing many of these algorithms together with a supervisory mechanism to select and activate the appropriate algorithms.

The supervision of these algorithms is non-trivial because they differ in their underlying representations, organizational structure, activation sequences, and parameter computation methods. The role of these algorithms ranges from feedback control, adaptive control, and identification to diagnosis and automatic reconfiguration. Their underlying representations are thus based on different branches of control theory including classical control, fuzzy control, and neuro-control. Their internal organizations are also different, where complex algorithms may consist of a number of sub-algorithms. In addition, the sequence in which these algorithms are activated differs as well. For example, an auto-tuning sequence may involve the switching to relay control, followed by the determination of tuning parameters that will be used to design the feedback controller. Another activation sequence is required for self-tuning regulation, in which model identification is first activated, followed by an online controller design. Furthermore, the proper operations of these algorithms require that certain pre-conditions be met. For example, an identification algorithm requires sufficient excitation to be injected into the controlled process. The methods employed to tune these algorithms differ greatly as well, from pole-placement to robust control techniques or even heuristics-based approaches. Consequently, the design of a supervisory scheme is a complex task, as one has to take into account the fundamental differences between these algorithms.

The system designer needs to systematically address and resolve the differences between algorithms, often asking questions such as:

- How can the supervisory mechanism coordinate the given algorithms that employ different computational techniques and activation sequences?
- How can the supervisory mechanism account for the different ways in which these algorithms interact to achieve the overall system objective?
- How can the algorithms and supervisory elements be organized so that the supervisory mechanism can coordinate them effectively?
- What decisions must the supervisory mechanism make? What are the criteria for these decisions and what are the bases for these criteria?

It is difficult to provide general answers to these questions due to the differences between the algorithms and the context in which they are employed. An alternative approach can be devised based on the observation that an experienced designer is often able to apply past knowledge of similar problems and their solutions to a new problem. Such problem–solution pairs are known as design patterns. This chapter proposes a formal approach, adapting design patterns to the supervision of multiple algorithms, resulting in a supervisory architectural description language known as SPADE.

7.1.1 Design Patterns

The concept of design patterns was introduced in software engineering to improve the design and reusability of object-oriented software systems.[18] A design pattern identifies, names, and abstracts the key aspects of a common design structure that make it useful for creating a reusable object-oriented design. Software developers have long struggled to achieve better reuse and flexibility in their software products and can greatly benefit from using design patterns that encapsulate the knowledge and experience underlying many design and engineering efforts. Design patterns facilitate the reuse and sharing of design knowledge by allowing software engineers to adapt general solutions to fit specific problems.[19,20]

The basic concept of a pattern-based design approach for supervised control systems is depicted in Figure VI.7.1. The input is a given control system comprised of a number of algorithms which constitute the supervisory problem. The output is comprised of the supervisory problem and an appropriate supervisory mechanism, which together are known as the supervisory control system. The design framework is enclosed by the dashed perimeter and is comprised of a predefined catalog of design patterns as well as a design methodology. In this particular context, design patterns are known as supervisory schemes, which consist of a description of both a problem and its corresponding solution. A catalog of such supervisory schemes is required as it is not known *a priori* which pattern the given control system will exhibit. The number of supervisory schemes in the catalog defines the wealth of experience captured within the design methodology, which consists of identifying similar patterns occurring within different control problems and associating known supervisory solutions to each of these patterns.

The design methodology involves a series of steps, which define how the catalog of supervisory schemes is systematically applied to solve a supervisory problem. The first step maps the given

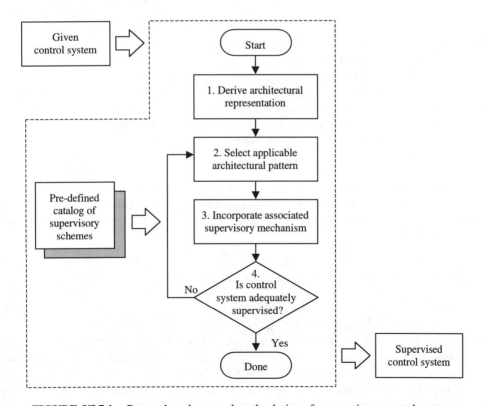

FIGURE VI.7.1 Pattern-based approach to the design of a supervisory control system.

control system into an abstract representation to facilitate the selection of a matching supervisory scheme in the second step. The third step incorporates the supervisory mechanism associated with the selected supervisory scheme. This process continues incrementally until the given control system is adequately supervised. Thus, the approach transforms the design process into a matching procedure that compares the system architecture against a library of patterns and applies the associated solutions.

The fact that terms such as direct, indirect, and adaptive control are familiar to a system designer indicates that the concept of patterns is not new. Existing work addressing the supervisory problem can be treated as providing patterns. For example, expert control (EC) can be perceived as a pattern for auto-tuning algorithms,[21,22] integrated process supervision (IPS) can be perceived as patterns for direct adaptive algorithms,[23–26] and multiple-model-multiple-controller (MMMC) can be viewed as a pattern for indirect adaptive algorithms.[27,28] However, discussions in these works lack formal treatment, and there has been little attempt to generalize and provide a mathematical representation that can be used for different existing supervisory schemes.

7.1.2 Architecture Description Languages

The effective application of a pattern-based approach requires a suitable representation of the given control system and the supervisory schemes. Such a representation must capture the organization of components — both algorithmic and supervisory ones — and their interactions. Existing works usually represent the interactions between the elements in a control system using informal diagrams. The specific role of each component, the permissible interactions between different components, and their contributions to the achievement of the overall system objective are often presented in a descriptive and informal manner. Such informal treatment severely limits the usefulness of the said diagrams, as implementers are forced to rely on their own interpretation when attempting to understand and apply these supervisory schemes. Furthermore, the lack of a formal treatment makes it difficult to enforce the principles advocated by a particular supervisory scheme.

A formal architectural description is a good candidate for an abstract representation because the initial design of a supervisory control system usually involves a high level decomposition of the system into smaller and more manageable units. In addition, it provides the basis for representing control configurations, which characterize the different problems faced by a system designer when attempting to design a supervisory mechanism. The architecture of a system specifies its computational components and the interactions between those components.[29] Architectural models thus clarify structural and semantic differences among components and interactions and are often composed incrementally to define larger systems. Ideally, the individual elements of an architectural model should be defined independently so that they can be reused in different contexts. The architecture only establishes the specifications of each individual element, which may in turn be defined as an architectural subsystem. Architectural models are primarily concerned with the following issues:[30]

- System Structure — an architectural description characterizes a system structure in terms of high-level computational elements and their interactions and provides the solution to a design problem in terms of a configuration of interacting components. It is not specifically about representing requirements (e.g., the abstract relationships between elements of a problem domain) or implementation details (e.g., algorithms and data structures).
- Abstract interactions — the interactions between architectural components can be many and complex, as reflected by the rich vocabulary usually available to system designers, and are often represented as labeled connections between components.
- Global properties — since the principal use of an architectural design is to reason about overall system behavior, it is typically concerned with a complete system.

The importance of capturing interactions between different components and being able to reason with such knowledge was highlighted in the early days of EC.[31,32] However, there was little attempt to provide a formal representation of this knowledge and relate it to overall system objectives. The closest attempts at formally representing the organization of a control system are within the realm of intelligent control architectures such as hierarchical distributed control (HDC),[33] nested hierarchical control (NHC),[34,35] and real-time control architecture (RCS),[36–38] each briefly described hereafter:

- HDC — a brief attempt to provide a formal representation of control system architectures was undertaken by Acar.[33] The author formulated a hierarchical representation of control nodes in an axiomatic manner and provided some informal guidelines as to how a control system involving multiple loops can be mapped to that representation. However, no attempt was made to formalize the representation of guidelines or the integration of multiple algorithms. There is no known continuation of this work.[33]
- NHC — this approach applies a recursive hierarchical structure whereby each lower level is a refinement of higher levels, and each level models a different operating resolution. NHC is primarily used for autonomous vehicles whereby the navigation of the vehicle within a large area is decomposed into smaller areas. NHC architecture has been gradually merged into RCS architecture[38] as described hereafter.
- RCS — developed over the course of two decades, and perhaps the most researched control system architecture so far, RCS has been applied to multiple and diverse systems.[36–39] It provides a reference architecture and an engineering methodology to aid the designers of complex control systems. Guidelines are provided for the decomposition of a problem into a set of hierarchically arranged control nodes. The hierarchical decomposition is guided by the principles of control theory, taking into account parameters such as system response times and factors such as planning horizons.

Despite the availability of a large amount of RCS literature, software libraries, as well as well-defined guidelines for the construction of control software for autonomous real-time systems, discussions of RCS have generally remained informal.[36–38] It is only recently that attempts at providing a formal description of RCS have been made through the use of architectural description languages (ADLs) from software engineering.[39,40] This exemplifies a growing interest in the control community for the application of software engineering techniques to the design and realization of control applications.[39–41,47]

ADL is a language that provides features for modeling a software system's conceptual architecture.[42–44] ADLs provide both a concrete syntax and a conceptual framework to characterize architectures. Some of the common terms used in architectural descriptions are:[43]

- A component is a unit of computation or data store.
- A connector is an interaction between two or more components.
- A port is an interface to a component.
- A role is an interface to a connector.
- A configuration is a set of components and connectors along with a set of attachments between ports of the components and roles of the connectors.

A survey and comparison of existing ADLs are available in the literature.[43,44] Figure VI.7.2 gives an example of how the interactions between components can be represented in the architectural description language known as ACME.[42] This sample model describes a client-server system connected via a remote procedure called `rpc`. The `client` has a port `send-request` from which it can send requests, and the `server` has a port `receive-request` from which it can receive requests. The `client` and `server` components as well as the `rpc` connector interact with each other when `attachments` are made.

```
System  simple_cs = {
  Component client = {Port send-request }
  Component server = {Port receive-request }
  Connector rpc = {Roles { caller, callee } }
  Attachments : {
       client.send-request to rpc.caller;
       server.receive-request to rpc. callee}
  }
}
```

FIGURE VI.7.2   Example of architectural description language.

7.1.3   Objective and Overview

The objective of the work presented in this chapter is to develop a supervisory architectural description language known as SPADE to provide a formal and generic representation of system architectures and architectural patterns. Such a high level of abstraction permits the mathematical manipulation, comparison, combination, and analysis of system and pattern descriptions and thereby provides a systematic application of the pattern-based design approach.

The development of SPADE adopts a number of concepts from software architecture research, especially in the area of architectural description languages. However, efforts in these areas apply to software systems whereas our work is within the realm of control systems, specifically process supervision. It is inevitable, therefore, that the original concepts have to be tailored and refined. First, the concepts of ports and connectors can be represented at a higher level of abstraction because the interactions between components in a control system are not as diverse as the communication protocols of a software system. For example, the input, output, and state vector elements are explicitly used in place of ports. This is beneficial to control engineers who are more familiar with vector representations, which can be easily mapped to behavioral representations based on systems theory.[45] Second, more concise and more general configuration formulae are used in place of connectors. This representation is more amenable to mathematical manipulation and, consequently, simplifies analysis.

To exemplify the essential concepts in SPADE, we introduce a typical supervisory control system, denoted S_{SMRAS} and depicted in Figure VI.7.3. S_{SMRAS} employs a PI controller C_{PI}, a model reference adaptive scheme (MRAS) C_{MRAS}, a performance monitor C_{PM}, a reference model C_{REF}, and a process supervisor C_{SMRAS}. The controlled process is denoted by C_{PLANT} and the signal source C_{SOURCE} provides the command signal. Components and interactions are represented in Figure VI.7.3 as boxes and connecting arcs, respectively. The latter model different types of information flow and, consequently, the relationships between interacting components, as elaborated on later in this chapter. The supervisory objective of S_{SMRAS} is to ensure that the closed-loop performance is maintained within a predefined level. The process supervisor C_{SMRAS} assesses system performance through the performance monitor C_{PM} that evaluates the closed-loop performance against that of the reference model C_{REF}. When a violation of the system objective is detected, the process supervisor activates the adaptive mechanism, in this case C_{MRAS}. Once it is ascertained that the performance specification is achieved, C_{SMRAS} turns off the adaptive mechanism C_{MRAS}.

The remainder of this chapter is organized as follows. Section 7.2 introduces SPADE and describes how an architectural description of a given control system can be derived. This realizes the first step in the pattern-based approach summarized in Figure VI.7.1. Section 7.3 describes the remaining steps by detailing how a supervisory scheme can be incorporated into a given control system. Section 7.4 describes how a catalog of supervisory schemes can be used. Finally, Section 7.5 concludes by highlighting the benefits of SPADE as a supervisory architectural description as well as current work that aims at extending SPADE beyond the architectural representation.

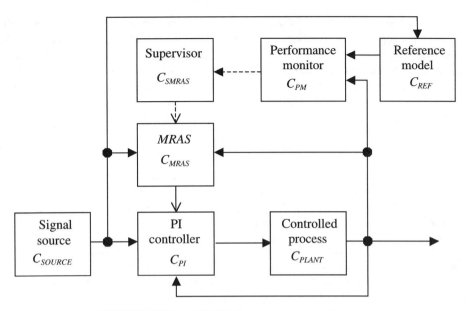

FIGURE VI.7.3   PI/MRAS supervisory control system.

7.2   Supervisory Architectural Description

SPADE provides a concise representation of a system architecture that realizes step 1 in the pattern-based approach (see Figure VI.7.1). SPADE formally represents a system denoted by $\mathcal{S}$ in Equation VI.7.1.

$$\mathcal{S} = \langle \mathcal{C}^*, \mathcal{L}^*, \mathcal{X}^* \rangle \qquad \text{(VI.7.1)}$$

The terms in Equation VI.7.1 are defined as follows. $\mathcal{C}^* = \{\mathcal{C}_1, \mathcal{C}_2, \ldots, \mathcal{C}_i, \ldots\}$ represents the set of components within the system. Each component has a transfer function, which usually involves some numeric computations to map inputs and other parameters to computational outputs. In addition, the transfer function is subject to supervisory actions that affect the component's mode of operation. $\mathcal{L}^* = \{\mathcal{L}_1, \mathcal{L}_2, \ldots, \mathcal{L}_i, \ldots\}$ represents a set of configuration formulae describing the interactions between components in $\mathcal{C}^*$. Each configuration formula denoted by $\mathcal{L} \in \mathcal{L}^*$ describes how adjacent components interact with each other. $\mathcal{X}^* = \{\mathcal{X}_1, \mathcal{X}_2, \ldots, \mathcal{X}_i, \ldots\}$ represents a set of configuration constraints that must be satisfied by the components when the system $\mathcal{S}$ is realized. Constraints provide the means to check the consistency of a system architecture against an implementation. The remainder of this section will further elaborate on the three terms in Equation VI.7.1.

7.2.1   Components

Components are the fundamental building blocks of a system. The set of components $\mathcal{C}^*_{SMRAS}$ in the PI/MRAS supervisory control system shown in Figure VI.7.3 is identified in Equation VI.7.2:

$$\mathcal{C}^*_{SMRAS} = \{\mathcal{C}_{SOURCE}, \mathcal{C}_{PI}, \mathcal{C}_{PLANT}, \mathcal{C}_{MRAS}, \mathcal{C}_{REF}, \mathcal{C}_{PM}, \mathcal{C}_{SMRAS}\} \qquad \text{(VI.7.2)}$$

SPADE distinguishes between two types of component behaviors, namely, computational and supervisory behaviors. Computational behaviors are used to represent the transfer functions of each component, whereas supervisory behaviors are used to represent different modes of operations within each component. Supervisory behaviors also represent the sequential activation and mode changes of different components to maintain closed-loop system performance. This deliberate separation of behavior types allows the system designer to concentrate on different processing requirements one at a time. SPADE formally represents a component as stated in Equation VI.7.3, where the three terms Σ, Π, and $\mathcal{R}^*$ represent the computational behavior, the supervisory behavior, and the roles played by a component, respectively. A set of components is denoted by $\mathcal{C}^*$.

$$\mathcal{C} = \langle \Sigma, \Pi, \mathcal{R}^* \rangle \qquad (\text{VI.7.3})$$

The computational behavior of a component, denoted by Σ, can be represented in several ways. However, as a general rule, it must be defined to facilitate the description of how a component interacts with other components. In SPADE, the computational behavior of a component $\mathcal{C}_i$, denoted by Σ_i, must be defined to include the following elements:

- The computational input vector space U_i, which represents the set of all possible values that the computational input vector $u_i(t)$ can take at any particular instant of time t
- The computational state vector space X_i, which represents the set of all possible values that the computational state vector $x_i(t)$ can take at any particular instant of time t
- The computational output vector space Y_i, which represents the set of all possible values that the computational output vector $y_i(t)$ can take at any particular instant of time t
- The parameter vector space W_i, which represents the set of all possible values the parameters $\boldsymbol{w}_i(t)$ of the transfer function of Σ_i can take at any instant in time t

Most mathematical system representations support this requirement.[45] SPADE extends the standard formalism with the inclusion of W_i to parameterize the transfer function of the component. The computational parameters are also used to describe the degradation of controlled processes and the effect of remedial actions that modify the transfer function of a component. The supervisory behavior of a component $\mathcal{C}_i$ is denoted by Π_i. It must be defined to include the following elements:

- The supervisory input Boolean vector space E_i, which represents the set of all possible values the supervisory input vector $\boldsymbol{e}_i(t) \in E_i^*$ can take at time t. The kth element in the vector $\boldsymbol{e}_i(t) = [e_i^{[1]}(t)\, e_i^{[2]}(t) \ldots e_i^{[k]}(t) \ldots]^T$ is a Boolean value indicating whether a specific input event is occurring at time t, i.e., $e_i^{[k]}(t) \in \{0, 1\}$.
- The supervisory output Boolean vector space F_i, which represents the set of all possible values the supervisory output vector $\boldsymbol{f}_i(t) \in F_i^*$ can take at time t. The kth element in the vector $\boldsymbol{f}_i(t) = [f_i^{[1]}(t) f_i^{[2]}(t) \ldots f_i^{[k]}(t) \ldots]^T$ is a Boolean variable indicating whether a specific output event is generated at time t, i.e., $f_i^{[k]}(t) \in \{0, 1\}$.

The supervisory input vector elements are the channels through which each component receives commands and status reports about system conditions and the basis on which they modify their computational transfer function. Supervisory components evaluate system conditions and issue commands and status reports accordingly to other components. These requirements for representing supervisory behaviors are supported both by finite state machines and discrete event systems.[46]

One of the drawbacks of existing supervisory schemes is their informal description of how their constituent parts are organized. This makes it difficult to separate the roles and responsibilities assigned to different components in the system architecture and results in a monolithic implementation. On the contrary, SPADE separates component roles as supervisory, control, and monitoring roles and associates distinct responsibilities to the components playing each role. This is achieved by first semantically partitioning supervisory input and output events into two sets, namely, commands and responses, denoted respectively by *Commands$_{Pr}$* and *Responses$_{Pr}$*. Commands consist of instructions such as activation and deactivation and can only originate from a supervisory component. Responses consist of status reports such as the failure of an algorithm or the completion of a task.

In SPADE, an arbitrary vector element within a component C_i is denoted by ε_i. The set of semantics assigned to an arbitrary element ε_i is denoted by $[\![\varepsilon_i]\!]$. The semantics of supervisory input events are partitioned into commands and responses, as formally expressed in Equations VI.7.4 and VI.7.5:

$$\forall e_i^{[j]}, [\![e_i^{[j]}]\!] \cap Commands_{Pr} = \{\ \} \wedge [\![e_i^{[j]}]\!] \cap Responses_{Pr} \neq \{\ \} \tag{VI.7.4}$$

$$\forall e_i^{[j]}, [\![e_i^{[j]}]\!] \cap Commands_{Pr} \neq \{\ \} \wedge [\![e_i^{[j]}]\!] \cap Responses_{Pr} = \{\ \} \tag{VI.7.5}$$

Equations VI.7.4 and VI.7.5 indicate that the semantics assigned to a supervisory input event must be mutually exclusive. The semantics of supervisory output events are partitioned into commands and responses in the same manner, as expressed in Equations VI.7.6 and VI.7.7:

$$\forall f_i^{[j]}, [\![f_i^{[j]}]\!] \cap Commands_{Pr} = \{\ \} \wedge [\![f_i^{[j]}]\!] \cap Responses_{Pr} \neq \{\ \} \tag{VI.7.6}$$

$$\forall f_i^{[j]}, [\![f_i^{[j]}]\!] \cap Commands_{Pr} \neq \{\ \} \wedge [\![f_i^{[j]}]\!] \cap Responses_{Pr} = \{\ \} \tag{VI.7.7}$$

When assigning the semantics of supervisory input and output events for a particular supervisory problem, the actual elements within the sets $Commands_{Pr}$ and $Responses_{Pr}$ are immaterial. However, the mutual exclusivity criterion formalized in Equation VI.7.8 must be satisfied.

$$Commands_{Pr} \cap Responses_{Pr} = \{\ \} \tag{VI.7.8}$$

Once the events are partitioned, the primary roles assigned to each component can be mathematically defined. A set of primary component roles $Roles_{Pr}$ is defined in SPADE, as shown in Equation VI.7.9:

$$Roles_{Pr} = \{supervisor, algorithm, monitor\} \tag{VI.7.9}$$

The set of roles played by a component C_i is denoted by the set $\mathcal{R}_i^* = \{\mathcal{R}_{i,1}, \mathcal{R}_{i,2}, \ldots, \mathcal{R}_{i,j}, \ldots\}$. The jth role of the component is denoted $\mathcal{R}_{i,j} \in \mathcal{R}_i^*$ and consists of a character string describing the purpose of including that component within the control system, i.e., its role.

Definition 1: Component Roles

A component plays the role of an algorithm if it has no supervisory outputs, i.e., it is purely computational. A component plays the role of supervisor if it sends command events. A component plays the role of a monitor if it sends response events. These roles are expressed formally in Equations VI.7.10–VI.7.12, respectively:

$$\mathcal{R}_i^* \cap Roles_{Pr} = \{algorithm\} \Leftrightarrow \dim(F_i) = 0 \tag{VI.7.10}$$

$$\mathcal{R}_i^* \cap Roles_{Pr} = \{supervisor\} \Leftrightarrow \dim(F_i) \neq 0 \wedge (\exists k, [\![f_i^{[k]}]\!] \cap Commands_{Pr} \neq \{\ \}) \tag{VI.7.11}$$

$$\mathcal{R}_i^* \cap Roles_{Pr} = \{monitor\} \Leftrightarrow \dim(F_i) \neq 0 \wedge (\forall k, [\![f_i^{[k]}]\!] \cap Responses_{Pr} = \{\ \}) \tag{VI.7.12}$$

It can be seen from the above equations that the primary roles are mutually exclusive, thus enforcing the requirement of a mathematical separation of component roles within a system. It must be noted that a supervisor can send response events to other, higher-level supervisors, thus allowing for the construction of a hierarchy of supervisors. However, a supervisor must also be responsible for sending commands to subordinate components. A monitor, on the other hand, cannot send any commands and consequently must be subordinate to some supervisor.

7.2.2   Configuration Formulae

Component interactions are abstractly represented in SPADE in terms of configuration formulae, denoted by $\mathcal{L}$. The concept of configuration formulae inherits from the concept of *attachment*, which describes information flowing from one specific vector element of a component to another. *Compositional operations* describe combinations of information flows that are permissible between two components. Configuration formulae provide an abstract representation that hides the details of the specific attachments involved, which is a distinguishing characteristic of SPADE compared to existing work in software architecture research. Configuration formulae also permit a mathematical comparison of two different architectures and thereby facilitates the matching of supervisory schemes and the incorporation of a supervisory mechanism (steps 2 and 3 in Figure VI.7.1).

7.2.2.1   Attachments

Two components interact directly with each other when they are connected together. For example, the outputs of a component are typically attached to the inputs of another. When such an attachment is made, the behaviors of both components are constrained such that the mathematical equations or laws governing their behaviors have to be simultaneously satisfied. An attachment between two components $\mathcal{C}_i$ and $\mathcal{C}_j$ is denoted by the pair $\langle \varepsilon_i, \varepsilon_j \rangle$, where ε_i and ε_j are arbitrary elements within any vector in $\mathcal{C}_i$ and $\mathcal{C}_j$, respectively. For the attachment $\langle \varepsilon_i, \varepsilon_j \rangle$ to be valid, the elements within the pair must have the same semantics, as expressed in Equation VI.7.13:

$$[\![\varepsilon_i]\!] = [\![\varepsilon_j]\!] \tag{VI.7.13}$$

SPADE assigns semantics to supervisory input vector elements as either commands or responses, which will be used subsequently to describe different types of attachments. The set of all attachments between $\mathcal{C}_i$ and $\mathcal{C}_j$ is denoted by $\mathcal{C}_i \triangle \mathcal{C}_j$. If $\mathcal{C}_i \triangle \mathcal{C}_j = \{\ \}$, there are no attachments between the two components and their behaviors are not constrained by one another. To uphold the principle of role separation within a system architecture, the supervisory elements of one component must never be attached to the computational elements of another, as expressed in Equations VI.7.14 and VI.7.15:

$$\Sigma_i \triangle \Pi_j = \{\ \} \tag{VI.7.14}$$

$$\Sigma_j \triangle \Pi_i = \{\ \} \tag{VI.7.15}$$

The attachments between any two components can either be between computational elements or between supervisory elements, as expressed in Equation VI.7.16:

$$\mathcal{C}_i \triangle \mathcal{C}_j = (\Sigma_i \triangle \Sigma_j) \cup (\Pi_i \triangle \Pi_j) \tag{VI.7.16}$$

Both computational and supervisory interactions can be refined as follows. Equation VI.7.17 states that the computational outputs of component $\mathcal{C}_i$ can be attached to the computational inputs, states, outputs, and parameters of another component $\mathcal{C}_j$ and vice-versa. Equation VI.7.18 states that the supervisory output events of a component $\mathcal{C}_i$ can be attached to the supervisory input events of another component $\mathcal{C}_j$ and vice-versa.

$$\Sigma_i \triangle \Sigma_j = (Y_i \triangle U_j) \cup (Y_i \triangle X_j) \cup (Y_i \triangle Y_j) \cup (Y_i \triangle W_j) \tag{VI.7.17}$$
$$\cup (Y_j \triangle U_i) \cup (Y_j \triangle X_i) \cup (Y_j \triangle Y_i) \cup (Y_j \triangle W_i)$$

$$\Pi_i \triangle \Pi_j = (F_i \triangle E_j) \cup (F_j \triangle E_i) \tag{VI.7.18}$$

SPADE provides a diagrammatic representation of attachments, summarized in Table VI.7.1, where the first and last two rows represent computational and supervisory attachments, respectively. Each attachment $\langle \varepsilon_i, \varepsilon_j \rangle$ in Table VI.7.1 is expressed in terms of its permissible type of elements

TABLE VI.7.1   Diagrammatic Representation of Attachments

Row	Notation	Element Type of ε_i	Element Type of ε_j	Semantics Constraints
1	$\longrightarrow$	computational output	computational input	—
2	$\longrightarrow$	computational output	computational parameter	—
3	$- - - - - \rightarrow$	supervisory output	supervisory input	response
4	$- - - - - \twoheadrightarrow$	supervisory output	supervisory input	command

for ε_i and ε_j as well as constraints on the semantics assigned to them. SPADE does not impose any semantic restrictions on computational attachments. Rows 1 and 2 represent the existence of an attachment from a computational output to a computational input and a computational parameter, respectively. Rows 3 and 4 represent the existence of an attachment from a supervisory output to a supervisory input having response and command semantics, respectively.

7.2.2.2   Composition Operations

Attachments describe the connections between adjacent components but not the exact relationships between them, which are typically quite detailed. Defining and using a more general and abstract representation that captures the essence of component interactions makes it possible to analyze system properties in an abstract manner. This is especially useful in the early stages of design when specific details are not yet known. SPADE provides such an abstraction through the concept of composition operations, which describe the relationship between adjacent components where attachments exist. They also describe how adjacent components can be combined to form a composite component in a manner that completely hides the interactions between them, i.e., where all attachments are accounted for. SPADE defines two categories of composition operations, which are detailed in the following paragraphs, namely:

- Compositions describing computational interactions
- Compositions describing supervisory interactions

Composition operations describing computational interactions between two components C_j and C_k are applicable when there are no supervisory attachments between them, i.e., $\Pi_j \bigtriangleup \Pi_k = \{\ \}$. For a computational composition to be valid, C_j must play the role of an algorithm and C_k can be either an algorithm or monitor. These conditions are expressed in Equations VI.7.19 and VI.7.20:

$$\|\mathcal{R}_j^* \cap \{\text{algorithm}\}\| = 1 \tag{VI.7.19}$$

$$\|\mathcal{R}_k^* \cap \{\text{algorithm, monitor}\}\| = 1 \tag{VI.7.20}$$

The supervisory behavior of the generated component C_i is defined such that each term in Π_i is a Cartesian product of the corresponding terms within C_j and C_k, as shown in Equations VI.7.21 and VI.7.22:

$$E_i = E_j \times E_k \tag{VI.7.21}$$

$$F_i = F_j \times F_k \tag{VI.7.22}$$

There are three computational compositions in SPADE, namely:

- The signal composition, denoted by $\twoheadrightarrow$
- The loop composition, denoted by $\circlearrowleft$
- The update composition, denoted by $\looparrowright$

A signal composition $\twoheadrightarrow$ occurs when the computational outputs of one component are attached to some computational inputs of another component. For example, the signal source C_{SOURCE} in Figure VI.7.1 outputs to the PI controller, C_{PI}. Both C_{SOURCE} and C_{PI} can be combined through a signal composition, formally expressed in Definition 2.

Definition 2: Signal Composition

A component C_i is a valid signal composition of the components C_j and C_k denoted by $C_i = C_j \twoheadrightarrow C_k$ if the conditions expressed in Equation VI.7.23 are satisfied.

$$valid(C_j \twoheadrightarrow C_k) \Leftrightarrow Y_j \vartriangle U_k \neq \{\ \} \wedge C_j \vartriangle C_k = Y_j \vartriangle U_k$$

$$\wedge \|\mathcal{R}_j^* \cap \{\text{algorithm}\}\| = 1 \wedge \|\mathcal{R}_k^* \cap \{\text{algorithm, monitor}\}\| = 1 \quad \text{(VI.7.23)}$$

The first condition on the right-hand side of Equation VI.7.23 states that there must be at least one attachment from the computational output of C_j to the computational input of C_k. The second condition states that permissible attachments are only those from the computational output vector elements of C_j to the computational input vector elements of the C_k. The third and fourth conditions respectively indicate that C_j must be an algorithm and that C_k can either be a monitor or algorithm. The resulting component C_i has a computational behavior Σ_i with terms derived as shown in Equations VI.7.24–VI.7.27:

$$U_i = U_j \times U_k / (Y_j \vartriangle U_k) \quad \text{(VI.7.24)}$$

$$X_i = X_j \times X_k \times (Y_j \vartriangle U_k) \quad \text{(VI.7.25)}$$

$$Y_i = Y_j \times Y_k \quad \text{(VI.7.26)}$$

$$W_i = W_j \times W_k \quad \text{(VI.7.27)}$$

Equation VI.7.24 states that the resulting input vector space U_i is a concatenation of the input vector spaces of U_j and U_k with the elements in $Y_j \vartriangle U_k$ removed. Equation VI.7.25 states that the resulting state vector space X_i is a concatenation of the input vector spaces of X_j and X_k with the elements in $Y_j \vartriangle U_k$ concatenated as well. The symbols $\times$ and / respectively denote the concatenation and removal operations, where the former is a generalization of the standard Cartesian product to handle the concatenation of attachments.

A loop composition $\circlearrowleft$ occurs when the output of one component is attached to the input of another component and vice versa. It may seem at first that a loop composition is equivalent to and might be defined instead as two reciprocal signal compositions, that is, two components C_j and C_k in a loop can be represented by $C_j \twoheadrightarrow C_k$ and $C_k \twoheadrightarrow C_j$. However, this is not a sound composition because the purpose of composition operations is to encapsulate all interactions between two components to produce a single composite component. The signal composition $C_j \twoheadrightarrow C_k$ only accounts for all attachments in one direction and is therefore inadmissible in this case. The same argument applies for $C_k \twoheadrightarrow C_j$. Thus, a separate representation is required, i.e., the loop composition, formally expressed in Definition 3.

Definition 3: Loop Composition

A component C_i is a valid loop composition of the components C_j and C_k, denoted by $C_i = C_j \circlearrowleft C_k$, if the conditions expressed in Equation VI.7.28 are satisfied.

$$valid(C_j \circlearrowleft C_k) \Leftrightarrow Y_j \vartriangle U_k \neq \{\ \} \wedge Y_k \vartriangle U_j \neq \{\ \} \wedge C_j \vartriangle C_k = Y_j \vartriangle U_k \cup Y_k \vartriangle U_j$$

$$\wedge \|\mathcal{R}_j^* \cap \{\text{algorithm}\}\| = 1 \wedge \|\mathcal{R}_k^* \cap \{\text{algorithm}\}\| = 1 \quad \text{(VI.7.28)}$$

The first two conditions on the right-hand side of Equation VI.7.23 state that there must be at least one attachment from the computational output vector elements of C_j to the computational input vector elements of C_k and, conversely, at least one attachment from the computational output vector elements of C_k to the computational input vector elements of C_j. The third condition states that the only permissible attachments between C_j and C_k are attachments from the computational output vector elements of C_j to the computational input vector elements of C_k and, conversely, attachments from the computational output vector elements of C_k to the computational input vector elements of C_j. The last two conditions indicate that both C_j and C_k can only play the role of an algorithm. The individual terms of the resulting computational behavior Σ_i are listed in Equations VI.7.29–VI.7.32:

$$U_i = U_j \times U_k / (Y_j \bigtriangleup U_k) \tag{VI.7.29}$$

$$X_i = X_j \times X_k \times (Y_j \bigtriangleup U_k) \tag{VI.7.30}$$

$$Y_i = Y_j \times Y_k \tag{VI.7.31}$$

$$W_i = W_j \times W_k \tag{VI.7.32}$$

An update composition $\leftrightsquigarrow$ occurs when two components are combined in sequence such that the first updates the parameters of the other. A typical example is that of an adaptive algorithm updating a controller. The update composition is expressed formally in Definition 4.

Definition 4: Update Composition

A component C_i is a valid update composition of the components C_j and C_k, denoted by $C_i = C_j \leftrightsquigarrow C_k$, if the conditions expressed in Equation VI.7.33 are satisfied.

$$valid(C_j \leftrightsquigarrow C_k) \Leftrightarrow Y_j \bigtriangleup W_k \neq \{\ \} \wedge C_j \bigtriangleup C_k = Y_j \bigtriangleup W_k \cup Y_j \bigtriangleup X_k \cup Y_k \bigtriangleup U_j$$

$$\wedge \| \mathcal{R}_j^* \cap \{\text{algorithm}\} \| = 1 \wedge \| \mathcal{R}_k^* \cap \{\text{algorithm monitor}\} \| = 1 \tag{VI.7.33}$$

The first condition on the right-hand side of Equation VI.7.33 states that there should be at least one attachment between the output vector elements of C_j to the computational parameter vector elements of C_k. The second condition specifies which attachments are permissible in an update composition, as follows. Attachments from the computational output vector elements of C_j to the computational input vector elements of C_k are permissible and there must be at least one such attachment. Attachments from the computational output vector elements of C_j to the computational state vector elements of C_k are optional. They occur when C_j needs to perform some reset action on C_k such as part of an adaptive algorithm. Finally, attachments from the computational output vector elements of C_k to the computational state vector elements of C_j are admissible as well. They occur when C_j needs to compute the parameters of C_k by observing the outputs of C_k. The last two conditions in Equation VI.7.33 indicate that C_j must be an algorithm and that C_k can be either a monitor or an algorithm. The individual terms of the resulting computational behavior Σ_i are listed in Equations VI.7.34–VI.7.37:

$$U_i = U_j \times U_k \tag{VI.7.34}$$

$$X_i = X_j \times X_k \times (Y_j \bigtriangleup W_k) \tag{VI.7.35}$$

$$Y_i = Y_j \times Y_k \tag{VI.7.36}$$

$$W_i = W_j \times W_k / (Y_j \bigtriangleup W_k) \tag{VI.7.37}$$

Composition operations describing supervisory interactions between two components C_j and C_k are applicable when there are no computational attachments between them, i.e., $\Sigma_j \bigtriangleup \Sigma_k = \{\ \}$. The computational behavior in the generated component C_i is such that each term in Σ_i is a Cartesian product of the corresponding terms within C_j and C_k. The resulting component has a supervisory input and output vector space, respectively defined in Equations VI.7.38 and VI.7.39:

$$E_i = E_j \times E_k/(F_j \bigtriangleup E_k)/(F_k \bigtriangleup E_j) \tag{VI.7.38}$$

$$F_i = F_j \times F_k/(F_j \bigtriangleup E_k)/(F_k \bigtriangleup E_j) \tag{VI.7.39}$$

The rest of the terms are a Cartesian product of the source components C_j and C_k. In addition, C_j must play the role of a supervisor, as indicated in Equation VI.7.40:

$$\|\mathcal{R}_j^* \cap \{\text{supervisor}\}\| = 1 \tag{VI.7.40}$$

There are two compositions describing supervisory interactions, namely:

- The monitoring composition, denoted by $\leftarrow$
- The supervisory composition, denoted by $\rightsquigarrow$

Definition 5: Monitoring Composition

A component C_i is a valid monitoring composition of two components C_j and C_k, denoted as $C_i = C_j \leftarrow C_k$, if C_k sends supervisory responses to C_j, as expressed in Equation VI.7.41:

$$valid(C_j \leftarrow C_k) \Leftrightarrow (\forall \langle \varepsilon_k, \varepsilon_j \rangle \in F_k \bigtriangleup E_j \Rightarrow [\![\varepsilon_k]\!] \cap Responses_{Pr} \neq \{\ \}) \wedge F_k \bigtriangleup E_j \neq \{\ \}$$

$$\wedge \forall \langle \varepsilon_j, \varepsilon_k \rangle \in F_j \bigtriangleup E_k \Rightarrow [\![\varepsilon_j]\!] \cap Commands_{Pr} \neq \{\ \} \wedge C_j \bigtriangleup C_k = \Pi_j \bigtriangleup \Pi_k$$

$$\wedge \|\mathcal{R}_j^* \cap \{\text{supervisor}\}\| = 1 \wedge \|\mathcal{R}_k^* \cap \{\text{monitor, supervisor}\}\| = 1 \tag{VI.7.41}$$

The first condition on the right-hand side of Equation VI.7.41 states that all attached output events from C_k to C_j must be responses and there must be at least one such attachment. The second and third conditions state that all attached output events from C_j to C_k, if any, must be commands, and that no other attachments are permissible. Finally, the last two conditions indicate that C_j must play the role of a supervisor and C_k must be either a monitor or supervisor. This implies that supervisors do not monitor algorithms directly.

Definition 6: Supervisory Composition

A component C_i is a valid supervisory composition of two components C_j and C_k, denoted as $C_i = C_j \rightsquigarrow C_k$, if C_j output commands to and receives responses from C_k, as expressed in Equation VI.7.42:

$$valid(C_j \rightsquigarrow C_k) \Leftrightarrow (\forall \langle \varepsilon_j, \varepsilon_k \rangle \in F_j \bigtriangleup E_k \Rightarrow [\![\varepsilon_j]\!] \cap Commands_{Pr} \neq \{\ \}) \wedge F_j \bigtriangleup E_k \neq \{\ \}$$

$$\wedge C_j \bigtriangleup C_k = F_j \bigtriangleup E_k \wedge \|\mathcal{R}_j^* \cap \{\text{supervisor}\}\| = 1 \tag{VI.7.42}$$

The first condition on the right-hand side of Equation VI.7.42 states that all attached output events from C_j to C_k must be commands, and there must be at least one such attachment. The second condition states that all attachments are supervisory and not computational. Finally, the third condition indicates that C_j must play the role of a supervisor. However, there is no constraint on the role of C_k.

A diagrammatic representation of valid composition operations is illustrated in Table VI.7.2 and can be used as follows. If any cell in Table VI.7.2 can visually represent the attachments between two components, then they produce the valid composition indicated at the bottom of each cell.

7.2.2.3 Configuration Formulae

This section extends the use of composition operations to abstractly represent the interaction between two components to a more general case involving multiple components through configuration formulae. Configuration formulae use composition operations as primitives to describe the interaction

TABLE VI.7.2   Diagrammatic Representation of Valid Composition Operations

$\mathcal{C}_i \longrightarrow \mathcal{C}_j$ valid signal composition, $\mathcal{C}_i \rightsquigarrow \mathcal{C}_j$	$\mathcal{C}_i \longleftarrow \mathcal{C}_j$ valid loop composition, $\mathcal{C}_i \circlearrowleft \mathcal{C}_j$
$\mathcal{C}_i \rightleftarrows \mathcal{C}_j$ valid update composition, $\mathcal{C}_i \twoheadrightarrow \mathcal{C}_j$	$\mathcal{C}_i \longrightarrow \mathcal{C}_j$ valid update composition, $\mathcal{C}_i \twoheadrightarrow \mathcal{C}_j$
$\mathcal{C}_i \dashrightarrow \mathcal{C}_j$ valid supervisory composition $\mathcal{C}_i \rightsquigarrow\!\!\!\!\rightsquigarrow \mathcal{C}_j$	$\mathcal{C}_i \dashleftarrow \mathcal{C}_j$ valid monitoring composition $\mathcal{C}_i \leftarrow \mathcal{C}_j$	$\mathcal{C}_i \dashleftarrow \mathcal{C}_j$ valid monitoring composition $\mathcal{C}_i \leftarrow \mathcal{C}_j$

between two or more components. A configuration formula $\mathcal{L}$ is a string of symbols that provide a partial description of component interactions within a control system.

Definition 7: Well-Formed Configuration Formula

A configuration formula $\mathcal{L}$ is well formed if it obeys the following rules:

- $\mathcal{C}$ is a well-formed formula if $\mathcal{C} \in \mathcal{C}^*$.
- If $\mathcal{L}$ is a well-formed formula, so are $\mathcal{C} \rightsquigarrow (\mathcal{L})$, $\mathcal{C} \leftarrow (\mathcal{L})$, $\mathcal{C} \twoheadrightarrow (\mathcal{L})$, $\mathcal{C} \circlearrowleft \mathcal{L}$, and $\mathcal{C} \twoheadrightarrow (\mathcal{L})$.
- If $\mathcal{L}$ is a well-formed formula, so are $(\mathcal{L}) \rightsquigarrow \mathcal{C}$, $(\mathcal{L}) \leftarrow \mathcal{C}$, $(\mathcal{L}) \twoheadrightarrow \mathcal{C}$, $(\mathcal{L}) \circlearrowleft \mathcal{C}$, and $(\mathcal{L}) \twoheadrightarrow \mathcal{C}$.
- No other strings or symbols are well-formed formulae.

It must be noted that a well-formed configuration formula, which expresses a composition hierarchy amongst components as per Definition 7, may not be necessarily valid. The validity of a configuration formula $\mathcal{L}$ is expressed by the predicate *valid* $(\mathcal{L})$ and is true if $\mathcal{L}$ has the semantics of being able to generate a composite component that is valid in $\mathcal{S}$. The parentheses in the above rules are introduced because the validity criteria defined previously are, in general, not commutative. To facilitate easy manipulation of configuration formulae, two operations Υ and Θ are defined. The unary operation $\Upsilon(\mathcal{L})$ converts a formula $\mathcal{L}$ to a set representation, as defined in Equation VI.7.43:

$$\Upsilon(\mathcal{L}) = \begin{cases} \{\mathcal{L}\} & \text{if } valid(\mathcal{L}) \\ \{\ \} & \text{otherwise} \end{cases} \tag{VI.7.43}$$

The operation Θ applies multiple compositional operations and converts the result into a set representation, as defined in Equation VI.7.44:

$$\begin{aligned}
\mathcal{L}_i \Theta \mathcal{L}_j =\ & \Upsilon(\mathcal{L}_i \rightsquigarrow \mathcal{L}_j) \cup \Upsilon(\mathcal{L}_j \rightsquigarrow \mathcal{L}_i) \\
& \cup \Upsilon(\mathcal{L}_i \leftarrow \mathcal{L}_j) \cup \Upsilon(\mathcal{L}_j \leftarrow \mathcal{L}_i) \\
& \cup \Upsilon(\mathcal{L}_i \twoheadrightarrow \mathcal{L}_j) \cup \Upsilon(\mathcal{L}_j \twoheadrightarrow \mathcal{L}_i) \\
& \cup \Upsilon(\mathcal{L}_i \circlearrowleft \mathcal{L}_j) \cup \Upsilon(\mathcal{L}_j \circlearrowleft \mathcal{L}_i) \\
& \cup \Upsilon(\mathcal{L}_i \twoheadrightarrow \mathcal{L}_j) \cup \Upsilon(\mathcal{L}_j \twoheadrightarrow \mathcal{L}_i)
\end{aligned} \tag{VI.7.44}$$

This can be extended to the case when the operands are both sets of configuration formulae, $\mathcal{L}_i^*$ and $\mathcal{L}_j^*$, as shown in Equation VI.7.45:

$$\mathcal{L}_i^* \Theta \mathcal{L}_i^* = \bigcup_{\mathcal{L}_i \in \mathcal{L}_i^*} \left(\bigcup_{\mathcal{L}_j \in \mathcal{L}_j^*} \mathcal{L}_i \Theta \mathcal{L}_j \right) \tag{VI.7.45}$$

Definition 8: Components Within a Configuration Formula

The set of constituent components within a configuration formula $\mathcal{L}$ is denoted by $?\mathcal{L}$. The number of components within a configuration formula $\mathcal{L}$ is known as its length and is denoted by $\|?\mathcal{L}\|$. For example, if $\mathcal{L} = \mathcal{C}_A \twoheadrightarrow (\mathcal{C}_B \leftrightsquigarrow \mathcal{C}_C)$, then $?\mathcal{L} = \{\mathcal{C}_A, \mathcal{C}_B, \mathcal{C}_C\}$ and $\|?\mathcal{L}\| = 3$.

The second term $\mathcal{L}_i^*$ in the formal representation of system architecture $\mathcal{S}_i = \langle \mathcal{C}_i^*, \mathcal{L}_i^*, \mathcal{X}_i^* \rangle$ refers to a list of all configuration formulae of length 2 and can be derived using $\mathcal{L}_i^* = \mathcal{C}_i^* \Theta \mathcal{C}_i^*$. This has the semantics of examining all attachments between adjacent components in $\mathcal{S}_i$ and determining which composition operation describes them, as illustrated in Example 1 below.

Example 1: Configuration Formulae for PI/MRAS Supervisory Control System

This example derives the configuration formulae for the PI/MRAS supervisory control system, the components of which are listed in Equation VI.7.46. The configuration formulae are superimposed onto the component diagram, as shown in Figure VI.7.4.

$$\mathcal{C}_{SMRAS}^* = \{\mathcal{C}_{SMRAS}, \mathcal{C}_{PM}, \mathcal{C}_{REF}, \mathcal{C}_{SOURCE}, \mathcal{C}_{PLANT}, \mathcal{C}_{PI}, \mathcal{C}_{MARS}\} \tag{VI.7.46}$$

The set of configuration formulae in term $\mathcal{L}_{SMRAS}^*$ is derived by examining all the attachments in $\mathcal{S}_{SMRAS}$. Table VI.7.3 lists the attachments between the different components and the corresponding configuration formulae. The first row indicates that the only attachment between the signal source and the PI controller is $\mathcal{C}_{SOURCE} \triangle \mathcal{C}_{PI} = \{\langle y_{SOURCE}, u_{PI}^{[ref]} \rangle\}$. This produces a signal composition expressed by the formula $(\mathcal{C}_{SOURCE} \twoheadrightarrow \mathcal{C}_{PI})$ in accordance with the validity checks listed in the

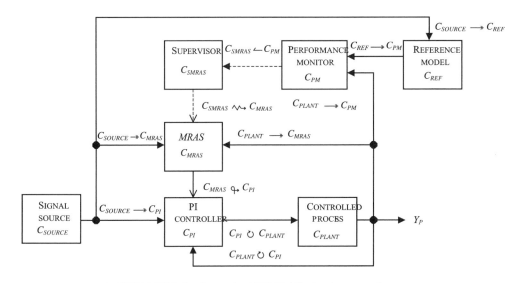

FIGURE VI.7.4   Supervised PI/MRAS adaptive control system.

TABLE VI.7.3 Attachments and Configuration Formulae in PI/MRAS Supervisory Control System

No.	Attachments	Configuration Formulae
1	$\mathcal{C}_{SOURCE} \Delta \mathcal{C}_{PI} = \{\langle y_{SOURCE}, u_{PI}^{[refl]} \rangle\}$	$(\mathcal{C}_{SOURCE} \twoheadrightarrow \mathcal{C}_{PI})$
2	$\mathcal{C}_{SOURCE} \Delta \mathcal{C}_{MRAS} = \{\langle y_{SOURCE}, u_{MRAS}^{[refl]} \rangle\}$	$(\mathcal{C}_{SOURCE} \twoheadrightarrow \mathcal{C}_{MRAS})$
3	$\mathcal{C}_{SOURCE} \Delta \mathcal{C}_{REF} = \{\langle y_{SOURCE}, u_{REF} \rangle\}$	$(\mathcal{C}_{SOURCE} \twoheadrightarrow \mathcal{C}_{REF})$
4	$\mathcal{C}_{PI} \Delta \mathcal{C}_{PLANT} = \{\langle y_{PI}, u_{PLANT} \rangle, \langle y_{PLANT}, u_{PI}^{[fdb]} \rangle\}$	$(\mathcal{C}_{PI} \bigcirc \mathcal{C}_{PLANT}), (\mathcal{C}_{PLANT} \bigcirc \mathcal{C}_{PI})$
5	$\mathcal{C}_{PLANT} \Delta \mathcal{C}_{MRAS} = \{\langle y_{PLANT}, u_{MRAS}^{[fdb]} \rangle\}$	$(\mathcal{C}_{PLANT} \twoheadrightarrow \mathcal{C}_{MRAS})$
6	$\mathcal{C}_{PLANT} \Delta \mathcal{C}_{PM} = \{\langle y_{PLANT}, u_{PM}^{[fdb]} \rangle\}$	$(\mathcal{C}_{PLANT} \twoheadrightarrow \mathcal{C}_{PM})$
7	$\mathcal{C}_{MRAS} \Delta \mathcal{C}_{PI} = \{\langle y_{MRAS}^{[kp]}, w_{PI}^{[kp]} \rangle, \langle y_{MRAS}^{[ki]}, w_{PI}^{[ki]} \rangle\}$	$(\mathcal{C}_{MRAS} \leftrightsquigarrow \mathcal{C}_{PI})$
8	$\mathcal{C}_{REF} \Delta \mathcal{C}_{PM} = \{\langle y_{REF}, u_{PM}^{[refl]} \rangle\}$	$(\mathcal{C}_{REF} \twoheadrightarrow \mathcal{C}_{PM})$
9	$\mathcal{C}_{SMRAS} \Delta \mathcal{C}_{PM} = \{\langle f_{PM}^{[fail\ Controller]}, e_{SMRAS}^{[unacceptable]} \rangle\}$ $\cup \{\langle f_{PM}^{[acceptable]}, e_{SMRAS}^{[complete\ Adapt]} \rangle, \langle f_{PM}^{[fail\ Adapt]}, e_{SMRAS}^{[fail]} \rangle\}$	$(\mathcal{C}_{SMRAS} \leftarrow \mathcal{C}_{PM})$
10	$\mathcal{C}_{SMRAS} \Delta \mathcal{C}_{MRAS} = \{\langle f_{SMRAS}^{[activate\ Adapt]}, e_{MRAS}^{[activate]} \rangle\}$ $\cup \{\langle f_{SMRAS}^{[deactivate\ Adapt]}, e_{MRAS}^{[deactivate]} \rangle\}$	$(\mathcal{C}_{SMRAS} \rightsquigarrow \mathcal{C}_{MRAS})$

preceding sections. The rest of Table VI.7.3 is derived in a similar manner, thus producing $\mathcal{L}_{SMRAS}^*$ as expressed in Equation VI.7.47:

$$\mathcal{L}_{SMRAS}^* = \{\mathcal{C}_{SOURCE} \twoheadrightarrow \mathcal{C}_{PI}, \mathcal{C}_{SOURCE} \twoheadrightarrow \mathcal{C}_{MRAS}, \mathcal{C}_{SOURCE} \twoheadrightarrow \mathcal{C}_{REF}, \mathcal{C}_{PI} \bigcirc \mathcal{C}_{PLANT}, \mathcal{C}_{PLANT} \bigcirc \mathcal{C}_{PI}\}$$
$$\cup \{\mathcal{C}_{PLANT} \twoheadrightarrow \mathcal{C}_{MRAS}, \mathcal{C}_{PLANT} \twoheadrightarrow \mathcal{C}_{PM}, \mathcal{C}_{MRAS} \leftrightsquigarrow \mathcal{C}_{PI}\}$$
$$\cup \{\mathcal{C}_{REF} \twoheadrightarrow \mathcal{C}_{PM}, \mathcal{C}_{SMRAS} \leftarrow \mathcal{C}_{PM}, \mathcal{C}_{SMRAS} \rightsquigarrow \mathcal{C}_{MRAS}\} \qquad (VI.7.47)$$

Configuration formulae of length 2 are sufficient to provide an abstraction of attachments within a system since attachments themselves connect only two components. Configuration formulae of length greater than two thus describe composition hierarchies.

7.2.3 Configuration Constraints

Configuration constraints are extensions to configuration formulae, which provide the means to describe important aspects of component interactions that configuration formulae cannot capture. Each configuration constraint is denoted by a predicate $\mathcal{X}$, and its validity is evaluated by the predicate *valid* $(\mathcal{X})$. $\mathcal{X}^*$ denotes the set of configuration constraints that have to be satisfied by individual or groups of components within a system.

For instance, the *role constraint* defined in Definition 9 is important, as sometimes components are indistinguishable in terms of their attachments alone and, thus, additional information is required to uniquely identify them within a configuration. This can be achieved by examining whether a component exhibits a particular role.

Definition 9: Role Constraint
A component $\mathcal{C}_i$ is said to play the role $\mathcal{R}$ if the predicate $\rho(\mathcal{C}_i, \mathcal{R})$ is true, as expressed in Equation VI.7.48:

$$\rho(\mathcal{C}_i, \mathcal{R}) \Leftrightarrow \{\mathcal{R}\} \subseteq \mathcal{R}_i^* \qquad (VI.7.48)$$

Examples of role constraints that are used quite frequently are $\rho(\mathcal{C}_i, \text{process})$ and $\rho(\mathcal{C}_i, \text{model})$, which indicate that the component $\mathcal{C}_i$ plays the role of a controlled process and a design model, respectively. In Example 1 above, the set of configuration constraints in $\mathcal{S}_{MRAS}$ is expressed in Equation VI.7.49:

$$\mathcal{X}_{SMRAS}^* = \{\rho(\mathcal{C}_{PLANT}, \text{process})\} \qquad (VI.7.49)$$

Another important constraint is the *orphaned supervisor constraint*. A supervisor is orphaned if it is not supervised by any other supervisor and, thus, appears as the root of the supervisory hierarchy. A hierarchical supervisory scheme is allowed only one orphaned supervisor, known as the root supervisor.

Definition 10: Orphaned Supervisor
A component C_i is an orphan supervisor denoted by the predicate $\vartheta(C_i)$ if it plays the role of a supervisor and no other component is supervising it, as expressed in Equation VI.7.50:

$$\vartheta(C_i) \Leftrightarrow (\mathcal{R}_i^* \cap \{\text{supervisor}\}) \wedge (\neg \exists C_j, C_j \rightsquigarrow C_i) \tag{VI.7.50}$$

7.2.4 Analyzing System Architectures

This section describes how a given system architecture $\mathcal{S} = \langle C^*, \mathcal{L}^*, \mathcal{X}^* \rangle$ can be analyzed through the comparison of two architectures and the identification of the properties they satisfy.

7.2.4.1 Architectural Comparison

This section provides a definition of subset and equivalence relationships between systems in Definitions 11 and 12, respectively. The union of two architectures is specified in Definition 13.

Definition 11: System Architecture Subset Relationship
A system architecture $\mathcal{S}_i = \langle C_i^*, \mathcal{L}_i^*, \mathcal{X}_i^* \rangle$ is a subset of $\mathcal{S}_j = \langle C_j^*, \mathcal{L}_j^*, \mathcal{X}_j^* \rangle$, denoted by $\mathcal{S}_i \subseteq \mathcal{S}_j$, if the criterion expressed in Equation VI.7.51 is satisfied.

$$\mathcal{S}_i \subseteq \mathcal{S}_j \Leftrightarrow C_i^* \subseteq C_j^* \wedge \mathcal{L}_i^* \subseteq \mathcal{L}_j^* \wedge \mathcal{X}_i^* \subseteq \mathcal{X}_j^* \tag{VI.7.51}$$

Definition 12: System Architecture Equivalence Relationship
Two system architectures $\mathcal{S}_i = \langle C_i^*, \mathcal{L}_i^*, \mathcal{X}_i^* \rangle$ and $\mathcal{S}_j = \langle C_j^*, \mathcal{L}_j^*, \mathcal{X}_j^* \rangle$ are equivalent, denoted by $\mathcal{S}_i = \mathcal{S}_j$, if the criterion expressed in Equation VI.7.52 is satisfied.

$$\mathcal{S}_i = \mathcal{S}_j \Leftrightarrow \mathcal{S}_i \subseteq \mathcal{S}_j \wedge \mathcal{S}_j \subseteq \mathcal{S}_i \tag{VI.7.52}$$

Definition 13: Union of System Architectures
The union of two system architectures $\mathcal{S}_i = \langle C_i^*, \mathcal{L}_i^*, \mathcal{X}_i^* \rangle$ and $\mathcal{S}_j = \langle C_j^*, \mathcal{L}_j^*, \mathcal{X}_j^* \rangle$ results in another system architecture, denoted by $\mathcal{S}_k = \langle C_k^*, \mathcal{L}_k^*, \mathcal{X}_k^* \rangle = \mathcal{S}_i \cup \mathcal{S}_j$, where the individual terms of $\mathcal{S}_k$ are defined in Equations VI.7.53–VI.7.55. In order for the union to be valid, the constraints in $\mathcal{X}_i^*$ and $\mathcal{X}_j^*$ must remain valid in $\mathcal{X}_k^*$. This is, in general, not the case, and further analysis is required.

$$C_k^* = C_i^* \cup C_j^* \tag{VI.7.53}$$

$$\mathcal{L}_k^* = \mathcal{L}_i^* \cup \mathcal{L}_j^* \tag{VI.7.54}$$

$$\mathcal{X}_k^* = \mathcal{X}_i^* \cup \mathcal{X}_j^* \tag{VI.7.55}$$

Example 2: Architectural Comparison
This example illustrates the inadequacy of architectural equivalence when attempting to compare two adaptive control systems. The first adaptive control system, denoted by $\mathcal{S}_{MRAS} = \langle C_{MRAS}^*, \mathcal{L}_{MRAS}^2 \rangle$ and illustrated in Figure VI.7.5a, consists of a plant, a PI controller, and a model reference adaptive mechanism, denoted by C_{PLANT}, C_{PI}, and C_{MRAS}, respectively. The second adaptive control configuration system, denoted by $\mathcal{S}_{CMAC} = \langle C_{CMAC}^*, \mathcal{L}_{CMAC}^*, \mathcal{X}_{CMAC}^* \rangle$ and illustrated in Figure VI.7.5b, is comprised of the same plant but employs a CMAC, denoted by C_{CMAC}, as the feedback controller and a learning algorithm, denoted by C_{ACMAC}, as the adaptive mechanism. It can be easily ascertained that $\mathcal{S}_{MRAS}$ and $\mathcal{S}_{CMAC}$ are different from a system viewpoint since they employ different components,

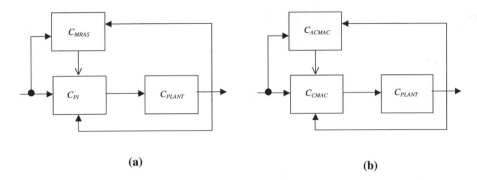

FIGURE VI.7.5 Comparing adaptive control system architectures: (a) PI/MRAS; (b) CMAC/ACMAC.

i.e., since $\mathcal{C}^*_{CMAC} \neq \mathcal{C}^*_{MRAS}$. However, it can be visually verified from Figure VI.7.5 that they exhibit the same interaction patterns. More specifically, they are both instances of a direct-adaptive configuration, and, thus, they can employ similar supervisory mechanisms.

7.2.4.2 Architectural Properties

The set of architectural properties exhibited by a system $\mathcal{S}_i = \langle \mathcal{C}^*_i, \mathcal{L}^*_i, \mathcal{X}^*_i \rangle$ is denoted by $\mathcal{H}^*_i = \mathcal{H}_{i,1}, \mathcal{H}_{i,2}, \ldots, \mathcal{H}_{i,j}, \ldots \}$, where each $\mathcal{H}_{i,j} \in \mathcal{H}^*_i$ is a property. The following paragraphs will describe several properties and discuss the context in which they are used.

A system architecture is *consistently represented* if its parts are not in conflict with each other, as expressed formally in Definition 14.

Definition 14: Consistent Property
A system $\mathcal{S}_i$ is represented consistently if the conditions expressed in Equation VI.7.56 are satisfied, that is, if all configuration formulae and constraints are valid.

$$\mathcal{H}^c \in \mathcal{H}^*_i \Leftrightarrow (\forall \mathcal{X} \in \mathcal{X}^*_i, valid(\mathcal{X})) \wedge (\forall \mathcal{L} \in \mathcal{L}^*_i, valid(\mathcal{L})) \tag{VI.7.56}$$

One of the requirements of SPADE is to provide an enforceable organization of components, which is achieved by clearly separating the roles of all system components. A system that achieves this objective is considered *neat*, as expressed formally in Definition 15.

Definition 15: Neat Property
A system $\mathcal{S}_i$ is neat if all its components play only one primary role and all attachments between components are represented by one and only one composition operation. This property is denoted by $\mathcal{H}^N \in \mathcal{H}^*_i$ and requires the conditions expressed in Equation VI.7.57 to be satisfied.

$$\mathcal{H}^N \in \mathcal{H}^*_i \Leftrightarrow \forall \mathcal{C}_i \in \mathcal{C}^*_i, \|\mathcal{R}^*_i \cap Roles_{PR}\| = 1$$
$$\wedge (\forall \mathcal{C}_{j,k} \in \mathcal{C}^*_i, \mathcal{C}_j \bigtriangleup \mathcal{C}_k \Leftrightarrow \|\Upsilon(\mathcal{C}_j \Theta \mathcal{C}_k)\| = 1) \tag{VI.7.57}$$

It is possible to devise a supervisory scheme that is not neat. However, this will result in a monolithic implementation, which does not facilitate ease of maintenance and analysis. Distinguishing between a control system and a supervisory control system can be achieved by examining the roles played by their components according to the following definitions.

Definition 16: Control System
A system denoted by $\mathcal{S}_P = \langle \mathcal{C}^*_P, \mathcal{L}^*_P, \mathcal{X}^*_P \rangle$ is a control system if it is comprised of neither supervisors nor monitors, as expressed in Equation VI.7.58:

$$\mathcal{H}^{ST} \in \mathcal{H}^*_S \Leftrightarrow \mathcal{C}^* = \{\mathcal{C}_i \in \mathcal{C}^*_P | \mathcal{R}^*_i \cap Roles_{PR} = \{algorithm\}\} \tag{VI.7.58}$$

Definition 17: Supervisory Control System
A system denoted by $\mathcal{S}_S = \langle \mathcal{C}_S^*, \mathcal{L}_S^*, \mathcal{X}_S^* \rangle$ is a supervisory control system if there is at least one component playing the role of supervisor, monitor, and algorithm, respectively, as expressed in Equation VI.7.59:

$$\mathcal{H}^{ST} \in \mathcal{H}_S^* \Leftrightarrow \{\mathcal{C}_i \in \mathcal{C}_S^* | \mathcal{R}_i^* \cap Roles_{PR} = \{algorithm\}\} \neq \{ \ \}$$

$$\wedge \ \{\mathcal{C}_i \in \mathcal{C}_S^* | \mathcal{R}_i^* \cap Roles_{PR} = \{monitor\}\} \neq \{ \ \}$$

$$\wedge \ \{\mathcal{C}_i \in \mathcal{C}_S^* | \mathcal{R}_i^* \cap Roles_{PR} = \{supervisor\}\} \neq \{ \ \} \qquad \text{(VI.7.59)}$$

The set of supervisory components within a system denoted by $\mathcal{S} = \langle C^*, \mathcal{L}^*, \mathcal{X}^* \rangle$ is referred to as $C^* - \{\mathcal{C}_i \in C^* | \{algorithm \in \mathcal{R}_i^*\}$ and is comprised of monitors and supervisors. The separation between a control system and a supervisory control system is important in this work as the former constitutes the supervisory problem and the latter constitutes the supervisory solution.

In a supervisory control system, algorithmic components are arranged according to their respective control configurations. On the other hand, supervisors are arranged hierarchically to manage the complexity of the supervisory task. The definition of *hierarchical supervision* thus requires the use of a transitive supervision relationship denoted by $\leadsto_T$ and recursively defined in Equation VI.7.60:

$$\mathcal{C}_i \leadsto_T \mathcal{C}_j \Leftarrow \mathcal{C}_i \leadsto \mathcal{C}_j$$

$$\mathcal{C}_i \leadsto_T \mathcal{C}_j \Leftarrow \exists \mathcal{C}_k, \mathcal{C}_i \leadsto \mathcal{C}_k \wedge \mathcal{C}_k \leadsto_T \mathcal{C}_j \qquad \text{(VI.7.60)}$$

Definition 18: Hierarchical Property
A control system $\mathcal{S}_i = \langle \mathcal{C}_i^*, \mathcal{L}_i^*, \mathcal{X}_i^* \rangle$ is hierarchical if all its supervisors are arranged in a hierarchical manner, as expressed in Equation VI.7.61:

$$\mathcal{H}^H \in \mathcal{H}_i^* \Leftrightarrow \|\{\mathcal{C} \in \mathcal{C}_i^* | \vartheta(\mathcal{C})\}\| = 1$$

$$\wedge \ \exists \mathcal{C}_j \in \mathcal{C}_i^*, (\vartheta(\mathcal{C}_j) \wedge (\forall \mathcal{C}_k \in \mathcal{C}_i^*, \mathcal{C}_j \neq \mathcal{C}_k \Leftrightarrow \mathcal{C}_j \leadsto_T \mathcal{C}_k))$$

$$\wedge \ \forall \mathcal{C}_j, \mathcal{C}_k \in \mathcal{C}_i^*, \mathcal{C}_j \leadsto_T \mathcal{C}_k \Rightarrow \neg \mathcal{C}_k \leadsto_T \mathcal{C}_j \qquad \text{(VI.7.61)}$$

The first condition on the right-hand side of Equation VI.7.61 states that there must be only one orphaned supervisor, known as the root supervisor. The second condition indicates that the root supervisor either directly or indirectly supervises all other components. Finally, the third condition states that there must not be any cyclic supervision relationship.

7.3   Supervisory Schemes

The mathematical formulation in SPADE lays the foundation for the representation and application of supervisory schemes comprised of both an architectural representation and a supervisory augmentation.

Definition 19: Architectural Pattern
An architectural pattern is formally expressed in Equation VI.7.62, where $\underline{\mathcal{C}}^*$, $\underline{\mathcal{L}}^*$, and $\underline{\mathcal{X}}^*$ represent a set of placeholders for components, a representative set of configuration formulae, and a set of constraints, respectively.

$$\underline{\mathcal{A}} = \langle \underline{\mathcal{C}}^*, \underline{\mathcal{L}}^*, \underline{\mathcal{X}}^* \rangle \qquad \text{(VI.7.62)}$$

The main difference between an architectural pattern and a system architecture $\mathcal{S} = \langle \mathcal{C}^*, \mathcal{L}^*, \mathcal{X}^* \rangle$ is that $\underline{\mathcal{A}}$ is a template and, hence, does not include any actual component. The underscore character is used to indicate a placeholder, rather than an actual component, within a system architecture. The placeholders are specified as follows:

- $\underline{\mathcal{C}}^* = \{\underline{\mathcal{C}}_1, \underline{\mathcal{C}}_2, \dots \}$ is the set of placeholders for components.
- $\underline{\mathcal{L}}^* = \{\underline{\mathcal{L}}_1, \underline{\mathcal{L}}_2, \dots \}$ is the set of configuration formulae that represent the key interactions that are sufficient to categorize a class of systems.
- $\underline{\mathcal{X}}^* = \{\underline{\mathcal{X}}_1, \underline{\mathcal{X}}_2, \dots, \underline{\mathcal{X}}_j, \dots \}$ is a representative set of constraints used to ensure consistency when performing architectural pattern matching.

In addition to characterizing different control configurations, architectural patterns are also useful in the design of the supervisory mechanism for a candidate control system. Because each control configuration applies the same activation principles and sequences, a supervisory mechanism can be designed for each configuration *a priori*. Given a candidate control system, the system designer has to identify the control configurations it can exhibit and apply the associated supervisory mechanism. Each supervisory mechanism is comprised of a set of supervisory and monitoring components that can be used for a particular control configuration.

Consequently, a design guideline for a given control system $\mathcal{S}_P$ can be conceptually represented as follows. If $\mathcal{S}_P$ exhibits an architectural pattern $\underline{\mathcal{A}}_F$, then the recommendation is to incorporate supervisory components represented by $\mathcal{B}_A$ that are known to be applicable for control problems characterized by $\underline{\mathcal{A}}_F$. $\mathcal{B}_A$ is known as a *supervisory augmentation* and is defined in Definition 20.

Definition 20: Supervisory Augmentation
A supervisory augmentation defines a set of supervisory components and how they should interact with a given control system, as expressed formally in Equation VI.7.63:

$$\mathcal{B}_i = \langle \underline{\mathcal{C}}_i^*, \underline{\mathcal{L}}_i^*, \underline{\mathcal{X}}_i^* \rangle \tag{VI.7.63}$$

The adornment under each term indicates that it contains a combination of placeholders and actual components. The first term $\underline{\mathcal{C}}_i^*$ can be decomposed as a union of placeholders denoted by $\underline{\mathcal{C}}_i^*$ and actual components $\mathcal{C}_i^*$, i.e., $\underline{\mathcal{C}}_i^* = \underline{\mathcal{C}}_i^* \cup \mathcal{C}_i^*$. The set of configuration formulae denoted by $\underline{\mathcal{L}}_i^*$ and the constraints denoted by $\underline{\mathcal{X}}_i^*$ are thus applied to both placeholders and actual components. $\underline{\mathcal{X}}_i^*$ represents the set of constraints that must be satisfied after incorporating the components into $\underline{\mathcal{C}}_i^*$. This provides the means to ensure that the components are properly incorporated.

A design guideline describes when a supervisory augmentation can be applied. This is expressed as an *architectural inference rule*, as follows:

$$\text{if } \mathcal{S}_P \sqsubseteq_C \underline{\mathcal{A}}_F \text{ then } \mathcal{S}_S = \mathcal{S}_P \uplus \mathcal{B}_A$$

An architectural inference rule is comprised of two steps. The first step involves matching $\mathcal{S}_P$ against a given architectural pattern $\underline{\mathcal{A}}_F$. The second step involves augmenting $\mathcal{S}_P$ with the supervisory and monitoring components defined in $\mathcal{B}_A$ to produce a supervisory control system $\mathcal{S}_S$. The symbol $\uplus$ used in the architectural inference rule refers to the incorporation of $\mathcal{B}_A$ into $\mathcal{S}_P$ and is formally described in Definition 28. An architectural inference rule can be abstracted as a pair of architectural pattern $\underline{\mathcal{A}}_F$ and its associated supervisory augmentation $\mathcal{B}_A$. Each pair is known as a pattern augmentation pair denoted by $\mathcal{P} = \langle \underline{\mathcal{A}}_F, \mathcal{B}_A \rangle$ and constitutes a mathematical representation of a design guideline in SPADE.

7.3.1 Applying Patterns

This section describes the mechanism to perform the actual incorporation of a supervisory mechanism into a candidate control system to produce a supervisory control system. Figure VI.7.6 illustrates the matching concept by attempting to match an architectural pattern $\underline{A}_i = \langle \underline{C}_i^*, \underline{L}_i^*, \underline{X}_i^* \rangle$ against a candidate system-architecture $S_l = \langle C_l^*, L_l^*, X_l^* \rangle$. The first step in the matching process is to substitute a set of bindings $\mathcal{K}_j^* = \{\mathcal{K}_1, \mathcal{K}_2, \dots\}$ into an architectural pattern $\underline{A}_i = \langle \underline{C}_i^*, \underline{L}_i^*, \underline{X}_i^* \rangle$ to produce a bounded architectural pattern $A_k = \langle C_k^*, L_k^*, X_k^* \rangle$. The individual terms of $A_k = \langle C_k^*, L_k^*, X_k^* \rangle$ are computed according to Definition 21. The second step compares the bounded architectural pattern against the candidate system-architecture.

Architectural augmentations define which additional supervisory components are required to provide adequate supervision for a given control configuration. They also identify, at a high level of abstraction, which interactions are required between the supervisory and the given components. Each placeholder within an architectural pattern must be substituted by a distinct configuration formula within a candidate system architecture before a comparison can be made between the architectural pattern and the candidate system architecture. This substitution resolves the different naming conventions used by the system architecture and the architectural pattern. The substitution operation is known as the binding of the placeholders and is defined as a pair $\mathcal{K}_i = \langle \mathcal{L}_i, \underline{C}_i \rangle$, comprised of:

- A configuration formula from a system-architecture, denoted by $\mathcal{L}_i$
- A placeholder from an architectural pattern, denoted by $\underline{C}_i$

The substitution of a binding $\mathcal{K}$ into a configuration formula $\underline{L}$ is denoted by $\underline{L} \Leftarrow \mathcal{K}$. This substitution operation can be repeated for a set of bindings $\mathcal{K}^* = \{\mathcal{K}_1, \mathcal{K}_2, \dots\}$ until all the placeholders within an architectural pattern are replaced. The substitution of a set of bindings into a set of configuration formulae is denoted by $\underline{L}^* \Leftarrow \mathcal{K}^*$. The substitution of a set of bindings $\mathcal{K}_j^*$ into an architectural pattern $\underline{A}_i$ results in a bounded architectural pattern denoted by $A_k = \underline{A}_i \Leftarrow \mathcal{K}_j^*$.

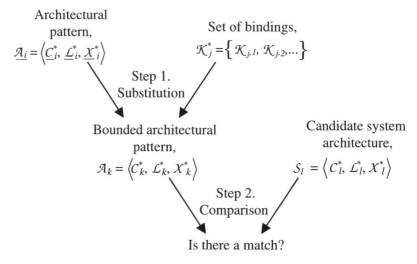

FIGURE VI.7.6 Substitution and matching process.

Definition 21: Substitution

The substitution of a set of bindings $\mathcal{K}_i^*$ into an architectural pattern $\underline{\mathcal{A}}_j = \langle \underline{\mathcal{C}}_j^*, \underline{\mathcal{L}}_j^*, \underline{\mathcal{X}}_j^* \rangle$ results in a bounded architectural pattern, $\mathcal{A}_k = \underline{\mathcal{A}}_j \Leftarrow \mathcal{K}_i^*$. The individual terms of $\mathcal{A}_k = \langle \mathcal{C}_k^*, \mathcal{L}_k^*, \mathcal{X}_k^* \rangle$ is computed as indicated in Equations VI.7.64–VI.7.66:

$$C_k^* = \{ \mathcal{C} | \mathcal{C} \in ?\mathcal{L}_m \wedge \langle \mathcal{L}_m, \underline{\mathcal{C}}_m \rangle \in \mathcal{K}_i^* \} \tag{VI.7.64}$$

$$\mathcal{L}_k^* = \underline{\mathcal{L}}_j^* \Leftarrow \mathcal{K}_i^* \tag{VI.7.65}$$

$$\mathcal{X}_k^* = \underline{\mathcal{X}}_j^* \Leftarrow \mathcal{K}_i^* \tag{VI.7.66}$$

The operation ? in Equation VI.7.64 extracts the components from a configuration formula and has been specified in Definition 8. $\mathcal{A}_k$ comprises actual components within the bindings as determined by Equation VI.7.64. The configuration formulae in $\mathcal{A}_k$ are the same as those in $\underline{\mathcal{A}}_j$ except that they are now bounded, as expressed in Equation VI.7.65. Finally, the configuration constraints in $\mathcal{A}_k$ are bound to actual components, as expressed in Equation VI.7.66. After substitution, the bounded architectural pattern $\mathcal{A}_k$ uses the same component labels as the candidate system architecture $\mathcal{S}_l = \langle \mathcal{C}_l^*, \mathcal{L}_l^*, \mathcal{X}_l^* \rangle$. However, comparisons may not be possible yet because $\mathcal{S}_l$ and $\mathcal{A}_k$ may have a different number of components. This is resolved through the concept of a configuration closure, which provides an exhaustive list of all possible chains of component interactions.

Definition 22: Configuration Closure

The configuration closure of a system $\mathcal{S}_i = \langle \mathcal{C}_i^*, \mathcal{L}_i^*, \mathcal{X}_i^* \rangle$, denoted by $\mathcal{L}_i^+ \cdot \mathcal{L}_i^+$, represents all possible compositions of components within $\mathcal{S}_i$ and is defined iteratively, as expressed in Equations VI.7.67–VI.7.71:

$$\mathcal{L}_i^0 = \{ \; \} \tag{VI.7.67}$$

$$\mathcal{L}_i^1 = \mathcal{C}_i^* \tag{VI.7.68}$$

$$\mathcal{L}_i^2 = \mathcal{L}_i^1 \odot \mathcal{L}_i^1 \tag{VI.7.69}$$

$$\mathcal{L}_i^3 = \mathcal{L}_i^1 \odot \mathcal{L}_i^2 \cup \mathcal{L}_i^2 \odot \mathcal{L}_i^1 \tag{VI.7.70}$$

$$\mathcal{L}_i^N = \bigcup_{I+J=N} \mathcal{L}_i^I \odot \mathcal{L}_i^J \tag{VI.7.71}$$

Equations VI.7.67 and VI.7.68 define the initial conditions and first iteration, respectively. As an example, Equations VI.7.69 and VI.7.70 define the second and third iterations. More generally, Equation VI.7.71 defines the Nth iteration, which generates the set of configuration formulae depicted by $\mathcal{L}_i^N$. The configuration closure $\mathcal{L}_i^+$ is defined as the union of all sets of configuration formulae of length $0 \leq J \leq \infty$ generated at each iteration, as expressed in Equation VI.7.72:

$$\mathcal{L}_i^+ = \bigcup_{J=0}^{\infty} \mathcal{L}_i^J \tag{VI.7.72}$$

The definition of $\mathcal{L}^+$ meets the requirements of a closure, as follows. For any $\mathcal{L}_1, \mathcal{L}_2 \in \mathcal{L}^+$, the application of any compositional operation results in a configuration formula, which also belongs to $\mathcal{L}^+$; in other words, $(\mathcal{L}_1 \odot \mathcal{L}_2) \in \mathcal{L}^+$.

Example 3: Substitution

The concept of substitution is illustrated in Figure VI.7.7, where an attempt is made at using the architectural pattern $\underline{\mathcal{A}}_1$, illustrated in Figure VI.7.7a, to capture the essence of two candidate system architectures $\mathcal{S}_{2a}$ and $\mathcal{S}_{2b}$, shown in Figure VI.7.7b and c, respectively. The symbols R, U, and Y

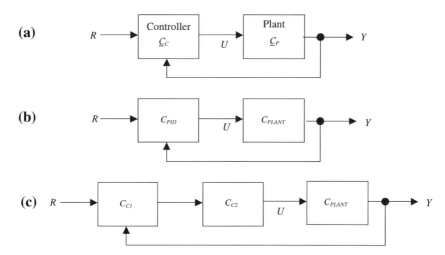

FIGURE VI.7.7 Matching example: (a) architectural pattern $\mathcal{A}_1$, and two candidate system architectures (b) $\mathcal{S}_{2a}$, and (c) $\mathcal{S}_{2b}$.

denote the reference signal, control effort, and system output, respectively. $\mathcal{A}_1$ clearly represents a feedback pattern, and both $\mathcal{S}_{2a}$ and $\mathcal{S}_{2b}$ are two particular instances of a feedback pattern.

The architectural pattern $\underline{\mathcal{A}}_1 = \langle \underline{\mathcal{C}}_1^*, \underline{\mathcal{L}}_1^*, \underline{\mathcal{X}}_1^* \rangle$ represents a feedback configuration, as described in Equations VI.7.74–VI.7.75, which state that the placeholders stand for a controlled process $\underline{\mathcal{C}}_P$ and a controller $\underline{\mathcal{C}}_C$, that the interaction between the two placeholders $\underline{\mathcal{C}}_C$ and $\underline{\mathcal{C}}_P$ is a loop composition, and that a process role constraint is placed on the placeholder $\underline{\mathcal{C}}_P$.

$$\underline{\mathcal{C}}_1^* = \{\underline{\mathcal{C}}_P, \underline{\mathcal{C}}_C\} \tag{VI.7.73}$$

$$\underline{\mathcal{L}}_1^* = \{\underline{\mathcal{C}}_C \circlearrowright \underline{\mathcal{C}}_P\} \tag{VI.7.74}$$

$$\underline{\mathcal{X}}_1^* = \{\rho(\underline{\mathcal{C}}_P, \text{process})\} \tag{VI.7.75}$$

The system architecture for $\mathcal{S}_{2a} = \langle C_{2a}^*, \mathcal{L}_{2a}^*, \mathcal{X}_{2a}^* \rangle$ is defined in Equations VI.7.76–VI.7.78 in terms of its constituent components C_{2a}^*, a list of component interactions $\mathcal{L}_{2a}^*$, and the constraints that it conforms to $\mathcal{X}_{2a}^*$.

$$C_{2a}^* = \{C_{PLANT}, C_{PID}\} \tag{VI.7.76}$$

$$\mathcal{L}_{2a}^* = \{C_{PID} \circlearrowright C_{PLANT}, C_{PLANT} \circlearrowright C_{PID}\} \tag{VI.7.77}$$

$$\mathcal{X}_{2a}^* = \{\rho(C_{PLANT}, \text{process})\} \tag{VI.7.78}$$

The system architecture for $\mathcal{S}_{2b} = \langle C_{2b}^*, \mathcal{L}_{2b}^*, \mathcal{X}_{2b}^* \rangle$ is defined similarly in Equations VI.7.79–VI.7.81:

$$C_{2b}^* = \{C_{PLANT}, C_{C1}, C_{C2}\} \tag{VI.7.79}$$

$$\mathcal{L}_{2b}^* = \{C_{C1} \twoheadrightarrow C_{C2}, C_{C2} \twoheadrightarrow C_{PLANT}, C_{PLANT} \twoheadrightarrow C_{C1}\} \tag{VI.7.80}$$

$$\mathcal{X}_{2b}^* = \{\rho(C_{PLANT}, \text{process})\} \tag{VI.7.81}$$

In order to match an architectural pattern against a candidate system architecture, each placeholder within the architectural pattern must be substituted by a distinct configuration formula within the candidate system architecture. In the case of $\mathcal{S}_{2a}$, the placeholders $\underline{\mathcal{C}}_C$ and $\underline{\mathcal{C}}_P$ must bound to C_{PID} and C_{PLANT}, respectively. The set of bindings is thus identified as expressed in Equation VI.7.82:

$$K_{2a}^* = \{\langle C_{PLANT}, \underline{\mathcal{C}}_P \rangle, \langle C_{PID}, \underline{\mathcal{C}}_C \rangle\} \tag{VI.7.82}$$

When the substitution is made, the bounded architectural pattern is computed as $\mathcal{A}_{3a} = \underline{\mathcal{A}}_{2a} \Leftarrow \mathcal{K}_{2a}^*$, the individual terms of the bounded architectural pattern are $\mathcal{A}_{3a} = \langle \mathcal{C}_{3a}^*, \mathcal{L}_{3a}^*, \mathcal{X}_{3a}^* \rangle$, and its individual terms are computed as expressed in Equations VI.7.83–VI.7.85:

$$\mathcal{C}_{3a}^* = \{\mathcal{C} | \mathcal{C} \in ?\mathcal{L}_m \wedge \langle \mathcal{L}_m, \underline{\mathcal{C}}_m \rangle \in \mathcal{K}_{2a}^* \} = \{\mathcal{C}_{PLANT}, \mathcal{C}_{PID}\} \subseteq \mathcal{C}_{2a}^* \tag{VI.7.83}$$

$$\mathcal{L}_{3a}^* = \underline{\mathcal{L}}_1^* \Leftarrow \mathcal{K}_{2a}^* = \{\mathcal{C}_{PID} \circlearrowleft \mathcal{C}_{PLANT}\} \subseteq \mathcal{L}_{2a}^* \tag{VI.7.84}$$

$$\mathcal{X}_{3a}^* = \underline{\mathcal{X}}_1^* \Leftarrow \mathcal{K}_{2a}^* = \{\rho(\mathcal{C}_{PLANT}, \text{process})\} \subseteq \mathcal{X}_{2a}^* \tag{VI.7.85}$$

The resulting bounded architectural pattern $\mathcal{A}_{3a}$ is a subset of $\mathcal{S}_{2a}$, which implies that $\mathcal{S}_{2a}$ exhibits a feedback pattern, represented by $\underline{\mathcal{A}}_1 = \langle \underline{\mathcal{C}}_1^*, \underline{\mathcal{L}}_1^*, \underline{\mathcal{X}}_1^* \rangle$. The case for $\mathcal{S}_{2b}$ is more complex because of the different number of components and placeholders, thus making direct bindings impossible. This is resolved by using the configuration closure of $\mathcal{S}_{2b}$, as expressed in Equation VI.7.86:

$$\mathcal{L}_{2b}^+ = \{\mathcal{C}_{PLANT}, \mathcal{C}_{C1}, \mathcal{C}_{C2}\} \cup \{\mathcal{C}_{C1} \twoheadrightarrow \mathcal{C}_{C2}, \mathcal{C}_{C2} \twoheadrightarrow \mathcal{C}_{PLANT}, \mathcal{C}_{PLANT} \twoheadrightarrow \mathcal{C}_{C1}\}$$

$$\cup \{(\mathcal{C}_{C1} \twoheadrightarrow \mathcal{C}_{C2}) \circlearrowleft \mathcal{C}_{PLANT}, (\mathcal{C}_{C2} \twoheadrightarrow \mathcal{C}_{PLANT}) \circlearrowleft \mathcal{C}_{C1}, (\mathcal{C}_{PLANT} \twoheadrightarrow \mathcal{C}_{C1}) \circlearrowleft \mathcal{C}_{C2}\}$$

$$\cup \{\mathcal{C}_{PLANT} \circlearrowleft (\mathcal{C}_{C1} \twoheadrightarrow \mathcal{C}_{C2}), \mathcal{C}_{C1} \circlearrowleft (\mathcal{C}_{C2} \twoheadrightarrow \mathcal{C}_{PLANT}), \mathcal{C}_{C2} \circlearrowleft (\mathcal{C}_{PLANT} \twoheadrightarrow \mathcal{C}_{C1})\} \tag{VI.7.86}$$

The placeholders $\underline{\mathcal{C}}_C$ and $\underline{\mathcal{C}}_P$ can now be bound to $\mathcal{C}_{C1} \twoheadrightarrow \mathcal{C}_{C2}$ and $\mathcal{C}_{PLANT}$ respectively, thus yielding the set of bindings $\mathcal{K}_{2b}^*$, expressed in Equation VI.7.87:

$$\mathcal{K}_{2b}^* = \{\langle \mathcal{C}_{PLANT}, \underline{\mathcal{C}}_P \rangle, \langle \mathcal{C}_{C1} \twoheadrightarrow \mathcal{C}_{C2}, \underline{\mathcal{C}}_C \rangle\} \tag{VI.7.87}$$

When the substitution is made, the bounded architectural pattern is denoted by $\mathcal{A}_{3b} = \langle \mathcal{C}_{3b}^*, \mathcal{L}_{3b}^*, \mathcal{X}_{3b}^* \rangle$, where the individual terms are computed as expressed in Equations VI.7.88–VI.7.90:

$$\mathcal{C}_{3b}^* = \{\mathcal{C} | \mathcal{C} \in ?\mathcal{L}_m \wedge \langle \mathcal{L}_m, \underline{\mathcal{C}}_m \rangle \in \mathcal{K}_{2b}^* \} = \{\mathcal{C}_{PLANT}, \mathcal{C}_{C1}, \mathcal{C}_{C2}\} \subseteq \mathcal{C}_{2b}^* \tag{VI.7.88}$$

$$\mathcal{L}_{3b}^* = \underline{\mathcal{L}}_1^* \Leftarrow \mathcal{K}_{2b}^* = \{(\mathcal{C}_{C1} \twoheadrightarrow \mathcal{C}_{C2}) \circlearrowleft \mathcal{C}_{PLANT}\} \subseteq \mathcal{L}_{2b}^+ \tag{VI.7.89}$$

$$\mathcal{X}_{3b}^* = \underline{\mathcal{X}}_1^* \Leftarrow \mathcal{K}_{2b}^* = \{(\rho(\mathcal{C}_{PLANT}, \text{process})\} \subseteq \mathcal{X}_{2b}^* \tag{VI.7.90}$$

The bounded architectural pattern $\mathcal{A}_{3b}$ is not a subset of $\mathcal{S}_{2b}$, as observed in Equation VI.7.89, which implies that to match $\mathcal{A}_{3b}$ against $\mathcal{S}_{2b}$, the configuration closure of $\mathcal{S}_{2b}$ has to be used instead. Using this technique, architectural patterns can be made more general. This example also shows that there exist several ways to define a match, which will be discussed in Section 7.3.1.1.

Based on the above example, the concept of a *valid binding* is formally described in Definition 23.

Definition 23: Valid Binding
A set of bindings $\mathcal{K}_k^*$ is a valid binding of a source architectural pattern $\underline{\mathcal{A}}_i = \langle \underline{\mathcal{C}}_i^*, \underline{\mathcal{L}}_i^*, \underline{\mathcal{X}}_i^* \rangle$ to a candidate system $\mathcal{S}_j = \langle \mathcal{C}_j^*, \mathcal{L}_j^*, \mathcal{X}_j^* \rangle$, having a configuration closure $\mathcal{L}_i^+$, if the criteria expressed in Equations VI.7.91–VI.7.95 are satisfied:

$$\forall \langle \mathcal{L}_m, \underline{\mathcal{C}}_m \rangle \in \mathcal{K}_k^*, \mathcal{L}_m \in \mathcal{L}_j^+ \tag{VI.7.91}$$

$$\forall \langle \mathcal{L}_m, \underline{\mathcal{C}}_m \rangle, \langle \mathcal{L}_n, \underline{\mathcal{C}}_n \rangle \in \mathcal{K}_k^*, m \neq n \Leftrightarrow \mathcal{L}_m \cap \mathcal{L}_n = \{\ \} \tag{VI.7.92}$$

$$(\mathcal{L}_i^* \Leftarrow \mathcal{K}_k^*) \subseteq \mathcal{L}_j^+ \tag{VI.7.93}$$

$$(\mathcal{X}_i^* \Leftarrow \mathcal{K}_k^*) \subseteq \mathcal{X}_k^* \tag{VI.7.94}$$

$$\{\underline{\mathcal{C}}_m | \langle \mathcal{L}_m, \underline{\mathcal{C}}_m \rangle \in \mathcal{K}_k^* \} = \underline{\mathcal{C}}_i^* \tag{VI.7.95}$$

Equation VI.7.91 states that the configuration formulae used to bind the placeholders must be members of the candidate system architecture. Equation VI.7.92 states that the configuration formulae from S_j which are used to bind the placeholders must not overlap, i.e., they must not contain the same actual components from S_j. Equation VI.7.93 states that the configuration formulae resulting from a substitution must be subsets of the candidate configuration closure. Equation VI.7.94 states that the set of configuration constraints resulting from a substitution is also a subset of the candidate set of constraints. Finally, Equation VI.7.95 states that all placeholders in the source architectural pattern must be taken into account by the set of valid bindings.

7.3.1.1 Architectural Matching

If a set of valid bindings exists to match an architectural pattern $\underline{A}_i = \langle \underline{C}_i^*, \underline{L}_i^*, \underline{X}_i^* \rangle$ against a candidate system architecture $S_j = \langle C_j^*, L_j^*, X_j^* \rangle$, the system architecture is deemed to exhibit the architectural pattern. This is mathematically denoted as $S_j \sqsubseteq \underline{A}_i$ and S_j is considered to belong to a class of system architectures that can be described by $\underline{A}_i$. In Example 3, the architectural pattern $\underline{A}_1 = \langle \underline{C}_1^*, \underline{L}_1^*, \underline{X}_1^* \rangle$ represents a feedback configuration. Since both S_{2a} and S_{2b} exhibit the architectural pattern $\underline{A}_1$, they are considered as belonging to the class of feedback control systems, which is expressed as $S_{2a} \sqsubseteq \underline{A}_1$ and $S_{2b} \sqsubseteq \underline{A}_1$.

Architectural matching is designed to be purely syntactical in accordance with whether the terms in a bounded architecture also appear in a candidate system architecture. There are different degrees of similarity between the bounded architectural pattern and the candidate system architecture, as listed below in decreasing order of similarity:

- Exhaustive match
- Comprehensive match
- Partial match
- Closure match

In the strongest case, there exists a set of valid bindings that, when substituted into the architectural pattern, produces a bounded architectural pattern that exhaustively matches the candidate system architecture, as expressed in Definition 24:

Definition 24: Exhaustive Matching
An architectural description $\underline{A}_i = \langle \underline{C}_i^*, \underline{L}_i^*, \underline{X}_i^* \rangle$ provides an exhaustive match against a candidate system architecture $S_j = \langle C_j^*, L_j^*, X_j^* \rangle$, denoted as $S_j \sqsubseteq_E \underline{A}_i$, if the conditions expressed in Equation VI.7.96 are satisfied, namely, that all components, configuration formulae, and constraints in S_j must be represented by $\underline{A}_i$.

$$\exists \mathcal{K}_k^* = \{\mathcal{K}_1, \mathcal{K}_2, \ldots\}, \{\mathcal{C} | \mathcal{C} \in ?\mathcal{L}_m \wedge \langle \mathcal{L}_m, \underline{\mathcal{C}}_m \rangle \in \mathcal{K}_k^*\} = \mathcal{C}_j^*$$

$$\wedge \; (\underline{\mathcal{L}}_i^* \Leftarrow \mathcal{K}_k^*) = \mathcal{L}_j^* \wedge (\underline{\mathcal{X}}_i^* \Leftarrow \mathcal{K}_k^*) = \mathcal{X}_j^* \tag{VI.7.96}$$

Although an exhaustive match provides the greatest accuracy, it may not be particularly useful since the architectural pattern then loses generality by being overly restrictive. This would defeat the purpose of having patterns and frameworks to achieve generality. Consequently, weaker matching conditions are more appropriate.

In a slightly weaker case of a comprehensive match, all components in the candidate system architecture are represented by the architectural pattern. However, only essential configuration formulae in the candidate system architecture are represented by the source architectural pattern. Comprehensive matching has been illustrated in the case of candidate control systems S_{2a} in Example 3, whereby the architectural pattern $\underline{A}_i$ comprises only one configuration formula.

Definition 25: Comprehensive Matching
An architectural description $\underline{\mathcal{A}}_i = \langle \underline{\mathcal{C}}_i^*, \underline{\mathcal{L}}_i^*, \underline{\mathcal{X}}_i^* \rangle$ provides a comprehensive match against a candidate system architecture $\mathcal{S}_j = \langle \mathcal{C}_j^*, \mathcal{L}_j^*, \mathcal{X}_j^* \rangle$, denoted as $\mathcal{S}_j \sqsubseteq_C \underline{\mathcal{A}}_i$, if the conditions expressed in Equation VI.7.97 are satisfied, namely, that the configuration formulae in the set of bindings must cover all components in the candidate system architecture.

$$\exists \mathcal{K}_k^* = \{\mathcal{K}_1, \mathcal{K}_2, \dots\}, \{\mathcal{C} | \mathcal{C} \in \, ?\mathcal{L}_m \wedge \langle \mathcal{L}_m, \underline{\mathcal{C}}_m \rangle \in \mathcal{K}_k^* \} = \mathcal{C}_j^*$$

$$\wedge \, (\underline{\mathcal{L}}_i^* \Leftarrow \mathcal{K}_k^*) \subseteq \mathcal{L}_j^* \wedge (\underline{\mathcal{X}}_i^* \Leftarrow \mathcal{K}_k^*) \subseteq \mathcal{X}_j^* \tag{VI.7.97}$$

In the case of a partial match, the architectural pattern describes the essential interactions between a subset of components within the candidate system architecture, as expressed in Definition 26.

Definition 26: Partial Matching
An architectural description $\underline{\mathcal{A}}_i = \langle \underline{\mathcal{C}}_i^*, \underline{\mathcal{L}}_i^*, \underline{\mathcal{X}}_i^* \rangle$ provides a partial match against a candidate system architecture $\mathcal{S}_j = \langle \mathcal{C}_j^*, \mathcal{L}_j^*, \mathcal{X}_j^* \rangle$, denoted as $\mathcal{S}_j \sqsubseteq_A \underline{\mathcal{A}}_i$, if the criterion expressed in Equation VI.7.98 is satisfied:

$$\forall \mathcal{K}_k^* = \{\mathcal{K}_1, \mathcal{K}_2, \dots\}, \{?\mathcal{L}_m | \langle \mathcal{L}_m, \underline{\mathcal{C}}_m \rangle \in \mathcal{K}_k^* \} \subset \mathcal{C}_j^*$$

$$\wedge \, (\underline{\mathcal{L}}_i^* \Leftarrow \mathcal{K}_k^*) \subset \mathcal{L}_j^* \wedge (\underline{\mathcal{X}}_i^* \Leftarrow \mathcal{K}_k^*) \subset \mathcal{X}_j^* \tag{VI.7.98}$$

A partial match indicates that multiple architectural patterns are necessary to fully describe a system architecture. For example, a control system may have an adaptive algorithm and a plant diagnosis algorithm. One architectural pattern may be used to represent the interaction between the adaptive algorithm and the controller, and another may be used to represent the interaction between the plant diagnosis algorithm and the plant. Consequently, both architectural patterns are needed to represent the entire interactions within the control system. Definition 31 specifies how architectural patterns can be combined to produce more sophisticated ones. In addition, multiple architectural patterns signify the need for multiple supervisory components, one for each pattern. Consequently, additional hierarchical supervisors will also be needed to coordinate these supervisors.

The weakest and most general case involves the matching of the configuration closure, as illustrated in the case of the candidate control system $\mathcal{S}_{2b}$ in Example 3. This is termed closure matching, as expressed in Definition 27.

Definition 27: Closure Matching
An architectural description $\underline{\mathcal{A}}_i = \langle \underline{\mathcal{C}}_i^*, \underline{\mathcal{L}}_i^*, \underline{\mathcal{X}}_i^* \rangle$ provides a closure match against a candidate system architecture $\mathcal{S}_j = \langle \mathcal{C}_j^*, \mathcal{L}_j^*, \mathcal{X}_j^* \rangle$, having a configuration closure $\mathcal{L}_j^+$, denoted as $\mathcal{S}_j \sqsubseteq_L \underline{\mathcal{A}}_i$, if the conditions expressed in Equation VI.7.99 are satisfied:

$$\exists \mathcal{K}_k^* = \{\mathcal{K}_1, \mathcal{K}_2, \dots\}, \{\mathcal{C} | \mathcal{C} \in \, ?\mathcal{L}_m \wedge \langle \mathcal{L}_m, \underline{\mathcal{C}}_m \rangle \in \mathcal{K}_k^* \} = \mathcal{C}_j^*$$

$$\wedge \, (\underline{\mathcal{L}}_i^* \Leftarrow \mathcal{K}_k^*) \subseteq \mathcal{L}_j^+ \wedge (\underline{\mathcal{X}}_i^* \Leftarrow \mathcal{K}_k^*) = \mathcal{X}_j^* \tag{VI.7.99}$$

These conditions differ from those in an exhaustive or comprehensive match through the use of a configuration closure. This is useful for handling cases where placeholders in the architectural patterns are substituted by configuration formulae of arbitrary length, as in the case of $\mathcal{S}_{2b}$ in Example 3. The use of a closure match must be handled with care since the number of possible matches can be very large. In subsequent discussions, only bindings from one component to one placeholder will be considered, as this does not limit the generality and the applicability of SPADE.

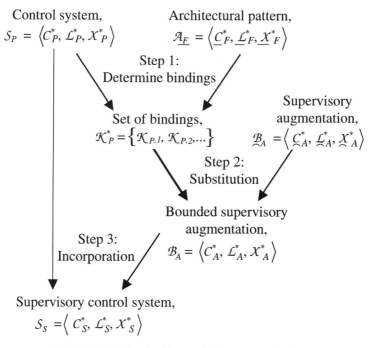

FIGURE VI.7.8   Architectural inference mechanism.

7.3.1.2 Incorporation of Supervisory Components

The incorporation of a supervisory augmentation into a candidate control system is depicted in Figure VI.7.8 and is comprised of three steps.

- **Step 1: Determine a set of bindings** — this step determines a matching architectural pattern $\underline{\mathcal{A}}_F$ that describes the essential interactions in the control system $\mathcal{S}_P$. The matching procedure produces a set of valid bindings denoted by $\mathcal{K}_P^*$. The degree of match can be exhaustive, comprehensive, or partial. If an exhaustive or a comprehensive match occurs, the incorporation of a single supervisory augmentation suffices. If there is a partial match, multiple supervisory augmentations have to be incorporated to ensure that all components are supervised. This produces a hierarchical supervisory scheme and will be elaborated on in subsequent sections.
- **Step 2: Bind supervisory augmentation** — each architectural pattern $\underline{\mathcal{A}}_F$ is associated with a supervisory augmentation denoted by $\underline{\mathcal{B}}_A$. A supervisory augmentation describes how additional supervisory components are to be incorporated into a control system. In order for the supervisory augmentation to be incorporated, the naming convention between the supervisory augmentation and the control system must be made consistent. This is achieved by binding the placeholders in $\underline{\mathcal{B}}_A$ with $\mathcal{K}_P^*$ in this step, which results in a bounded supervisory augmentation $\mathcal{B}_A$.
- **Step 3: Incorporate bounded supervisory augmentation** — the final step performs the actual incorporation of supervisory components, which results in a supervisory control system denoted by $\mathcal{S}_S$. The third step is formalized in Definition 28.

Definition 28: Incorporation of Supervisory Augmentation

The incorporation of a supervisory augmentation $\mathcal{B}_i = \langle \mathcal{C}_i^*, \mathcal{L}_i^*, \mathcal{X}_i^* \rangle$ under a set of bindings system $\mathcal{K}_j^*$ into a candidate system $\mathcal{S}_k = \langle \mathcal{C}_k^*, \mathcal{L}_k^*, \mathcal{X}_k^* \rangle$ results in another system denoted by $\mathcal{S}_l = \langle \mathcal{C}_l^*, \mathcal{L}_l^*, \mathcal{X}_l^* \rangle$.

This operation is expressed as $\mathcal{S}_l = \mathcal{S}_k \cup (\mathcal{B}_i \Leftleftarrows \mathcal{K}_j^*)$, and the terms of the resulting system are expressed in Equations VI.7.100–VI.7.102:

$$C_l^* = (C_i^* \Leftleftarrows \mathcal{K}_j^*) \cup C_k^* \tag{VI.7.100}$$

$$\mathcal{L}_l^* = (\mathcal{L}_i^* \Leftleftarrows \mathcal{K}_j^*) \cup \mathcal{L}_k^* \tag{VI.7.101}$$

$$\mathcal{X}_l^* = (\mathcal{X}_i^* \Leftleftarrows \mathcal{K}_j^*) \cup \{\mathcal{X} \in \mathcal{X}_k^* | valid(\mathcal{X})\} \tag{VI.7.102}$$

Equation VI.7.100 states that the resultant system consists of the components within the original system C_j^* and the supplementary components C_i^*, defined within the supervisory augmentation. Equation VI.7.101 states that the resulting set of configuration formulae is a union of those in the original system $\mathcal{L}_k^*$ and those in the bounded supervisory augmentation. It must be noted that the constraints in the candidate system $\mathcal{S}_k = \langle C_k^*, \mathcal{L}_k^*, \mathcal{X}_k^* \rangle$ may not be valid after the incorporation of $\mathcal{B}_i$, hence, the first term in Equation VI.7.102 states that all constraints specified in the supervisory augmentation must also be valid in the resulting system $\mathcal{S}_i$. This provides a means to ascertain the incorporation step as expressed in Equation VI.7.103.

$$\forall \mathcal{X} \in (\mathcal{X}_i^* \Leftleftarrows \mathcal{K}_j^*), valid(\mathcal{X}) \tag{VI.7.103}$$

In addition, the second term in Equation VI.7.102 indicates that only those constraints that remain valid continue to appear in the resulting system $\mathcal{S}_l$. The incorporation of a supervisory augmentation $\mathcal{B}_i$ into a system $\mathcal{S}_k$ with a set of bindings $\mathcal{K}_j^*$ produces a new system denoted by $\mathcal{S}_l = \mathcal{S}_k \cup (\mathcal{B}_i \Leftleftarrows \mathcal{K}_j^*)$. This expression explicitly indicates the set of bindings used. A more compact expression can be written as $\mathcal{S}_l = \mathcal{S}_k \uplus \mathcal{B}_i$. In this case, there is no mention of any set of bindings and they are implicit. This permits a simple way to express an architectural inference rule as follows:

$$\text{if } \mathcal{S}_P \sqsubseteq_C \mathcal{A}_F \text{ then } \mathcal{S}_S = \mathcal{S}_P \uplus \mathcal{B}_A$$

This compact representation is similar to the rules within a typical knowledge base. Bindings are explicitly mentioned in such rules because they are only available when rules are applied. The above rule reads "if a system $\mathcal{S}_P$ is comprehensively described by an architectural pattern $\mathcal{A}_F$, then a supervisory control system can be derived by incorporating the supervisory augmentation $\mathcal{B}_A$ into $\mathcal{S}_P$." Example 4 illustrates how a direct adaptive configuration can be augmented with a supervisor using a pattern-based approach.

Example 4: Incorporation of Supervisory Components
This example illustrates the incorporation of supervisory components into an adaptive control system by demonstrating the step-by-step derivation of the PI/MRAS realization. Four diagrams are used in this example, namely:

- The candidate control system denoted by $\mathcal{S}_{CMRAS}$ in Figure VI.7.9
- The architectural pattern denoted by $\mathcal{A}_{FDA}$ in Figure VI.7.10
- The supervisory augmentation denoted by $\mathcal{B}_{ADA}$ in Figure VI.7.11
- The resulting supervisory control system denoted by $\mathcal{S}_{SMRAS}$ in Figure VI.7.12

The candidate control system is formally expressed as $\mathcal{S}_{CMRAS} = \langle C_{CMRAS}^*, \mathcal{L}_{CMRAS}^*, \mathcal{X}_{CMRAS}^* \rangle$, where the individual terms of $\mathcal{S}_{CMRAS}$ are defined in Equations VI.7.104–VI.7.106:

$$C_{CMRAS}^* = \{C_{SOURCE}, C_{PLANT}, C_{PI}, C_{MRAS}\} \tag{VI.7.104}$$

$$\mathcal{L}_{CMRAS}^* = \{C_{SOURCE} \twoheadrightarrow C_{PI}, C_{SOURCE} \twoheadrightarrow C_{MRAS}, C_{PI} \circlearrowright C_{PLANT}, C_{PLANT} \circlearrowleft C_{PI}\}$$
$$\cup \{C_{MRAS} \rightsquigarrow C_{PI}, C_{PLANT} \twoheadrightarrow C_{MRAS}\} \tag{VI.7.105}$$

$$\mathcal{X}_{CMRAS}^* = \{\rho(C_{PLANT}, \text{process})\} \tag{VI.7.106}$$

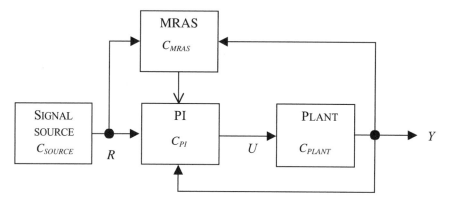

FIGURE VI.7.9   Candidate control system.

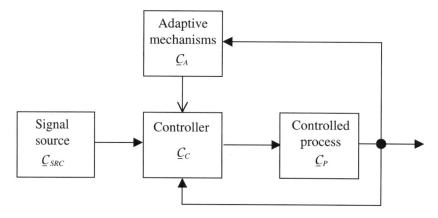

FIGURE VI.7.10   Architectural pattern.

The architectural pattern is expressed formally as $\underline{\mathcal{A}}_{FDA} = \langle \mathcal{C}^*_{FDA}, \mathcal{L}^*_{FDA}, \mathcal{X}^*_{FDA} \rangle$, where the individual terms of $\underline{\mathcal{A}}_{FDA}$ are defined in Equations VI.7.107–VI.7.109:

$$\mathcal{C}^*_{FDA} = \{ \underline{\mathcal{C}}_{SRC}, \underline{\mathcal{C}}_P, \underline{\mathcal{C}}_C, \underline{\mathcal{C}}_{DA} \} \tag{VI.7.107}$$

$$\mathcal{L}^*_{FDA} = \{ \underline{\mathcal{C}}_{SRC} \twoheadrightarrow \underline{\mathcal{C}}_C, \underline{\mathcal{C}}_C \circlearrowleft \underline{\mathcal{C}}_P, \underline{\mathcal{C}}_A \leftrightarrow \underline{\mathcal{C}}_C \} \tag{VI.7.108}$$

$$\mathcal{X}^*_{FDA} = \{ \rho(\underline{\mathcal{C}}_P, \text{process}) \} \tag{VI.7.109}$$

In Equation VI.7.107, the elements in $\mathcal{C}^*_{FDA}$ are placeholders for a signal source $\mathcal{C}^*_{SRC}$, a controlled process $\underline{\mathcal{C}}_P$, a controller $\underline{\mathcal{C}}_C$, and a direct-adaptive mechanism $\underline{\mathcal{C}}_A$. The interactions between these placeholders are depicted in Figure VI.7.10 and are formalized in Equation VI.7.108. The only constraint in $\underline{\mathcal{A}}_{FDA}$ is the process designation constraint expressed in Equation VI.7.109.

- **Step 1: Determine a set of bindings** — the first step in the architectural inference procedure is to determine a matching architectural pattern for the candidate control system, $\mathcal{S}_{CMRAS}$. One such architectural pattern is $\underline{\mathcal{A}}_{FDA}$. A set of valid bindings that matches $\underline{\mathcal{A}}_{FDA}$ against $\mathcal{S}_{CMRAS}$ is identified in Equation VI.7.110:

$$\mathcal{K}^*_{CMRAS} = \{ \langle \mathcal{C}_{SOURCE}, \underline{\mathcal{C}}_{SRC} \rangle, \langle \mathcal{C}_{PLANT}, \mathcal{C}_P \rangle, \langle \mathcal{C}_{PI}, \underline{\mathcal{C}}_C \rangle, \langle \mathcal{C}_{MRAS}, \underline{\mathcal{C}}_A \rangle \} \tag{VI.7.110}$$

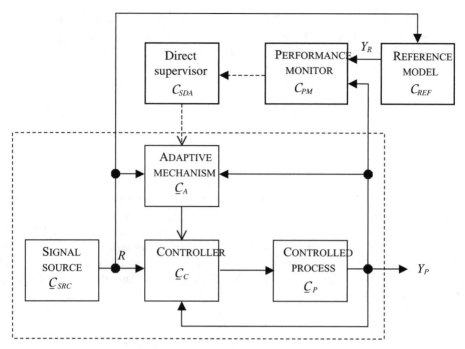

FIGURE VI.7.11　Supervisory augmentation.

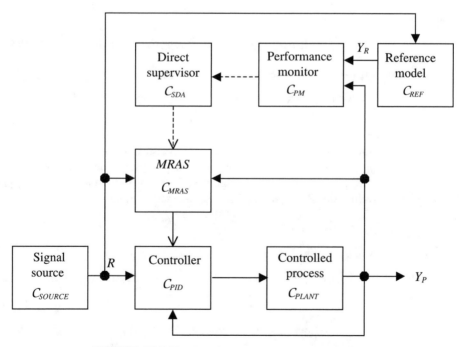

FIGURE VI.7.12　Resulting supervisory control system.

- **Step 2: Bind supervisory augmentation** — the second step substitutes the bindings into the supervisory augmentation associated with $\underline{\mathcal{A}}_{FDA}$, denoted $\mathcal{B}_{ADA} = \langle \mathcal{C}^*_{ADA}, \mathcal{L}^*_{ADA}, \mathcal{X}^*_{ADA} \rangle$, which defines the additional components that are to provide adequate supervision. The individual terms of $\mathcal{B}_{ADA}$ are defined in Equations VI.7.111–VI.7.113:

$$\mathcal{C}^*_{ADA} = \{\mathcal{C}_{SDA}, \mathcal{C}_{PM}, \mathcal{C}_{REF}, \underline{\mathcal{C}}_{SRC}, \underline{\mathcal{C}}_P, \underline{\mathcal{C}}_A\} \tag{VI.7.111}$$

$$\mathcal{L}^*_{ADA} = \{\mathcal{C}_{SDA} \leftharpoonup \mathcal{C}_{PM}, \mathcal{C}_{REF} \twoheadrightarrow \mathcal{C}_{PM}\}$$

$$\cup \{\mathcal{C}_{SDA} \rightsquigarrow \underline{\mathcal{C}}_A\} \cup \{\underline{\mathcal{C}}_{SRC} \twoheadrightarrow \mathcal{C}_{REF}, \underline{\mathcal{C}}_P \twoheadrightarrow \mathcal{C}_{PM}\} \tag{VI.7.112}$$

$$\mathcal{X}^*_{ADA} = \{\vartheta(\mathcal{C}_{SDA})\} \tag{VI.7.113}$$

Equation VI.7.111 states that the augmentation is comprised of three additional components, namely, a supervisor $\mathcal{C}_{SDA}$, a performance monitor $\mathcal{C}_{PM}$, and a reference model $\mathcal{C}_{REF}$. The three terms in Equation VI.7.112 list the interactions among the three additional components, between the supervisor and the placeholders, and between other additional components and the placeholders supporting the supervisor, respectively. The reference model $\mathcal{C}_{REF}$ is included to permit the comparison of the transfer function of the feedback control system $\underline{\mathcal{C}}_C \circlearrowright \underline{\mathcal{C}}_P$ against that of the reference model $\mathcal{C}_{REF}$. The difference between their respective outputs Y_R and Y_P is usually used as a measure of system performance, the value of which is computed by the performance monitor $\mathcal{C}_{PM}$. Based on the recommendation of $\mathcal{C}_{PM}$, the supervisor $\mathcal{C}_{SDA}$ decides whether the adaptive mechanism bounded to $\underline{\mathcal{C}}_A$ has to be activated. Equation VI.7.113 states that after augmentation, the supervisor $\mathcal{C}_{SDA}$ is an orphaned supervisor.

Upon substituting the set of bindings, $\mathcal{K}^*_{CMRAS}$, the resulting bounded supervisory augmentation denoted by $\mathcal{B}_{FSDA} = \langle \mathcal{C}^*_{FSDA}, \mathcal{L}^*_{FSDA}, \mathcal{X}^*_{FSDA} \rangle$ is described by Equations VI.7.114–VI.7.116:

$$\mathcal{C}^*_{FSDA} = \{\mathcal{C}_{SDA}, \mathcal{C}_{PM}, \mathcal{C}_{REF}, \mathcal{C}_{SOURCE}, \mathcal{C}_{PLANT}, \mathcal{C}_{MRAS}\} \tag{VI.7.114}$$

$$\mathcal{L}^*_{FSDA} = \{\mathcal{C}_{SDA} \leftharpoonup \mathcal{C}_{PM}, \mathcal{C}_{REF} \twoheadrightarrow \mathcal{C}_{PM}\}$$

$$\cup \{\mathcal{C}_{SDA} \rightsquigarrow \mathcal{C}_{MRAS}\} \cup \{\mathcal{C}_{SOURCE} \twoheadrightarrow \mathcal{C}_{REF}, \mathcal{C}_{PLANT} \twoheadrightarrow \mathcal{C}_{PM}\} \tag{VI.7.115}$$

$$\mathcal{X}^*_{FSDA} = \{\vartheta(\mathcal{C}_{SDA})\} \tag{VI.7.116}$$

- **Step 3: Incorporate bounded supervisory augmentation** — the incorporation of $\mathcal{B}_{FSDA}$ into $\mathcal{S}_{CMRAS}$ results in a supervisory control system denoted by $\mathcal{S}_{SMRAS} = \langle \mathcal{C}^*_{SMRAS}, \mathcal{L}^*_{SMRAS}, \mathcal{X}^*_{SMRAS} \rangle$ and depicted in Figure VI.7.12. The individual terms of $\mathcal{S}_{SMRAS}$ are computed using Definition 28 and the results are expressed in Equations VI.7.117–VI.7.119:

$$\mathcal{C}^*_{SMRAS} = \{\mathcal{C}_{SOURCE}, \mathcal{C}_{PLANT}, \mathcal{C}_{PI}, \mathcal{C}_{MRAS}, \mathcal{C}_{SDA}, \mathcal{C}_{PM}, \mathcal{C}_{REF}\} \tag{VI.7.117}$$

$$\mathcal{L}^*_{SMRAS} = \{\mathcal{C}_{SOURCE} \twoheadrightarrow \mathcal{C}_{PI}, \mathcal{C}_{SOURCE} \twoheadrightarrow \mathcal{C}_{MRAS}, \mathcal{C}_{PI} \circlearrowright \mathcal{C}_{PLANT}, \mathcal{C}_{PLANT} \circlearrowright \mathcal{C}_{PI}\}$$

$$\cup \{\mathcal{C}_{MRAS} \looparrowright \mathcal{C}_{PI}, \mathcal{C}_{SOURCE} \twoheadrightarrow \mathcal{C}_{MRAS}\} \cup \{\mathcal{C}_{SDA} \leftharpoonup \mathcal{C}_{PM}, \mathcal{C}_{REF} \twoheadrightarrow \mathcal{C}_{PM}, \mathcal{C}_{SDA} \rightsquigarrow \mathcal{C}_{MRAS}\}$$

$$\cup \{\mathcal{C}_{SOURCE} \twoheadrightarrow \mathcal{C}_{REF}, \mathcal{C}_{PLANT} \twoheadrightarrow \mathcal{C}_{PM}\} \tag{VI.7.118}$$

$$\mathcal{X}^*_{SMRAS} = \{\rho(\mathcal{C}_{PLANT}, \text{process}), \vartheta(\mathcal{C}_{SDA})\} \tag{VI.7.119}$$

Equations VI.7.117 and VI.7.118 list the components of the resulting supervisory control system as well as the resulting configuration formulae. The first two terms in Equation VI.7.118 are derived from the candidate control system in Equation VI.7.105, before the incorporation step, while the last two terms are derived from the bounded supervisory augmentation in Equation VI.7.115. Finally,

TABLE VI.7.4 Semantics of Vector Elements

Vector Element	Description	Semantics	Configuration Formulae
$u_{PM}^{[ref]}$	command signal to reference model	command signal	$\mathcal{C}_{SRC} \twoheadrightarrow \mathcal{C}_{REF}$
$u_{PM}^{[fdb]}$	feedback input to performance monitor	feedback	$\underline{\mathcal{C}}_{P} \twoheadrightarrow \mathcal{C}_{PM}$
$f_{SDA}^{[activate\ Adapt]}$	command to activate adaptation	command, activate	$\mathcal{C}_{SDA} \rightsquigarrow \underline{\mathcal{C}}_{A}$
$f_{SDA}^{[deactivate\ Adapt]}$	command to deactivate adaptation	command, deactivate	$\mathcal{C}_{SDA} \rightsquigarrow \underline{\mathcal{C}}_{A}$

Equation VI.7.119 defines the constraints to be satisfied by $\mathcal{S}_{SMRAS}$. The resulting supervisory control system is the PI/MRAS supervisory control system depicted in Figure VI.7.12. This example demonstrates the specific steps involved in the construction of a system architecture using a pattern-based approach.

7.3.1.3 Component Attachments

The previous section describes how supervisory components are incorporated and operate at a high level of abstraction. At a finer level of detail, specific attachments have to be made between existing and incorporated components. Although the supervisory augmentation does not explicitly state how attachments are made, it does state which components should interact. The semantics assigned to the vector elements of a component guides how these attachments can be identified. Section 7.2.2.1 has established that attachments are valid only if the semantics between the interacting elements ε_i and ε_j is the same. In other words, the attachment $\langle \varepsilon_i, \varepsilon_j \rangle$ is valid only when $[\![\varepsilon_i]\!] = [\![\varepsilon_j]\!]$. This provides the criterion to determine how component attachments are made after the incorporation step by:

- Associating the vector elements within the additional components of the supervisory augmentation with their corresponding configuration formulae
- Using the semantics assigned to these vector elements to search for the appropriate ones in the bounded configuration after the incorporation step

Example 5 illustrates how component attachments are identified using the semantics assigned to the vector elements.

Example 5: Component Attachments
In Example 4, the supervisory augmentation $\mathcal{B}_{ADA}$ specifies that the performance monitor $\mathcal{C}_{PM}$ must interact with the controlled process $\underline{\mathcal{C}}_P$, that the reference model $\mathcal{C}_{REF}$ must receive input from the command signal source $\underline{\mathcal{C}}_{SRC}$, and that the supervisor must interact with the adaptive mechanism $\underline{\mathcal{C}}_A$. The semantics of the vector elements within the performance monitor $\mathcal{C}_{PM}$ and the adaptation supervisor $\mathcal{C}_{SDA}$ are specified in Table VI.7.4. The first two columns list the vector element along with a brief description, the third column indicates its associated semantics, and the last column identifies the configuration formula that defines which other component it should be attached to.

A specific attachment is made when a placeholder in the configuration formula is bounded. For example, the first row of Table VI.7.4 indicates that the signal source $\underline{\mathcal{C}}_{SRC}$ has to be attached to the performance monitor $\mathcal{C}_{PM}$. In Example 4, $\underline{\mathcal{C}}_{SRC}$ is bounded to $\mathcal{C}_{SOURCE}$, and the only vector element within $\mathcal{C}_{SOURCE}$ that fulfills the configuration formula $\underline{\mathcal{C}}_{SRC} \twoheadrightarrow \mathcal{C}_{PM}$ and the element semantics is $y_{SOURCE}^{[ref]}$. Consequently, the first row of Table VI.7.4 generates an attachment $\langle y_{SOURCE}^{[ref]}, u_{PM}^{[ref]} \rangle$. Other attachments are derived using the same principle.

7.3.2   Relating Architectural Patterns

Several relationships can be defined between architectural patterns. However, this requires that the naming convention used in the placeholders must be made consistent. This can be achieved using a set of bindings similar to the case of attempting to match an architectural pattern against a system architecture. The substitution of a set of bindings $\underline{\mathcal{K}}_k^*$ into an architectural pattern $\mathcal{A}_i = \langle \underline{\mathcal{C}}_i^*, \underline{\mathcal{L}}_i^*, \underline{\mathcal{X}}_i^* \rangle$ produces yet another architectural pattern $\underline{\mathcal{A}}_j = \langle \underline{\mathcal{C}}_j^*, \underline{\mathcal{L}}_j^*, \underline{\mathcal{X}}_j^* \rangle$ where the individual terms are defined in Equations VI.7.120–VI.7.122:

$$\underline{\mathcal{C}}_j^* = \{ \mathcal{C} | \mathcal{C} \in ?\mathcal{L}_m \wedge \langle \mathcal{L}_m, \underline{\mathcal{C}}_m \rangle \in \underline{\mathcal{K}}_k^* \} \tag{VI.7.120}$$

$$\underline{\mathcal{L}}_j^* = \underline{\mathcal{L}}_i^* \Leftarrow \mathcal{K}_k^* \tag{VI.7.121}$$

$$\underline{\mathcal{X}}_j^* = \underline{\mathcal{X}}_i^* \Leftarrow \underline{\mathcal{K}}_k^* \tag{VI.7.122}$$

The above equations are similar to that in Definition 21, which describes the binding of components into an architectural pattern to produce a system architecture. The difference in this case is that the binding is from placeholders to placeholders. Using this concept, the subset and equivalence relationships are expressed in Definitions 29 and 30.

Definition 29:   Architectural Pattern Subset Relationship

An architectural pattern $\mathcal{A}_i = \langle \underline{\mathcal{C}}_i^*, \underline{\mathcal{L}}_i^*, \underline{\mathcal{X}}_i^* \rangle$ is a subset of another architectural pattern $\underline{\mathcal{A}}_j = \langle \underline{\mathcal{C}}_j^*, \underline{\mathcal{L}}_j^*, \underline{\mathcal{X}}_j^* \rangle$, denoted by $\underline{\mathcal{A}}_i \subseteq \underline{\mathcal{A}}_j$, if the conditions expressed in Equation VI.7.123 are satisfied.

$$\underline{\mathcal{A}}_i \subseteq \underline{\mathcal{A}}_j \Leftrightarrow \exists \underline{\mathcal{K}}_k^*, \{ \mathcal{C} | \mathcal{C} \in ?\mathcal{L}_m \wedge \langle \mathcal{L}_m, \underline{\mathcal{C}}_m \rangle \in \underline{\mathcal{K}}_k^* \} \subseteq \underline{\mathcal{C}}_j^*$$

$$\wedge \, (\underline{\mathcal{L}}_i^* \Leftarrow \underline{\mathcal{K}}_k^*) \subseteq \underline{\mathcal{L}}_j^* \wedge (\underline{\mathcal{X}}_i^* \Leftarrow \underline{\mathcal{K}}_k^*) \subseteq \underline{\mathcal{X}}_j^* \tag{VI.7.123}$$

Equation VI.7.123 states that a subset relationship exists if there exists a set of bindings which, when substituted into $\underline{\mathcal{A}}_i$, produces another architectural pattern whereby each individual term is a subset of $\mathcal{A}_j$. If two architectural patterns use the same component naming convention, the set of bindings is no longer necessary and the above criteria for subset relationship simplify to that in Equation VI.7.124. That is, if $\mathcal{A}_i$ is a subset of $\mathcal{A}_j$ then all placeholders, configuration formulae, and configuration constraints in $\mathcal{A}_i$ must also appear in $\mathcal{A}_j$.

$$\underline{\mathcal{A}}_i \subseteq \underline{\mathcal{A}}_j \Leftrightarrow \underline{\mathcal{C}}_i^* \subseteq \underline{\mathcal{C}}_j^* \wedge \underline{\mathcal{L}}_i^* \subseteq \underline{\mathcal{L}}_j^* \wedge \underline{\mathcal{X}}_i^* \subseteq \underline{\mathcal{X}}_j^* \tag{VI.7.124}$$

The definition of this subset relation between architectural patterns permits the definition of an equivalence relationship, as follows.

Definition 30:   Architectural Pattern Equivalence Relationship

Two architectural patterns $\mathcal{A}_i = \langle \underline{\mathcal{C}}_i^*, \underline{\mathcal{L}}_i^*, \underline{\mathcal{X}}_i^* \rangle$ and $\underline{\mathcal{A}}_j = \langle \underline{\mathcal{C}}_j^*, \underline{\mathcal{L}}_j^*, \underline{\mathcal{X}}_j^* \rangle$ are equivalent, as denoted by $\underline{\mathcal{A}}_i = \underline{\mathcal{A}}_j$, if the criterion in Equation VI.7.125 is satisfied:

$$\underline{\mathcal{A}}_i = \underline{\mathcal{A}}_j \Leftrightarrow \underline{\mathcal{A}}_i \subseteq \underline{\mathcal{A}}_j \wedge \underline{\mathcal{A}}_j \subseteq \underline{\mathcal{A}}_i \tag{VI.7.125}$$

Definition 31:   Union of Architectural Patterns

The union of two architectural patterns $\mathcal{A}_i = \langle \underline{\mathcal{C}}_i^*, \underline{\mathcal{L}}_i^*, \underline{\mathcal{X}}_i^* \rangle$ and $\underline{\mathcal{A}}_j = \langle \underline{\mathcal{C}}_j^*, \underline{\mathcal{L}}_j^*, \underline{\mathcal{X}}_j^* \rangle$ produces another architectural pattern, denoted by $\underline{\mathcal{A}}_k = \underline{\mathcal{A}}_i \cup \underline{\mathcal{A}}_j$. The individual terms of $\underline{\mathcal{A}}_k = \langle \underline{\mathcal{C}}_k^*, \underline{\mathcal{L}}_k^*, \underline{\mathcal{X}}_k^* \rangle$ are determined in Equations VI.7.126–VI.7.128.

$$\underline{\mathcal{C}}_k^* = \underline{\mathcal{C}}_i^* \cup \underline{\mathcal{C}}_j^* \tag{VI.7.126}$$

$$\underline{\mathcal{L}}_k^* = \underline{\mathcal{L}}_i^* \cup \underline{\mathcal{L}}_j^* \tag{VI.7.127}$$

$$\underline{\mathcal{X}}_k^* = \underline{\mathcal{X}}_i^* \cup \underline{\mathcal{X}}_j^* \tag{VI.7.128}$$

The union operation can be used to progressively develop more sophisticated architectural patterns from simpler ones.

7.4 Supervisory Frameworks

In practice, it seldom happens that a single supervisory scheme is sufficient to describe a given control system. Usually, each pattern is documented to describe a certain part of a typical control system, and a number of these patterns are incrementally and hierarchically applied to produce the required supervisory control system.

This process is illustrated in Figure VI.7.13 where the candidate control system $\mathcal{S}_P$ exhibits two control configurations characterized by the architectural patterns $\underline{\mathcal{A}}_1$ and $\underline{\mathcal{A}}_2$, both of which only provide partial descriptions. Patterns $\underline{\mathcal{A}}_1$ and $\underline{\mathcal{A}}_2$ characterize configurations involving the components $\mathcal{C}_{A1}$ and $\mathcal{C}_{A2}$, respectively. The application of supervisory augmentations associated with $\underline{\mathcal{A}}_1$ and $\underline{\mathcal{A}}_2$ results in the inclusion of the two orphaned supervisors $\mathcal{C}_{S1}$ and $\mathcal{C}_{S2}$, which have to be coordinated by another supervisor. An architectural pattern $\underline{\mathcal{A}}_3$ is designed to identify where a hierarchical supervisor $\mathcal{C}_{S3}$ is to be incorporated. It does not require a full description of the system, as expressed by the non-overlapping patterns in Figure VI.7.13c. The incorporation of $\mathcal{C}_{S3}$ results in only one orphaned supervisor, which satisfies the hierarchical supervision property $\mathcal{H}^H$ defined in Section 7.2.4.2. Consequently, a number of pattern–augmentation pairs are required to cover a sufficiently large class of control configurations, which constitute a supervisory framework.

Definition 32: Supervisory Framework
A supervisory framework defines how different guidelines expressed as a set of pattern–augmentation pairs $\mathcal{P}^*$ are appropriately applied to produce a system with some desirable properties $\mathcal{H}^*$, as expressed in Equation VI.7.129.

$$\mathcal{F} = \langle \mathcal{P}^*, \mathcal{H}^* \rangle \tag{VI.7.129}$$

$\mathcal{P}^* = \{\mathcal{P}_1, \mathcal{P}_2, \ldots, \mathcal{P}_t, \ldots\}$ is the set of pattern–augmentation pairs, where each pair $\mathcal{P}_i = \langle \underline{\mathcal{A}}_{f,i}, \mathcal{B}_{A,i} \rangle$ comprises an architectural pattern $\underline{\mathcal{A}}_{F,i}$ and a supervisory augmentation $\mathcal{B}_{A,i}$. The former characterizes a given control problem whereas the latter characterizes the corresponding solution. Each pair constitutes a guideline to solve a given supervisory problem. $\mathcal{H}^* = \{\mathcal{H}_1, \mathcal{H}_2, \ldots, \mathcal{H}_i, \ldots\}$

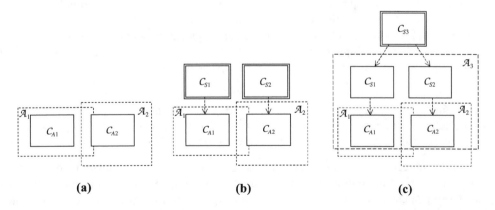

(a)   **(b)**   **(c)**

FIGURE VI.7.13   Hierarchical supervision: (a) candidate control system exhibiting two control configurations; (b) two supervisory augmentations incorporated; and (c) a hierarchical supervisor incorporated.

is the set of desirable properties that the framework produces. A possible list of these properties has been defined in Section 7.2.

7.4.1   Applying a Framework of Supervisory Schemes

Using the principles described in the preceding sections, a more general form of architectural inference mechanism can be devised, the inputs of which are:

- A framework $\mathcal{F}$ comprised of many pattern–augmentation pairs, where the ith pair is denoted by $\langle \mathcal{A}_{F,i}, \mathcal{B}_{A,i} \rangle$, rather than a single pair
- A candidate control system $\mathcal{S}_P$

The inference mechanism summarized in Figure VI.7.14 iteratively applies architectural inference rules until only one orphaned supervisor remains and only one component is not supervised, i.e., the root supervisor. The candidate control system, denoted by $\mathcal{S}_{PL}$, is initially assigned to $\mathcal{S}_P$. Each iteration consists of 5 steps:

- **Step 1: Find matching architectural inference rule** — this step searches available pattern–augmentation pairs $\langle \mathcal{A}_{F,i}, \mathcal{B}_{A,i} \rangle$ for one that either comprehensively or partially characterizes $\mathcal{S}_{PL}$.
- **Step 2: Apply architectural inference** — this step incorporates the selected supervisory augmentation $\mathcal{B}_{A,i}$ according to Definition 28, producing a system denoted by $\mathcal{S}_{SL} = \mathcal{S}_{PL} \uplus \mathcal{B}_{A,i}$.
- **Step 3: Check result** — this step checks whether the incorporation of supervisory components has completed, to ensure that there is only one orphaned supervisor, referred to as the root supervisor, and that all components other than the root supervisor are supervised or monitored by some supervisor.
- **Step 4: Prepare for next iteration** — if the conditions in Step 3 are not satisfied, $\mathcal{S}_{PL}$ is reassigned with $\mathcal{S}_{SL}$ in preparation for the next iteration.
- **Step 5: Final result** — if the conditions in Step 3 are satisfied, the final result $\mathcal{S}_S$ is the one produced after the augmentation in this step is applied, and, thus, $\mathcal{S}_S = \mathcal{S}_{SL}$.

The generalized architectural inference mechanism produces a supervisory control system that is hierarchical if it terminates successfully (step 4 in Figure VI.7.14). The hierarchical property has been defined earlier in Definition 18, the first condition of which is satisfied by being a termination criterion (step 3(s)a in Figure VI.7.1). Equation VI.7.61 states that there must be a transitive supervisory path from the root supervisor to all other components and that there must not be any cyclic supervisory relationship in the resulting supervisory control system. The satisfaction of these two conditions is demonstrated as follows. Consider an arbitrary component $\mathcal{C}_0$ before the execution of the inference mechanism. Suppose at a certain iteration a supervisor $\mathcal{C}_1$ is incorporated such that $\mathcal{C}_1 \rightsquigarrow \mathcal{C}_0 \vee \mathcal{C}_1 \leftarrow \mathcal{C}_0$. This is reasonable for two reasons, namely:

- The only permissible interaction between a supervisor and any other component must be based on either a supervisory composition or monitoring component.
- The arbitrary component $\mathcal{C}_0$ must be supervised before the final iteration, otherwise the inference mechanism cannot terminate according to Step 3(b).

At some subsequent iteration, another supervisor $\mathcal{C}_2$ will be incorporated such that $\mathcal{C}_2 \rightsquigarrow \mathcal{C}_1 \vee \mathcal{C}_2 \leftarrow \mathcal{C}_1$ according to the same argument. This implies that $\mathcal{C}_2$ transitively supervises $\mathcal{C}_0$, i.e., $\mathcal{C}_2 \rightsquigarrow_T \mathcal{C}_0$ and $\mathcal{C}_2 \rightsquigarrow_T \mathcal{C}_1$. Since supervisory augmentations are defined such that the newly included supervisor supervises or monitors existing components, there is no cyclic supervision. This can be affirmed by examining the relationship at the Nth subsequent iteration. At this iteration, the

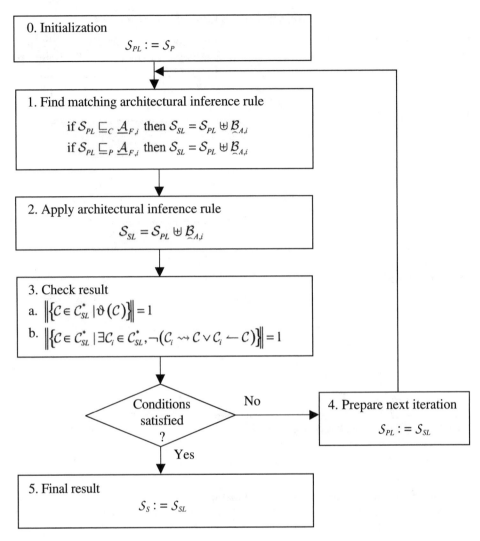

FIGURE VI.7.14 Generalized architectural inference mechanism.

supervisor $\mathcal{C}_N$ transitively supervises all preceding components as expressed in Equation VI.7.130, that is, there is no supervision relationship of the form $\mathcal{C}_i \rightsquigarrow_T \mathcal{C}_N, \forall 0 \leq i < N$.

$$\mathcal{C}_N \rightsquigarrow_T \mathcal{C}_i, \forall 0 \leq i < N \qquad (VI.7.130)$$

This demonstrates that the last condition in Equation VI.7.61 is satisfied. At the last iteration, $\mathcal{C}_N$ is the root supervisor, i.e., $\vartheta(\mathcal{C}_N)$. Since, $\mathcal{C}_0, \mathcal{C}_1, \ldots, \mathcal{C}_{N-1}$ are arbitrary components, it can be concluded that the root supervisor $\mathcal{C}_N$ transitively supervises all other components, which satisfies the second condition in Equation VI.7.61. Consequently, the generalized architectural inference mechanism produces a supervisory control system that is hierarchical provided that:

- All supervisory augmentations are defined such that no supervisor defined in the associated architectural pattern supervises newly incorporated supervisors.
- The original candidate control system has no cyclic supervision relationship in the first place.

7.4.2   Combining and Comparing Supervisory Frameworks

Different frameworks can be unified into a single supervisory framework. From the framework perspective, the unification of supervisory framework refers to the inclusion of different pattern–augmentation pairs from different frameworks into a larger framework. Existing frameworks such as expert control (EC), integrated process supervision (IPS), and multiple-model multiple-controller (MMMC) have been specifically demonstrated for only single control configurations and can be considered as single-scheme frameworks. EC has been demonstrated for auto-tuning control configurations, IPS for direct adaptive configurations, and MMMC for indirect-adaptive configurations. Consequently, the terms *supervisory scheme* and *supervisory framework* can be used interchangeably for frameworks that employ only one scheme. The inclusion of these schemes into a single framework implies that the resulting framework can be used to generate the supervisory mechanisms for control systems exhibiting one of the direct, indirect, and auto-tuning control configurations. This extends the coverage of a framework, but only in a limited sense since there is no provision for supervising a control system exhibiting more than one control configuration. Unification from the framework perspective is simply achieved by using the framework union operation formalized in Definition 35.

From a hierarchical perspective, the unification of supervisory schemes refers to their co-existence in a single supervisory control system. This provides the supervisory solution for a candidate control system involving multiple control configurations. When each supervisory scheme is applied, it produces an orphaned supervisor. As there can only be one orphaned supervisor in a system, a hierarchical supervisor is subsequently required to provide coordination between all the supervisors. Several relationships can be defined between frameworks, such as the subset, equivalence relationship, and the union operation.

Definition 33: Framework Subset Relationship

A framework $\mathcal{F}_i = \langle \mathcal{P}_i^*, \mathcal{H}_i^* \rangle$ is a subset of another framework $\mathcal{F}_i = \langle \mathcal{P}_j^*, \mathcal{H}_j^* \rangle$, denoted by $\mathcal{F}_i \subseteq \mathcal{F}_j$, if the criterion in Equation VI.7.131 is satisfied:

$$\mathcal{F}_i \subseteq \mathcal{F}_j \Leftrightarrow \mathcal{P}_i^* \subseteq \mathcal{P}_j^* \wedge \mathcal{H}_i^* \subseteq H_j^* \tag{VI.7.131}$$

Definition 34: Framework Equivalence Relationship

Two frameworks $\mathcal{F}_i = \langle \mathcal{P}_i^*, \mathcal{H}_i^* \rangle$ and $\mathcal{F}_i = \langle \mathcal{P}_j^*, \mathcal{H}_j^* \rangle$ are equivalent, denoted by $\mathcal{F}_i = \mathcal{F}_j$, if the criterion in Equation VI.7.132 is satisfied:

$$\mathcal{F}_i = \mathcal{F}_j \Leftrightarrow \mathcal{F}_i \subseteq \mathcal{F}_j \wedge \mathcal{F}_j \subseteq \mathcal{F}_i \tag{VI.7.132}$$

Definition 35: Union of Frameworks

The union of two frameworks $\mathcal{F}_i = \langle \mathcal{P}_i^*, \mathcal{H}_i^* \rangle$ and $\mathcal{F}_i = \langle \mathcal{P}_j^*, \mathcal{H}_j^* \rangle$ produces another framework, denoted by $\mathcal{F}_k = \mathcal{F}_i \cup \mathcal{F}_j$, where the individual terms of $\mathcal{F}_k = \langle \mathcal{P}_k^*, \mathcal{H}_k^* \rangle$ are computed as in Equations VI.7.133 and VI.7.134.

$$\mathcal{P}_k^* = \mathcal{P}_i^* \cup \mathcal{P}_j^* \tag{VI.7.133}$$

$$\mathcal{H}_k^* \subseteq \mathcal{H}_i^* \cup \mathcal{H}_j^* \tag{VI.7.134}$$

Equation VI.7.133 states that the new framework $\mathcal{F}_k$ is comprised of pattern–augmentation pairs from both $\mathcal{F}_i$ and $\mathcal{F}_j$. The properties produced are denoted by $\mathcal{H}_k^*$ in Equation VI.7.134. The specific properties in $\mathcal{H}_k^*$ require individual assessment. The union operation can be used to develop progressively more sophisticated frameworks from simpler ones and, consequently, provides the mechanism for the unification of different supervisory schemes.

7.5   Conclusion

A pattern-based approach to the design of a supervisory control system has been proposed. Such an approach allows sharing the knowledge of experienced system designers as documented within a catalog of supervisory schemes. The concept of patterns is different from the concept of software or component libraries, which are usually available in control system design packages, in that they represent the context in which a component or a group of components is employed in addition to describing their processing capability. This chapter has proposed a mathematical description for a system architecture, architectural patterns, and frameworks to facilitate the application of such a design approach.

7.5.1   Benefits

The use of SPADE permits architectural analysis and reasoning during system design, which, until now, has received limited attention. Architectural analysis allows characterizing high-level design considerations without being hampered by low-level behavioral characteristics. This concept constitutes the foundational basis for SPADE. This work demonstrates how the architectural representations can be used to perform further analysis and reasoning, such as characterizing different control configurations and performing architectural inference to construct a supervisory control system.

One of the distinguishing features of SPADE is the mapping of vector elements into ports and connectors into composition operations, which permits a direct mapping of dynamic systems formalism. Moreover, vector elements are familiar concepts for system designers. The use of composition operations abstracts the interaction between components and allows more sophisticated components to be composed. This is extended to the concept of configuration formulae, which describe interactions and provide an abstraction of the concept of connectors. In addition, architectural descriptions are represented as lists — components, configuration formulae, and constraints. The use of such techniques in abstracting architectures is an important contribution of this work as it allows the matching of system architectures to be greatly simplified. Other benefits of SPADE are discussed in the following paragraphs.

One of the drawbacks of existing supervisory schemes is their informal treatment of the system architecture and organization, which makes it difficult to separate the roles and responsibilities assigned to different components and results in a monolithic implementation. The requirement to provide an enforceable role separation is fulfilled through SPADE, which distinguishes between three primary component roles — algorithm, monitor, and supervisor. These roles define the responsibilities of a component and the interactions allowed with other components to fulfill these responsibilities. Specifically, a number of concepts have been mathematically formalized so their application can be facilitated, verified, and enforced, namely:

- The separation of computational and supervisory behaviors
- The naming of primary component roles — algorithm, monitor, and supervisor
- The definition of composition operations that restrict the type of interactions permitted between two components
- The definition of system properties to verify role separation — such as the consistent or neat properties
- The use of configuration formulae and architectural patterns to define incrementally more complex roles

One of the issues raised in expert control[22] is the fact that a purely rule-based representation of supervisory logic may be inadequate. Accordingly, the formalism in SPADE is devised to be independent of the underlying representation. Even though a discrete time representation is employed within the formulation of the computational and supervisory behaviors, it is not a requirement for the validity of SPADE, as attachments and configuration formulae, which are the fundamental concepts in SPADE, do not require the time dimension. Rather, they are formulated based on the semantics assigned to each element within SPADE. Behavioral formalism has its own representation of input, output, and state, and the concept of parameters is general as well. It is on such general grounds that the different compositional operations are defined, and, thus, generality across different underlying representations is achieved.

The improvement of system performance requires the parameters of some components to be adjusted. This implies that a viable behavioral representation must explicitly represent the parameters of algorithms and how they affect system conditions. In SPADE, the explicit representation of algorithm parameters is achieved by including the parameter vector within each component, allowing the separation of the update composition from the supervisory composition. That is, SPADE distinguishes between a component that actually updates or modifies the parameters of some other component from one that supervises such update activity. Consequently, the parameterization of components has the additional benefit of enforcing role separation between components. This is important as the existing literature often mingles these two activities and, consequently, employs monolithic components with multiple responsibilities. In addition, the representation of plant parameters provides the basis for describing system conditions and evaluating the effectiveness of different algorithms.

7.5.2 Future Work

In this chapter, a system is described from the architectural perspective. However, the behavioral and classification perspectives must be taken into account to provide a more complete representation of the system and patterns. The behavioral perspective, on the one hand, focuses on the dynamic aspects of each component and their interaction, i.e., how the states of each component evolve over time. One of the drawbacks of a purely architectural description is the lack of support for analyzing system behavior, which restricts the use of architectural descriptions to a specification language. Consequently, an ADL is typically comprised of a formal semantic theory as part of its underlying mechanism for characterizing architectures, to enhance its capability to model particular types of systems or particular aspects of a given system. By making SPADE independent of the underlying formalism, it is possible to incorporate a number of behavioral representations from systems theory. The classification perspective, on the other hand, focuses on the performance specification of the system and the remedial actions to be taken when these specifications are not met. The explicit introduction of monitoring components permits the encapsulation of classification logic so that work along this line can proceed independently. Separate publications currently under preparation will demonstrate how SPADE is able to support these two perspectives and integrate them into the discussed pattern-based approach to system design. Finally, the automation of the pattern matching process is also currently under investigation.

References

1. Lin, C.F., *Advanced Control Systems Design*, Prentice Hall, Upper Saddle River, NJ, 1994.
2. Isermann, R. and Lachmann, R., Parameter-adaptive control with configuration aids and supervision function, *Automatica*, 21(6), 1985.
3. Doyle, J.C., Francis, B.A., and Tennenbaum, A.R., *Feedback Control Theory*, Macmillan, New York, 1992.
4. Åström, K.J. and Wittenmark, B., *Adaptive Control*, Addison Wesley, Reading, MA, 1989.

5. Narendra, K.S. and Annaswarmy, A.M., *Stable Adaptive Systems*, Prentice Hall, Upper Saddle River, NJ, 1989.

6. Sastry, S. and Bodson, M., *Adaptive Control: Stability, Convergence, and Robustness*, Prentice Hall, Upper Saddle River, NJ, 1989.

7. Isermann, R., Lachmann, K.H., and Matko, D., *Adaptive Control Systems*, Prentice Hall, UK, 1992.

8. Scherer, W.T. and White, C.C., III, A survey of expert systems for equipment maintenance and diagnostics, in *Knowledge-Based System Diagnosis, Supervision and Control*, Tzafestas, S.G., Ed., Plenum Press, New York, 1989, 285.

9. Shen, Q., Fuzzy qualitative simulation and diagnosis of continuous dynamic systems, Ph.D. thesis, Heriot-Watt University, Edinburgh, 1991.

10. Isermann, R., Supervision, fault-detection and fault diagnosis methods—an introduction, in *Control Engineering Practice*, Vol. 5, No. 5, Elsevier Science, England, 1997, 639.

11. Cembrano, G. and Wells, G., Neural networks for control, in *Application of Artificial Intelligence In Process Control*, Boullart, L., Krijgsman, A., and Vingerhoeds, R.A., Eds., Pergamon Press, Oxford, UK, 1993, 388.

12. Jager, R., Adaptive fuzzy control, in *Application of Artificial Intelligence In Process Control*, Boullart, L., Krijgsman, A., and Vingerhoeds, R.A., Eds., Pergamon Press, Oxford, UK, 1993, 223.

13. Schiffman, W.H. and Geffers, H.W., Adaptive control of dynamic systems by backpropagation networks, *Neural Networks*, 6, 517, 1993.

14. Harris, C.J., Editor's introduction, in *Advances in Intelligent Control*, Harris, C.J., Ed., Taylor and Francis, London, UK, 1994, 1.

15. Kim, J.W., Moon, Y.K., and Zeigler, B.P., Designing fuzzy net controllers using genetic algorithms, *Control Syst.*, 15(3), 66, 1995.

16. Passino, K.M. and Yurkovich, S., *Fuzzy Control*, Addison Wesley, Reading, MA, 1998.

17. Eryurek, E. and Upadhyaya, B.R., Fault-tolerant control and diagnosis for large-scale systems, *Control Syst.*, 15(5), 34, 1995.

18. Gamma, E. et al., *Design Patterns: Elements of Reusable Object-Oriented Software*, Addison Wesley, Reading, MA, 1995.

19. Devedzic, V., Ontologies: borrowing from software patterns, *Intelligence New Vision AI Prac.*, 10(3), 122, 1999.

20. Srinivasan, S., Design patterns in object-oriented frameworks, *IEEE Comput.*, 32(2), 1999.

21. Årzén, K.E., Realization of expert system based feedback control, Ph.D. thesis, LUTFD2/ (TFRT-029)/1-199/(1987), Lund Institute of Technology, Sweden, 1987.

22. Årzén, K.E. and Åström, K.J., Expert control and fuzzy control, in *AI Symposium Series*, Davis, J.F., Stephanopoulos, G., and Venkatasubramanian, V., Eds., Vol. 92, No. 312, American Institute of Chemical Engineers, 1996, 47.

23. Quek, H.C., The application of artificial intelligence techniques to the integrated control of complex dynamic physical systems, Ph.D. thesis, Herriot-Watt University, Edinburgh, 1990.

24. Leitch, R. and Quek, H.C., A behavior classification for integrated process supervision, *IEE Proc. 3rd Int. Conf. Control*, 1, 127, 1991.

25. Leitch, R. and Quek, H.C., Architecture for integrated process supervision, *IEE Proc.-D Control Theory Appl.*, 139(3), 317, 1992.

26. Quek, H.C. and Leitch, R., Direct method for model reference adaptive PI controller using the gradient approach, *Proc. IEEE Region 10 Int. Conf. Tencon 93*, 4, 447, 1993.

27. Narendra, K.S., Balakrishnan, J., and Kemal Ciliz, M., Adaptation and learning using multiple models, switching and tuning, *Control Syst.*, 15(3), 1995.

28. Murray-Smith, R. and Johansen, T.A., *Multiple Model Approaches to Modelling and Control*, Taylor and Francis, London, UK, 1997.

29. Shaw, M. and Garlan, D., *Software Architecture: Perspectives on an Emerging Discipline*, Prentice Hall, Upper Saddle River, NJ, 1996.

30. Monroe, R.T. et al., Architectural styles, design patterns, and objects, *IEEE Software*, 14(1), 1997.

31. Voss, H., Architectural issues for expert systems in real-time control, IFAC Artificial Intelligence in Real-Time Control, Swansea, UK, 1989, 1.

32. Vepa, R., Monitoring and fault diagnosis in control engineering, in *Application of Artificial Intelligence In Process Control*, Boullart, L., Krijgsman, A., and Vingerhoeds, R.A., Eds., Pergamon Press, Oxford, UK, 1993, 456.

33. Acar, L. and Ozguner, U., Design of structure-based hierarchies for distributed intelligent control, in *An Introduction to Intelligent and Autonomous Control*, Antsakis, P.J., and Passino K.M., Eds., Kluwer Academic Publishers, Dordrecht, 1993, 79.

34. Meystel, A., Multi-resolutional recursive design operator for intelligent machines, in *Proceedings of the 1991 IEEE International Symposium on Intelligent Control*, Chicago, IL, August 1993, 42.

35. Meystel, A., Multi-resolutional feed-forward/feedback loops, Proceedings of the 1991 IEEE International Symposium on Intelligent Control, Chicago, IL, August 1993, 85.

36. Albus, J.S., A reference model architecture for intelligent systems design, in *An Introduction to Intelligent and Autonomous Control*, Antsakis, P.J., and Passino K.M., Eds., Kluwer Academic Publishers, Dordrecht, 1992, 27.

37. Huang, H.M. An architecture and a methodology for intelligent control, *IEEE Expert Intelligent Syst. Appl.*, 11(2), 28, 1996.

38. Huang, H.M. et al., Intelligent system control: a unified approach and applications, in *Gordon and Breach International Series in Engineering, Technology and Applied Science*, Volumes on Expert Systems Techniques and Applications, 1998.

39. Messina, E. et al., Representation of the RCS reference model architecture using an architectural description language, Proceedings of the 7th International Workshop on Computer Aided Systems Theory and Technology 1999, Vienna, Austria, Sept. 29–Oct. 2, 1999.

40. Dabrowski, C. et al., Formalizing the NIST D/RCS reference model architecture using an architectural description language, NISTIR 6443, National Institute of Standards and Technology, Gaithersburg, MD, December 1999.

41. Graves, A.R. and Czarnecki, C., Design patterns for behavior-based robotics, *IEEE Trans. Syst. Man Cybernetics—Part A Syst. Humans*, 30(1), 36, 2000.

42. Garlan, D., Monroe, R., and Wile, D., Acme: an architecture description interchange language, in Proceedings of CASCON 97, November 1997.

43. Medvidovic, N. and Taylor, R.N., A framework for classifying and comparing architecture description languages, Fifth ACM SIGSOFT Symposium on the Foundations of Software Engineering, Zurich, Switzerland, September 22–25, 1997, 60.

44. Medvidovic, N., A classification and comparison framework for software architecture description languages, Technical Report UCI-ICS-97-02, Department of Information and Computer Science, University of California, Irvine, February 1997.

45. Leigh, J.R., *Functional Analysis and Linear Control Theory*, Academic Press, UK, 1980.

46. Wonham, W.M., A control theory for discrete-event systems., in *Advanced Computing Concepts and Techniques in Control Engineering*, Denham, M.J. and Laub, A.J., Eds., Springer-Verlag, New York, 1988, 129.

47. Pasquier, M. and Quek, H.C., Integrated supervision: a generic component-based architecture for intelligent control systems, Proceedings of the IEEE 11th International Conference on System, Man and Cybernetics, Vol. 3, Beijing, China, 1996, 2333.

8

Feature-Based Integrated Design of Fuzzy Control Systems

8.1 Introduction VI-253
8.2 Synthesis of Fuzzy Control VI-254
8.3 Design Philosophy — A Feature-Based Hybrid
    Methodology VI-256
8.4 A Qualitative Approach VI-257
    Phase Plane Techniques • Nominal Rule Base
8.5 A Quantitative Approach VI-259
    Mathematical Model of Rule Base • Mathematical Model
    of Fuzzy Control • Dual Features • Design by Sliding •
    Design by PID — One Step Toward Autotuning
8.6 A Computer Simulation VI-271
    Dual Features of FLC • One Step Toward Autotuning of
    FZ-PID
8.7 Conclusions VI-281
Acknowledgment VI-281
References VI-281

Han-Xiong L
City University of Hong Kong

Guanrong Chen
City University of Hong Kong

8.1 Introduction

The basic idea of fuzzy logic control (FLC) was introduced by Zadeh in 1973[34] and then applied by Mamdani[15] in an attempt to control systems that are structurally too complex to model. Since then, FLC has become one of the most active and fruitful research areas in fuzzy mathematics and fuzzy systems theory, with many industrial applications reported.[6,10] The experience gained over the past three decades has shown that FLC may offer a preferable method for designing controllers for dynamic systems, even in most cases where traditional methods can be used.[16]

FLC can be classified into many different types.[30] The error feedback control is one of the most popular types studied extensively, which is named conventional FLC in this chapter. This conventional FLC usually uses the PID type structures for low-level control.[4] Just like its linear counterpart, there are PD-type FLC (FZ-PD),[20] PI-type FLC (FZ-PI),[33] and various PID-type FLC (FZ-PID),[17,21,27] aiming at different applications.

Although systematic analysis and design for FLCs are still considered premature, significant progress has been made in this pursuit. The design and analysis of conventional FLCs have been extensively studied to date, especially for the comparison between fuzzy control and conventional control.[1,4,7,8,19,23] However, many of these analyses and comparisons are qualitative or descriptive due to the lack of a universal mathematical model of the inference logic. A quantitative model for the max-min inference logic is first introduced by Ying[31] for quantitative analysis of the Mamdani-type FLC. A simplified model is presented later on in this chapter, to improve this kind of existing quantitative analysis.[12] Different features have been discovered through these quantitative approaches, notably including both the simple linear control feature and the variable structure control (VSC) feature. However, this sort of investigation is still incomplete, and there is a need for a thorough study of the subject, striving for a better understanding of the relationship between these features and the parameters of an FLC via a rigorous mathematical approach.

Autotuning a PID controller is already in the mature phase. However, tuning an FLC is, in most cases, performed on an experimental basis.[5] A common perception is that it is more difficult than tuning a conventional controller. Although extensive studies have been carried out in the last twenty years or so,[3,18,22,24,26,28,33] there is no systematic and simple tuning method for FLC today. The main reason is that there are too many parameters in an FLC that can affect the control performance.

This chapter aims to provide a more systematic, feasible, and application-oriented design methodology for fuzzy control systems by integrating different types of knowledge and methods from different applications. The methodology consists of two different approaches: nominal design and optimal tuning. Emphasis is placed on nominal design in this chapter, while the optimal tuning phase is still open for further research.

The nominal design phase employs a top-down approach. First, a qualitative method is introduced to construct the nominal rule base that is qualitative in nature. Using the standard triangle membership functions, the mathematical model of the linear rule base can be derived, followed by a module of FZ-PID control. The nominal design of FLC actually becomes a scaling gain design, and it proves that FZ-PID control is actually a quasi-sliding control with dual features from its linear counterpart and the sliding of the VSC. These features are adjustable by tuning the scaling gains and membership functions. As the sliding control is usually advantageous for complex processes while the linear control is better for simpler ones, the influence of these dual features on the FLC design and tuning is significant. Thus, nominal gains designs of FZ-PID can be based either on the well-tuned linear PID control or on VSC theory. The former design even leads to a possible autotuning.

A mathematical connection is studied between the lower-level scaling gains and the three higher-level proportional, integral, and derivative control action gains. A less-coupled gain structure is then designed to disclose the influence of each scaling gain to the different performance features. Although all these relationships are developed for FZ-PID around the system equilibrium state under certain approximations, practically, they can be applied to a global operational region. This is just like the classical control theory developed for linear systems that can practically work for many nonlinear processes. Thus, the conventional Ziegler–Nichols method for PID gain-tuning can be applied to the nominal scaling gain design and the fine-tuning thereafter. Simulations are included to demonstrate the dual features of FLCs and the feasibility of the autotuning strategy, thereby confirming the viability of the nominal design approach described in this chapter.

8.2   Synthesis of Fuzzy Control

In conventional two-term control, there are PD and PI controls, which can be expressed mathematically as

$$\mathbf{u}^{PD} = \mathbf{K_P}e + \mathbf{K_D}\dot{e} = \mathbf{K_P}(e + \mathbf{T_d}\dot{e})$$

$$\mathbf{u}^{PI} = \mathbf{K_P}e + \mathbf{K_I}\int edt = \mathbf{K_P}\left(e + \frac{1}{\mathbf{T_i}}\int edt\right) \tag{VI.8.1}$$

where e is the tracking error. The above control is a sum of different control actions. The proportional gain $\mathbf{K_P}$, integral gain $\mathbf{K_I}$, and derivative gain $\mathbf{K_D}$ represent the strengths of different control actions. The relationships between these control parameters are:

$$\mathbf{K_I} = \mathbf{K_P}/\mathbf{T_i}$$

$$\mathbf{K_D} = \mathbf{K_P}\mathbf{T_d}$$

where $\mathbf{T_i}$ and $\mathbf{T_d}$ are the integral and derivative times. Boldfaced symbols refer to conventional control parameters in this chapter, in contrast to the non-boldfaced symbols for fuzzy control parameters.

In fuzzy control, there is an analogous PD-type FLC (FZ-PD) and a PI-type FLC (FZ-PI). These two-term FLCs (i.e., FZ-PD/PI controllers) are well established and accepted in both scientific research and industrial applications. Their basic structures are shown in Figure VI.8.1, with $p = d/dt$ as the derivative operator, $1/p$ as the integral operator, K_e and K_d as input scaling gains, and K as the output scaling gain.

One of the most popular conventional controllers is the three-term PID control,[2] which can be expressed mathematically as

$$\mathbf{u}^{PID} = \mathbf{K_P}e + \mathbf{K_I}\int edt + \mathbf{K_D}\dot{e}$$

$$= \mathbf{K_P}\left(e + \frac{1}{\mathbf{T_i}}\int edt + \mathbf{T_d}\dot{e}\right) \qquad (VI.8.2)$$

Analogously, there are many versions of PID-type FLC (FZ-PID) control systems.[17] Since FZ-PI is good for the steady-state response and FZ-PD is good for the transient response, a combination

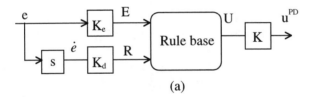

(a)

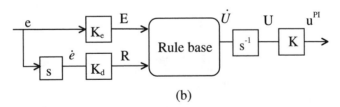

(b)

FIGURE VI.8.1 Structures of fuzzy two-term control: (a) FZ-PD; (b) FZ-PI.

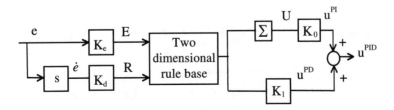

FIGURE VI.8.2 Structure of FZ-PID.

of these two should be able to achieve a good performance in both the transient and steady-state responses. For this reason, a simple two-input FZ-PID is proposed for the present discussion, as shown in Figure VI.8.2. This controller combines the features of both the FZ-PD and FZ-PI, with only one two-dimensional rule base.

8.3   Design Philosophy — A Feature-Based Hybrid Methodology

The main difficulty with fuzzy control is the absence of a systematic methodology for design and tuning due to the following reasons.

- Too many parameters exist in the knowledge base which come from different domains (either knowledge or control engineering) and possess different features (either quantitative or qualitative)
- Design and tuning are always related

Because the difficulty in design comes mainly from the coupling of parameters in the knowledge base, an FLC design can follow the hybrid design methodology by integrating both qualitative and quantitative approaches and both knowledge and control engineering. This is visualized in Figure VI.8.3. Two separate stages are suggested for design and tuning. The nominal design employs a top-down approach, from the qualitative level (higher level) to the quantitative level (lower level). It intends to find out the nominal model of the FLC, i.e., the initial parameters of the FLC. If the nominal model is not satisfactory, then fine-tuning can be used to explore finer parameters of the FLC through a bottom-up approach that continuously learns from the nominal model.

The nominal model of an FLC includes the nominal rule base and nominal database that contains membership functions (MFs) and scaling gains. The rule base forms qualitative control knowledge and, in this sense, conveys a general linguistic policy that needs to be determined beforehand and then sustained.[25] The nominal rule base should be designed qualitatively due to its qualitative nature. The database provides the necessary numerical calibration required for coping with more specific control situations. Obviously, scaling gains are of quantitative nature and, therefore, should be designed quantitatively. MFs bridging between rules and gains are of mixed nature. For simplicity, the nominal MFs are chosen as triangular. After determining the nominal rule base and MFs, all the design loads are shifted to the scaling gains that can be handled with various quantitative methods. Properly designed scaling gains are very critical to the nominal performance of the FLC.

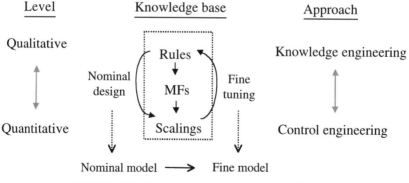

FIGURE VI.8.3   Hybrid methodology for FLC design.

8.4 A Qualitative Approach

Since the conventional FLC originates from control engineering rather than artificial intelligence, its knowledge base may be narrower than other types.[35] Actually, the rule base can be obtained independently of the process, and that is why a model-free FLC is possible. Based on some general knowledge about the dynamics (not necessarily the structure) of the process, a new methodology, the phase plane technique, is introduced here for systematic design of a two-dimensional rule base.

8.4.1 Phase Plane Techniques

The inputs to the rule base of the fuzzy two-term control are the error E and the change in error R. As shown in Figure VI.8.4, the time step response of the closed-loop system can be approximately classified into four areas, $A_1 \sim A_4$, and two sets of points, cross-over $\{b_1, b_2\}$ and peak-valley $\{c_1, c_2\}$.

The response area is defined as

$$A_1: E > 0 \text{ and } R < 0 \qquad A_2: E < 0 \text{ and } R < 0$$
$$A_3: E < 0 \text{ and } R > 0 \qquad A_4: E > 0 \text{ and } R > 0$$

and the cross-over points as

$$b_1 : \ E > 0 \to E < 0, \text{ when } R < 0 \qquad b_2 : \ E < 0 \to E > 0, \text{ when } R > 0$$

The peak-valley points are defined as

$$c_1 : \ R = 0, \ E < 0 \qquad c_2 : \ R = 0, \ E > 0$$

In classical control, phase plane is a useful tool for analyzing the stability of a nonlinear system. In fuzzy control, this error state space can also act as a bridge between the system performance and the rule base. The system equilibrium point is the origin of the phase plane. The rule base is actually constructed on the phase plane with E as the horizontal axis and R as the vertical axis. The rule base is then structured within the above four areas and with two sets of points, as shown in Figure VI.8.4.

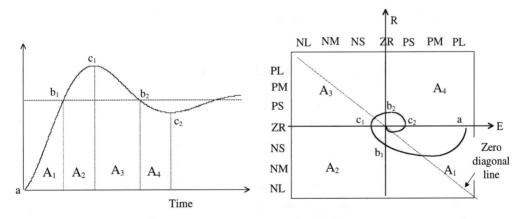

FIGURE VI.8.4   Mapping from time domain to rule base via state space.

8.4.2   Nominal Rule Base

The following metarules for rule base design can be obtained based on the fundamental features of R:

- Rule 1: If both E and R are zero, maintain the present control
- Rule 2: If E goes to zero at a satisfactory rate, maintain the present control
- Rule 3: If E is not self-correcting, more subrules are needed for the variation of the control output

By analyzing the phase-plane trajectory, one can decompose Rule 3 as:

- 3.1 Rules for cross-over points $\{b_1, b_2\}$ should prevent the overshoot in A_2/A_4:

  Since $E = 0$ and $R \neq 0$, non-zero R may cause the overshoot.
- 3.2 Rules for peak-valley points $\{c_1, c_2\}$ should speed up the response:

  Since $R = 0$, it cannot provide the self-correction.
- 3.3 Rules for area A_1 vary:

  - Above the diagonal line, R has less self-correction, so rules should speed up the response.
  - On the diagonal line, R has enough self-correction, so Rule 2 is used.
  - Below the diagonal line, R has over self-correction, so rules should prevent the possible overshoot in A_2.

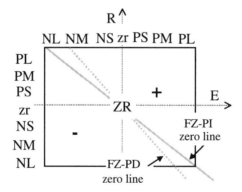

FIGURE VI.8.5   Rule base pattern of fuzzy two-term control.

R＼E	NL	NM	NS	ZR	PS	PM	PL
PL	zr	ps	pm	pl	pl	pl	pl
PM	ns	zr	ps	pm	pl	pl	pl
PS	nm	ns	zr	ps	pm	pl	pl
ZR	nl	nm	ns	zr	ps	pm	pl
NS	nl	nl	nm	ns	zr	ps	pm
NM	nl	nl	nl	nm	ns	zr	ps
NL	nl	nl	nl	nl	nm	ns	zr

u=0

FIGURE VI.8.6   Linear rule base.

- 3.4 Rules for area A_2 should decrease the overshoot around the peak:

Since $RE > 0$, R will enlarge the overshoot $|E|$.

- 3.5 Rules for area A_3 mirror the rules for area A_1:

  - Above the diagonal line, R has over self-correction, so rules should prevent the possible overshoot in A_4.
  - On the diagonal line, R has enough self-correction, so Rule 2 is used.
  - Below the diagonal line, R has less self-correction, so rules should speed up the response.

- 3.6 Rules for area A_4 should decrease the overshoot around the valley for the same reason as for Rule 3.4 (in area A_2).

These qualitative rules cover the complete space of the rule base and provide a consistent and continuous relationship between two adjacent locations. The pattern of the rule bases for FZ-PI and FZ-PD can be easily derived, as shown in Figure VI.8.5. The slight difference between these two rule bases is the switching line at which the sign of rules changes. Since the difference is not large, for simplicity, the linear rule base shown in Figure VI.8.6 could be used as the nominal rule base for both systems and for FZ-PID as well.

8.5 A Quantitative Approach

After designing the nominal rule base, the nominal MFs can be simply chosen to be the standard triangular. Then, all the design loads are shifted to scaling gains that can be handled by various quantitative methods. Properly designed scaling gains become very critical to the nominal performance of FLC. A quantitative method proposed by Ying[32] was used and modified[12] to obtain a mathematical model of the linear rule base, which will also help explore fundamental features of FLC and can be used in the gain design.

8.5.1 Mathematical Model of Rule Base

Generally, a rule base can be divided into many *inference cells* (ICs) with output rules on four corners, as shown in Figure VI.8.7. Any input data (E, R) to the rule base can be mapped from the local input $(e^*, \dot{e}^*)$ into the inference cell $IC(i, j)$ via the following formulas:

$$E = iA + e^*, \qquad R = jA + \dot{e}^*, \qquad (i, j = -M, \dots, -1, 0, 1 \dots, M)$$

Every IC region has four subregions, $(IC_1 \sim IC_4)$, as shown in Figure VI.8.8. Here, one of the diagonal lines is called an S-line, on which all the points have the same distance to the ZR diagonal line of the rule base. Fuzzification can be easily carried out on $IC(i, j)$, yielding a grade $(\mu_i, \mu_{i+1}, \mu_j, \mu_{j+1})$ for every input to the related MF on $(i, i+1, j, j+1)$.

The inference operation can also be carried out on the cell to create a crisp output from the rule base. The setup is as follows:

- Inference is based on Mamdani's max-min method
- The MFs of triangular shape are equally spread with width A for input MFs and width B for output MFs
- Defuzzification is based on the center-of-gravity method

Based on this setup, the crisp output U_l on region IC_l is obtained as follows:[12]

$$U_l = kB(1 - \gamma_l) + \frac{B}{A}\gamma_l S, \qquad (l = 1, 2, 3, 4) \tag{VI.8.3}$$

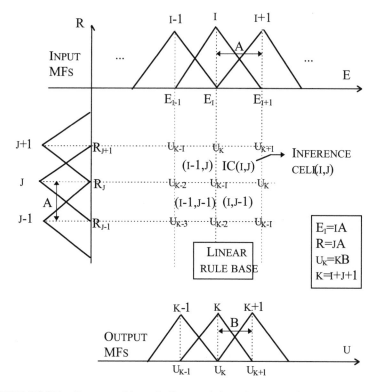

FIGURE VI.8.7   Decomposition of a linear rule base into a set of inference cells (ICs).

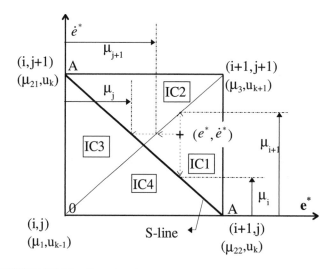

FIGURE VI.8.8   Functional composition of the inference cell $IC(i, j)$.

where

$$\gamma_1 = (1 + \mu_i)^{-1} = (2 - e^*/A)^{-1}$$

$$\gamma_2 = (1 + \mu_j)^{-1} = (2 - \dot{e}^*/A)^{-1}$$

$$\gamma_3 = (1 + \mu_{i+1})^{-1} = (1 + e^*/A)^{-1}$$

$$\gamma_4 = (1 + \mu_{j+1})^{-1} = (1 + \dot{e}^*/A)^{-1}$$

$$k = i + j + 1$$

$$S = K_d(\lambda e + \dot{e})$$

The gain γ_l is a nonlinear function in region $IC_l(2/3 \leq \gamma \leq 1)$.[12] For simplicity, the index l is omitted throughout this chapter. The model of the rule base will be briefed as

$$U = kB(1 - \gamma) + \frac{B}{A}\gamma S$$

$$= kB(1 - \gamma) + \frac{B}{A}\gamma K_d(\lambda e + \dot{e}) \tag{VI.8.4}$$

8.5.2 Mathematical Model of Fuzzy Control

The mathematical model of the fuzzy PD/PI controller shown in Figure VI.8.1 can be derived approximately as

$$u^{PD} = KU = KkB(1 - \gamma) + K\frac{B}{A}\gamma K_d(\lambda e + \dot{e}) \tag{VI.8.5}$$

$$u^{PI} = K\int U dt = KkB\int(1 - \gamma)dt + K\frac{B}{A}\gamma K_d\left(e + \int \lambda e dt\right) \tag{VI.8.6}$$

These models show that FZ-PD/PI control is a combination of a nonlinear PD/PI control component and a multi-level relay component, $kb(1 - \gamma)$, $k = 0, \pm 1, \pm 2, \ldots$.

Based on the expression from Equation VI.8.4, the mathematical model of the three-term FZ-PID in Figure VI.8.2 can be expressed as

$$u^{PID} = \{kB(1 - \gamma) + \frac{B}{A}\gamma S\}(K_0 + K_1 p)\frac{1}{p}$$

$$= u_r + u_3 \tag{VI.8.7}$$

with

$$p = \frac{d}{dt}, \quad \frac{1}{p} = \int dt$$

where the multi-level relay component u_r is

$$u_r = kB\{K_0 - (\gamma K_0 + \dot{\gamma} K_1)\}\frac{1}{p}$$

and the nonlinear three-term control component u_3 is

$$u_3 = \frac{B}{A}\{(\gamma K_0 + \dot{\gamma} K_1)S + \gamma K_1 \dot{S}\}\frac{1}{p}$$

By replacing $S = E + R = K_d(\lambda e + \dot{e})$, and letting $\beta = \frac{K_1}{K_0}$, u_3 is actually a nonlinear PID:

$$u_3 = \frac{B}{A}\{(\gamma K_0 + \dot{\gamma} K_1)E + [(\gamma K_0 + \dot{\gamma} K_1) + \gamma K_1]R + \gamma K_1 \dot{R}\}\frac{1}{p}$$

$$= \frac{B}{A}\gamma K_d K_0 \left\{[(1 + \frac{\dot{\gamma}}{\gamma}\beta) + \beta]e + \lambda \int (1 + \frac{\dot{\gamma}}{\gamma}\beta)edt + \beta \dot{e}\right\}$$

Strictly speaking, the FZ-PID control is a nonlinear control, which consists of a nonlinear PID control component u_3 and a nonlinear multi-level relay component u_r.

8.5.3   Dual Features

Practically, the linear rule base has a limitation, as shown in Figure VI.8.6. The output is saturated when it is larger than the maximum value allowed. It has a linear control feature and also conveys relay features when the output is saturated. Based on the VSC theory, this type of rule base actually defines the following switching function:

$$S = E + R = K_d(\lambda_e + \dot{e}) \tag{VI.8.8}$$

with the output rules $U = S$.

The diagonal line of the rule base, where all output rules are zero, is actually a sliding surface $S = 0$. The rest of the rules give the linguistic distances away from the sliding surface, like $S = PS$, etc. The rule base in Figure VI.8.6 contains some linguistic features from both linear control and sliding features.

8.5.3.1   Feature of Linear Control

Equations VI.8.5 and VI.8.6 consist of a relay term plus a nonlinear PD/PI control, while Equation VI.8.7 has a relay term plus a nonlinear PID control. When $k = 0$, the input $(e^*, \dot{e}^*)$ falls into the IC on the ZR diagonal line of the rule base.[12] FLCs in Equations VI.8.5, VI.8.6, and VI.8.7 become nonlinear PD/PI and PID controls as shown in Equation VI.8.9 below. They will be closer to their linear counterparts for a slow process where the variation of γ is small. They become linear PD/PI/PID controls when the inputs $(e^*, \dot{e}^*)$ are on the IC border, where $\gamma = 1$, because there are no overlapped MFs.[12]

$$u^{PD} = K\frac{B}{A}S = K\frac{B}{A}K_d(\lambda e + \dot{e})$$

$$u^{PI} = K\frac{B}{A}S\frac{1}{p} = K\frac{B}{A}K_d\left(e + \lambda \int edt\right) \tag{VI.8.9}$$

$$u^{PID} = K_0\frac{B}{A}K_d\left\{(\lambda + 1)e + \lambda \int edt + \dot{e}\right\}$$

8.5.3.2   Feature of Sliding Control

A slight transform of Equation VI.8.4 yields a model of a two-dimensional rule base as

$$U = kB + \frac{B}{A}\gamma(S - kA) \tag{VI.8.10}$$

The rule base output U is a monotonically increasing from U_{min} through U_{mid} to U_{max},[12] as follows:

$$U_{min} = (k - 1)B \quad \text{at } S = (k - 1)A$$
$$U_{mid} = kB \quad\quad\quad \text{at } S = kA$$
$$U_{max} = (k + 1)B \quad \text{at } S = (k + 1)A$$

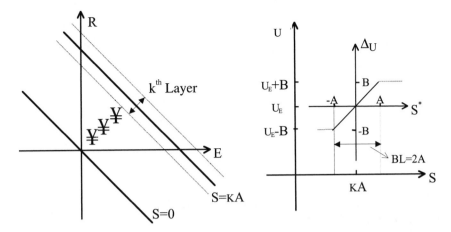

FIGURE VI.8.9   Sliding feature at the kth layer on the two-dimensional phase plane.

If the control outside the layer is approximated by $U_{\min}/U_{\max}$ in the cell IC, then the crisp output of FZ-PD/PI in Equations VI.8.5 or VI.8.6 can be written in a more general form as

$$u^{PD} = K^*U, \qquad U^{PI} = K^*U/p \qquad 1/p \text{ is the integrator} \qquad \text{(VI.8.11)}$$

where $U = u_e + u_2$ and

$$u_e = kB$$
$$u_2 = B\gamma \ \text{sat}\left(\frac{S^*}{A}\right)$$

with

$$S^* = S - kA$$
$$\text{sat}\ (S^*/A) = \begin{cases} \text{sgn}\ (S^*) & |S^*| \geq A \\ S^*/A & |S^*| \leq A \end{cases}$$

Just like the conventional VSC or sliding control,[9,29] u_e can be approximated as an equivalent control, which forces the state toward the sliding surface, and the function $\text{sat}(S^*/A)$ is actually the continuous approximation of the function $\text{sgn}(S^*)$ by introducing a boundary layer $2A$. As long as layer A is small enough, the control in Equation VI.8.11 can be viewed as a quasi-sliding control with a local sliding surface $S^* = 0$ at each layer, which is explained graphically in Figure VI.8.9. The global sliding surface for the system is still $S = 0$, and the local sliding surface is only a pseudo surface.

After defining the two-dimensional switching function as

$$S_2 = \lambda S + \dot{S} \qquad \text{(VI.8.12)}$$

the model of FZ-PID can be expressed similarly as

$$u^{PID} = K_0(u_e + u_3)\frac{1}{p} \qquad \text{(VI.8.13)}$$

where

$$u_e = kB$$

$$u_3 = B\gamma \, \text{sat} \left(\frac{S_2^*}{A_2} \right)$$

with

$$S_2^* = S_2 - kA_2, \quad A_2 = \lambda A$$

$$\text{sat} \, (S_2^*/A_2) = \begin{cases} \text{sgn} \, (S_2^*) & |S_2^*| \geq A_2 \\ S_2^*/A_2 & |S_2^*| \leq A_2 \end{cases}$$

8.5.3.3 Adjustment of Dual Features

Both qualitative and quantitative analyses show that fuzzy-PID combines the features from linear control and sliding of the VSC. These features come from the compensation term u_e and the local switching control term u_s. For simplicity, FZ-PD control is selected for illustration with detailed analysis.

Based on Equation VI.8.4, the ratio of dual features ρ of FZ-PD can be defined as

$$\rho = \frac{u_s}{u_e + u_s} \tag{VI.8.14}$$

with

$$u_e = kB \qquad\qquad \text{a compensation relay term}$$

$$u_s = B\gamma \frac{S - kA}{A} \qquad \text{a local linear (PD) control term}$$

Here, the larger ρ, the weaker the sliding effects and the stronger the PD effects. In general, when one part is stronger, the other part must be weaker and vice versa.

8.5.3.3.1 Dual Features and S (or Layer Index k)

When S is reduced, inputs are connected to a lower layer, and the compensation term u_e is smaller because of a smaller k. As the effect of this local linear control u_s is the same for every k, ρ is inversely proportional to S. Therefore, the sliding effect becomes stronger for a larger S and weaker for a smaller S.

When inputs are on the layer of the switching surface $(k = 0)$ $S \leq A$, fuzzy-PID becomes a nonlinear PID because u_e disappears. On the other hand, when S is beyond the maximum range of the rule base, fuzzy-PID becomes a pure VSC for the largest undesirable dynamics because, in this case, the linear control term disappears.

8.5.3.3.2 Dual Features and the Input IC Size A

The size of ICs is inversely proportional to the number of layers. For the same input range, a larger size A will have a smaller number of layers. Thus,

- If the input *IC* size A is large, ρ will be large because k will be small for the same S, which causes stronger linear control effects and weaker sliding effects
- If the input *IC* size A is small, ρ will be small because the relay u_e will be large for the same S, which causes weaker linear control effects and stronger sliding effects

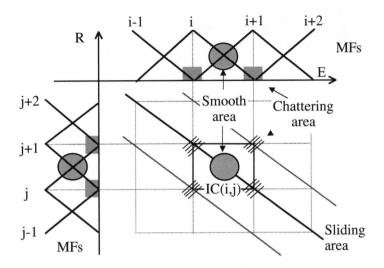

FIGURE VI.8.10 Dual features of the two-dimensional rule base.

8.5.3.3.3 *Dual Features and the Number of MFs (N)*

For the same input range, a larger number of MFs, N, will have smaller input IC size, A, which causes stronger sliding control and weaker linear control effects. On the other hand, a smaller N will have a larger A, which causes weaker sliding control and stronger linear control effects. Another well-known feature is that the overall control is more linear for a larger N because there will be more linear points with $\gamma = 1$.[12] As this linear feature only affects the switching gain γ, it should not have much influence on the sliding feature.

As the layer width varies from corners to the center, the dual features also vary, as shown in Figure VI.8.10. The central area of the IC is usually smoother due to the fact that the layers are thicker, whereas the four corners may have more chattering because the layers are thinner. This is due to the proper overlapping of input MFs, which usually provide a smooth control (e.g., input E varies smoothly between MF_i and MF_{i+1}). The area where three MFs interact may easily cause chattering (e.g., input E can easily jump from MF_i, over MF_{i+1} to MF_{i+2}).

8.5.3.3.4 *Dual Features and Input Scaling Gains*

Increasing the input scaling gain is equivalent to reducing S or using a smaller IC size (i.e., finer MFs), which, in turn, increases the linear control effects and reduces the sliding effects. Increasing the input scaling gain makes the saturation of S more easy. If S saturates, the FLC becomes a pure VSC because the linear control term disappears.

Decreasing the input scaling gain is equivalent to decreasing S or using a larger IC size (i.e., coarser MFs), which, in turn, reduces the sliding effects and increases the linear control effects. Reducing the input scaling gain makes the saturation of S more difficult. If the input is on the layer of the switching surface, the fuzzy-PID becomes a pure PID control because the VSC relay part disappears.

8.5.3.3.5 *Dual Features and Design*

The dual features of FLC provide a good guide for FLC design and tuning. It is well known that a linear controller is good for a simpler process, while the sliding control could achieve a robust performance for a more complex process with unknown disturbance. However, too much sliding effect may cause a large chattering effect. Because of the dual features, an FLC could be designed to have a robust performance for either of the above processes without much chattering.

For a simpler process, sliding control may not be very helpful. Therefore, the comparative gain-design methodology[13,33] be used to design and tune the nominal (initial) gains of the FLC, which is based on its well-tuned linear counterpart. For a complex process, the sliding effect is useful for suppressing the unknown disturbance. Thus, the nominal gains of FLC can be designed and tuned based on the VSC theory demonstrated below.

8.5.4 Design by Sliding

For a second order system,

$$\ddot{x} = f(x, t) + u \tag{VI.8.15}$$

with x_d representing a desired signal. For simplicity, only the FZ-PD in Equation VI.8.5 is discussed here.

8.5.4.1 Stability Design

Suppose the upper bound of the undesirable dynamics exists:

$$|\lambda \dot{e} + \ddot{x}_d - f|_{\max} = F \tag{VI.8.16}$$

From Equation VI.8.8, one has

$$\dot{S} = K_d(\lambda \dot{e} + \ddot{e}) = K_d\{\lambda \dot{e} + \ddot{x}_d - f - u\} \le K_d\{F - KU\} \tag{VI.8.17}$$

A Lyapunov function is chosen as

$$V(S, t) = \frac{1}{2K_d} S^2 \tag{VI.8.18}$$

giving

$$\dot{V} = \frac{1}{K_d} \dot{S} S \le S(F - KU)$$

To meet the stability requirement

$$\dot{V} < 0 \tag{VI.8.19}$$

one should have

$$K(S^*U) \ge S^*F$$

i.e.,

$$K \ge \frac{S^*F}{S^*U} \ge \frac{S^*F}{|S^*U|_{\min}} = \frac{F}{|U|_{\min}}, \qquad S^*U > 0$$

It is clear that $k = 0$ makes U in Equation VI.8.4 smaller. According to VSC theory, the relay term sgn (S), where S is the distance to the switching surface, is usually replaced by its continuous approximation S/A. In the FLC, S is also the distance to the switching surface on the rule base plane. When $k = 0$, S is in the layer where the switching surface is located. Thus, S/A can be approximated as the relay term sgn (S) when $k = 0$. Thus,

$$U_{\min} = |\gamma B \text{sgn } (S)|_{\min} = B\gamma_{\min}$$

As $\gamma_{\min} = 2/3$,[12] then $U_{\min} = 2B/3$, and

$$K \ge 1.5\frac{F}{B} \tag{VI.8.20}$$

The quasi-sliding control effect is achieved if the layer width A is sufficiently small. Larger output K will help suppress the undesirable dynamics F and enhance the stability.

8.5.4.2 Performance Design

From VI.8.8, one has

$$e = \frac{S}{K_d(p + \lambda)} \tag{VI.8.21}$$

When the input does not saturate, the input is within the layer. From Equation VI.8.17

$$\dot{S} \leq K_d \left\{ F - K(1 - \gamma)u_e - K\frac{B}{A}\gamma S \right\}$$

$$e \leq \frac{A}{KK_e B\gamma} \frac{F - K(1 - \gamma)u_e}{(1 + \frac{\rho}{\lambda})(1 + \frac{\lambda A}{KK_e B\gamma}p)} \tag{VI.8.22}$$

These performance criteria, Equations VI.8.21 and VI.8.22, can be used as guidelines for optimal tuning:

- A larger λ or KK_e can lead to the good tracking error, i.e., good performance.
- Keeping KK_e unchanged will result in the same performance.

8.5.5 Design by PID — One Step Toward Autotuning

The classical PID control in Equation VI.8.2 can be applied to process control, as shown in Figure VI.8.11.

The transfer function can be expressed as

$$\mathbf{u}(s)^{PID} = \frac{\mathbf{K_p}}{\mathbf{T_i}} \frac{(s + \mathbf{a})(s + \mathbf{b})}{s}$$

$$\mathbf{a}, \mathbf{b} = \frac{\mathbf{T_i}}{2} \left(1 \pm \sqrt{1 - \frac{4\mathbf{T_d}}{\mathbf{T_i}}} \right) \tag{VI.8.23}$$

On the root-locus plane, PID control offers two zeros, **a** and **b**, and one pole at the origin. The condition for real zeros is $\mathbf{T_i} \geq 4\mathbf{T_d}$. For a reasonably large gain, the closed-loop poles will be close to zero. Imaginary zeros will cause less damping and, hence, more oscillations than real zeros, as the resulting closed-loop poles would more likely be complex.

Proportional action can reduce the steady-state error, but too much of it can deteriorate the stability. Integral action will eliminate the steady-state error. Derivative action will improve the closed-loop stability. The parameters, $\mathbf{K_P}$, $\mathbf{T_i}$, and $\mathbf{T_d}$ are usually used for tuning by the Ziegler–Nichols frequency response method.[2] The PID parameters in terms of the ultimate gain $\mathbf{K_c}$ and the ultimate period $\mathbf{t_c}$ are presented in Table VI.8.1.

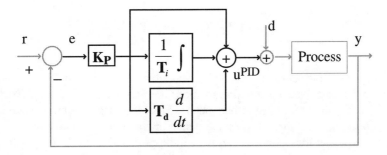

FIGURE VI.8.11   PID controller in the closed loop.

TABLE VI.8.1   PID Parameters According to
Ziegler–Nichols Frequency Response Method

K_P	T_i	T_d
$0.6\,\mathbf{K_c}$	$0.5\,\mathbf{t_c}$	$0.12\,\mathbf{t_c}$

8.5.5.1   Model Approximation by Feature Enhancement

The FZ-PID actually has two levels of gains. The scaling gains (K_e, K_d, K_0, K_1) are at the lower level; the fuzzy proportional, integral, and derivative gains (K_p, K_I, K_D) are at the higher level. These fuzzy gains represent the strengths of different control actions and directly affect the system performance. By tuning the scaling gains at the lower level, these fuzzy gains will be affected, as will the performance. As the control actions are fuzzily coupled, the contribution of each scaling gain (K_e, K_d, K_0, K_1) to different control action is still not very clear, which makes the actual design and tuning process rather difficult.

From Equation VI.8.5, it is known that FZ-PID is composed of a nonlinear relay term, u_r, and a nonlinear three-term control component, u_3. By replacing $S = E + \dot{E} = K_e(e + \alpha\dot{e})$ and letting $\beta = \frac{K_1}{K_0}$ and $C = \frac{B}{A}\gamma$, one has

$$u_3 = \frac{B}{A}\{(\gamma K_0 + \dot{\gamma} K_1)E + [(\gamma K_0 + \dot{\gamma} K_1) + \gamma K_1]\dot{E} + \gamma K_1 \ddot{E}\}\frac{1}{p}$$

$$= C K_e K_0 \left\{\left[\alpha\left(1 + \frac{\dot{\gamma}}{\gamma}\beta\right) + \beta\right]e + \int\left(1 + \frac{\dot{\gamma}}{\gamma}\beta\right)e + \alpha\beta\dot{e}\right\}$$

The three-term control has, thus, been explicitly represented. Strictly speaking, the FZ-PID control is a nonlinear control, which consists of a nonlinear three-term control component, u_3, and a nonlinear multi-level relay component, u_r. The control becomes a purely nonlinear three-term control u_3 as the nonlinear relay term u_r disappears on the switching surface $k = 0$.[12] The smaller the value of k, the more dominant the three-term control becomes.

8.5.5.1.1   *Approximations Around the Equilibrium State*
- In the area close to the equilibrium state, k can be approximated to be zero.
- When A is large or in the area close to the equilibrium state, the derivative $\dot{\gamma}$ can be assumed to be much smaller than γ itself.

The above approximations lead to $1 + \frac{\dot{\gamma}}{\gamma}\beta \approx 1$, which enhances the PID features of the FZ-PID, as

$$u^{PID} \approx u_3 \approx C K_e K_0 \left\{(\alpha + \beta)e + \int edt + \alpha\beta\dot{e}\right\} \qquad (VI.8.24)$$

Since the fuzzy system is very nonlinear, it is impossible to derive a precise and rigorous mathematical approach directly. The above approximations are very necessary for the analysis of fuzzy systems and the determination of initial parameters. First, the fuzzy system is studied in the area close to the equilibrium state, where the fuzzy system is approximated as a kind of linear system. The approach developed under these approximations will be extended to the global situation. Although this extension cannot be strictly proved by theory, its effectiveness can be confirmed practically via computer simulations.

Under the above approximations and conditions, the fuzzy-PID can be compared directly with the conventional PID control. The corresponding gains (K_p, K_I, K_D) of the three-term fuzzy control can be expressed as

$$K_p = CK_eK_0(\alpha + \beta)$$

$$K_I = CK_eK_0 \qquad \text{(VI.8.25)}$$

$$K_D = CK_eK_0\alpha\beta$$

The fuzzy integral time T_i and derivative time T_d can be obtained as

$$T_i \approx \alpha + \beta \qquad \text{(VI.8.26)}$$

$$T_d \approx \frac{\alpha\beta}{\alpha + \beta}$$

The scaling gains can be solved from Equations VI.8.25 and VI.8.26:

$$\alpha, \beta = \frac{T_i}{2}\left(1 \pm \sqrt{1 - \frac{4T_d}{T_i}}\right) \qquad \text{(VI.8.27)}$$

$$K_0 = \frac{K_p}{CK_e(\alpha + \beta)} = \frac{K_p}{CK_eT_i} \qquad \text{with } K_e = \text{constant}$$

There are two important issues in the above solutions:

1. Less oscillation: The real solutions for α and β require $T_i \geq 4T_d$. As the scaling gains are real, the condition $T_i \geq 4T_d$ is always maintained, which is equivalent to maintaining real zeros for the conventional PID. This implies that FZ-PID generally leads to more damping and, hence, less oscillations than its conventional counterparts.
2. Coupled gains: The influence of the scaling gains on the proportional, integral, or derivative action is still not clear because of their coupling.

8.5.5.2 Gain Decoupling Strategy

By applying $K_d = \alpha K_e$, the fuzzy gains in Equation VI.8.25 can be transformed into the less coupled expressions in Equation VI.8.28 as

$$K_p = CK_dK_0\left(1 + \frac{b}{a}\right)$$

$$K_I = CK_dK_0\frac{1}{a} \qquad \text{(VI.8.28)}$$

$$K_D = CK_dK_0\beta$$

With K_d held constant, the coupling effects disappear from the integral and derivative gains. The PID actions can be clearly seen in Equation VI.8.29, obtained by letting $K_p' = CK_dK_0$, so that

$$u_3 = K_p'\left\{\left(1 + \frac{\beta}{\alpha}\right)e + \frac{1}{\alpha}\int edt + \beta\dot{e}\right\} \qquad \text{(VI.8.29)}$$

The only difference between the FZ-PID Equation VI.8.29 and conventional PID control Equation VI.8.2 is the coupling effect in the proportional gain. Practically, K_p' can be approximated

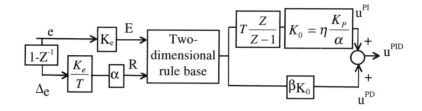

FIGURE VI.8.12   FZ-PID control with the less coupled gain structure.

TABLE VI.8.2   Parameters of PID Control and FZ-PID

	Proportional Gain	Integral Time	Derivative Time
PID	$\mathbf{K_P}$	$\mathbf{T_i}$	$\mathbf{T_d}$
FZ-PID	$K_P(K_0)$	α	$\beta(K_1)$

by K_P, and then one can obtain the approximate relationship between the scaling gains (α, β, K_0, K_1) and the more familiar three-term control parameters (K_P, T_i, T_d):

$$\alpha = T_i$$

$$b = T_d$$

$$K_0 \approx \frac{K_P}{CK_d} = \eta\frac{K_p}{\alpha} \tag{VI.8.30}$$

$$K_1 = \beta K_0$$

with $\eta = (CK_e)^{-1}$.

It is seen that α corresponds to the integral time and β to the derivative time; K_0 contributes significantly to the proportional action (K_P), and K_1 contributes more to the derivative action due to β. FZ-PID with this less coupled gain structure shown in Figure VI.8.12 can provide a clear connection between scaling gains and the more familiar three-term control parameters (K_P, K_I/T_i, K_D/T_d). In practice, the three parameters (K_P, α, β) are usually enough for design and tuning purposes. The parameter comparison with the conventional PID is shown in Table VI.8.2.

The following special cases exist:

- If the input and the ouput variables have the same numbers of MFs and the same spreads, i.e., $A = B$, then γ varies between 2/3 and 1,[12] and its average value is about 5/6.
- In practice, K_e is often chosen as unity to avoid input saturation.

Under these conditions, one obtains $\eta = 1.2$, and the output scaling K_0 can be simplified as

$$K_0 = 1.2\frac{K_p}{\alpha}$$

8.5.5.3   Gains Tuning

It is still difficult to choose initial gains for the FZ-PID. The FZ-PID in Figure VI.8.3 can be represented in a format equivalent to Figure VI.8.13, where the multi-level relay component u_r is

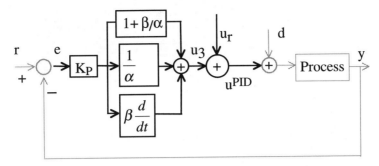

FIGURE VI.8.13   Equivalent format of FZ-PID control in the closed loop.

considered as an external input to be included into the process. Thus, only the three-term component, which is actually the nonlinear PID control, needs to be designed.

One approximate but convenient approach is to directly apply the Ziegler–Nichols method in Table VI.8.1 for obtaining the initial fuzzy gains. This is equivalent to using the well-tuned parameters $(\mathbf{K_P}, \mathbf{T_i}, \mathbf{T_d})$ of the conventional PID as the initial fuzzy parameters (K_P, T_i, T_d) of FZ-PID. Then the initial scaling gains can be calculated from Equations VI.8.27 or VI.8.30. Gain tuning should be based on the relationship between the scaling gains and the three control actions shown in Equation VI.8.30. The auto-tuning algorithm for conventional PID control[2] can be extended and developed for application to the fine-tuning of the FZ-PID.

8.6   A Computer Simulation

The aim of this computer simulation is to show the dual features of the fuzzy-PID for a process with large uncertainties and to compare it with the conventional PID on the performance of autotuning. The simulation period is 15 seconds, and the sampling step is 0.01 second. The quantitative criteria for measuring the performance are chosen as IAE and ITAE, defined by

$$IAE = \int |e| dt \qquad\qquad (VI.8.31)$$

$$ITAE = \int t|e| dt$$

8.6.1   Dual Features of FLC

The FLC is simulated to compare with the classical PD and VSC. A second-order process is chosen as in Equation VI.8.15, where the disturbance is supposed to be $f(t) = 10 \sin(t)$, and the constant desired signal (command signal) is denoted x_d. As the process has varying disturbance, the design is based on the VSC theory.

8.6.1.1   Performance of the Classical Control
8.6.1.1.1   *PD Control*
The simulation result for PD is summarized in Table VI.8.3 and illustrated in Figure VI.8.14. PD control cannot achieve a robust performance in the steady state because of the strongly varying disturbance.

8.6.1.1.2   *The Classical VSC*
As the process is a second-order system, the switching function should be first order.

$$S = \lambda e + \dot{e}$$

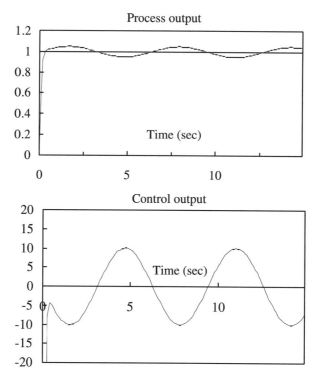

FIGURE VI.8.14   Performance of the PD.

TABLE VI.8.3   Performance Parameters of the PD

K	T_d	IAE/ITAE
200	0.1	0.59/3.74

According to Equation VI.8.16, the uncertainty bound of the undesirable dynamics is estimated as

$$F = |(\lambda - 2)\dot{e} - f|_{\max} = |(\lambda - 2)\dot{e} - 10 \sin t|_{\max} = 10 \qquad \text{(VI.8.32)}$$

Here, $\lambda = 2$ is chosen for a small value of F and a reasonably fast transient response. The control output is chosen as

$$u = (\lambda - 2)\dot{e} - \hat{f} + K \, \text{sgn}\left(\frac{S}{\phi}\right)$$

with boundary layer $2\phi = 0.1$, $K > 0$ to keep stability, and $\hat{f}$ is the estimate of the disturbance f.
Two different cases are examined:

- Case 1: $\hat{f} = f$, the disturbance is 100% known
- Case 2: $\hat{f} = 0.5f$, the disturbance is 50% known

The simulation results of VSC are summarized in Table VI.8.4 and shown in Figure VI.8.15. The gain K is chosen by trial-and-error.

TABLE VI.8.4 Performance of the VSC

Case	K	IAE/ITAE	Chattering
1	10	0.62/0.46	No
2	10	0.74/1.73	Yes

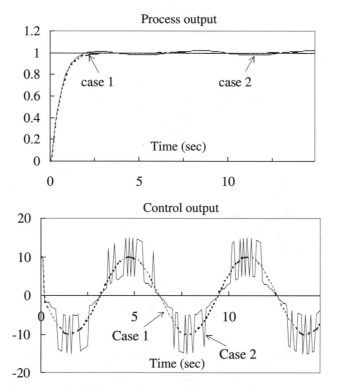

FIGURE VI.8.15 Performance of the VSC.

8.6.1.2 Performance of Fuzzy Control

Similar to the conventional VSC, the switching function is chosen as Equation VI.8.8 for FZ-PD and Equation VI.8.12 for FZ-PID. The disturbance $f(t)$ is assumed to be unknown.

8.6.1.2.1 *Fuzzy Control with More MFs: N = 7*

As the inputs and the output are normalized into $[-1, 1]$, seven triangular MFs with equal spread result in $A = 1/3$ and $B = 1/3$. The nominal gains were chosen as

$$K_e = 1$$

$$\lambda = 2 \quad \text{(from Equation VI.8.32)}$$

$$K \geq 1.5\frac{F}{B} = 4.5F = 45, \text{ and } K = 50 \text{ was chosen (from Equation VI.8.20)}$$

Optimal gains can be obtained through a minor adjustment from the nominal gains, following the performance criteria in Equations VI.8.21 and VI.8.22. In general, increasing the gain K or K_e

TABLE VI.8.5   Performance of FZ-PD with $N = 7$

No.	K_e	λ	K	IAE/ITAE	KK_e	Perf. Change	Chattering
1*	1	2	50	1.9/12.15	50	Stable	No
2	2	2	100	0.82/3	200	Better	No
3*	3	2	100	0.72/2	300	Better	Yes

TABLE VI.8.6   Performance of FZ-PD with $N = 7$

No.	K_e	λ	K_0	IAE/ITAE	$K_e K_0$	Perf. Change	Chattering
1*	1	2	50	1.7/10.95	50	Stable	No
2	2	2	200	0.58/1.38	400	Better	No
3*	3	2	200	0.51/0.86	600	Better	Yes

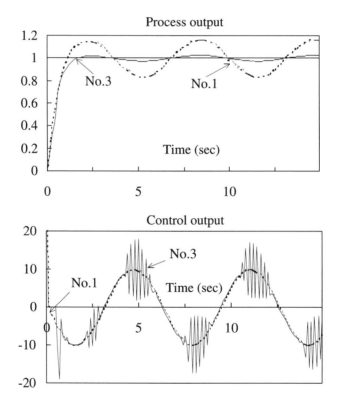

FIGURE VI.8.16   Performance of FZ-PD with more MFs.

will help with the suppression of the disturbance because it increases the VSC effects. For a larger gain, the chattering may appear in the control output. The simulation results of FZ-PD are shown in Table VI.8.5 (the one marked with * is plotted) and Figure VI.8.16. Following a similar procedure, the simulation results of FZ-PID are shown in Table VI.8.6 and Figure VI.8.17.

8.6.1.2.2 *FZ-PD with Less MFs:* $N = 3$

Only three MFs are used for each input and output variable. The simulation results under the same $K_e K$ are shown in Table VI.8.7 and Figure VI.8.18.

TABLE VI.8.7 Performance of VSC type FZ-PD with $N = 3$

No.	K_e	λ	K	IAE/ITAE	KK_e	Chattering
1*	1	2	50	2.42/16.39	50	No
3*	3	2	100	0.72/2	300	No

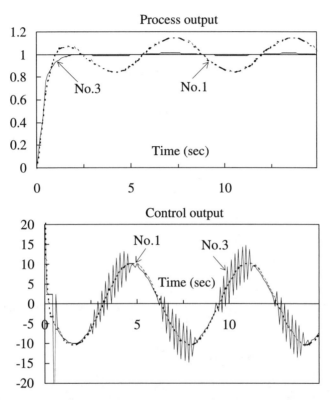

FIGURE VI.8.17 Performance of FZ-PID with more MFs.

8.6.1.3 Performance Analysis

In general, it is difficult for a simple PD controller to keep a robust performance for a complex process. Usually, the sliding control can achieve good robust performance, but it requires more knowledge about the process: otherwise, large chattering may arise from a large gain K. On the contrary, FLC needs less knowledge of the process because it does not need much detail about the uncertainty.

As previously discussed, FLC combines effects from its linear counterpart and the sliding control. Therefore, the performance of FLC is similar, yet superior, to both of them in controlling a complex process. In a situation with strong uncertainties, FLC is closer to the sliding control because a strong sliding effect is generated by its large gain. Due to its dual features, FLC can achieve good performance without too much chattering.

The dual features of FLC can be tuned through parameter adjustment. The less knowledge of the process, the larger the gain $K_e K$ is needed to suppress the larger uncertainty, which makes FLC closer to sliding control. However, a too-large $K_e K$ may cause large chattering, which should be

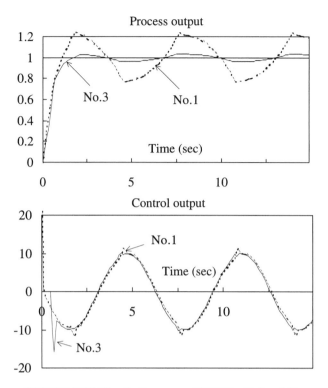

FIGURE VI.8.18   Performance of FZ-PD with less MFs.

avoided. Less MFs can have less sliding effect and also less chattering but make the system more nonlinear because a smaller amount of uncertainty can be suppressed.

8.6.2 One Step Toward Autotuning of FZ-PID

The performance of the fuzzy PID control in Figure VI.8.12, together with its autotuning features, is now compared with that of the conventional PID in Equation VI.8.2 for both selected linear and nonlinear processes.

8.6.2.1 Performance Comparison for Linear Process

A linear process with unknown load disturbance d is chosen as in Equation VI.8.33. The process output is y; the control output is u; the command signal is a step function applied at $t = 0$; the load disturbance d is applied at $t = 6$ sec.

$$y(s) = \frac{2.2}{1 + 0.5s + s^2}[u(s) + d(s)] \qquad (VI.8.33)$$

$$d(s) = \frac{5}{s}e^{-6s}$$

By using the frequency response method, the ultimate gain and period, $\mathbf{K_c} \approx 45$ and $\mathbf{t_c} \approx 0.63$, are obtained for the linear second-order process Equation VI.8.33. Parameters of PID control are calculated according to Table VI.8.1 and are shown, together with the performance indices obtained, in Table VI.8.8.

Inputs to the fuzzy system are normalized to $[-1, 1]$ by dividing the setpoint value. Thus, the maximum gain for the unsaturated E is $K_e = 1$ in order to avoid possible input saturation. The

TABLE VI.8.8   Performance of the Conventional PID

K_P	T_i	T_d	IAE	ITAE
27	0.3	0.1	0.67	1.14

TABLE VI.8.9   FZ-PID for the Linear Process

K_P	α	β	IAE	ITAE
27	0.3	0.1	0.34	0.37

TABLE VI.8.10   Integral Effect of α for Linear Process

K_P	α	β	IAE	ITAE
27	0.2	0.1	0.3	0.34
27	0.3	0.1	0.34	0.37
27	0.4	0.1	0.45	0.5

TABLE VI.8.11   Derivative Effect of β for Linear Process

K_P	α	β	IAE	ITAE
27	0.3	0.1	0.34	0.37
27	0.3	0.2	0.35	0.37
27	0.3	0.3	0.35	0.38

higher-level fuzzy gains (K_P, T_i, T_d) are initially derived directly using the Ziegler–Nichols method in Table VI.8.1, i.e., they are assumed to be the same as the optimally tuned PID gains (K_P, T_i, T_d) in Table VI.8.8. The scaling gains of the FZ-PID, easily derived from Equation VI.8.30, are shown in Table VI.8.9. The corresponding FZ-PID performance is also summarized in Figure VI.8.19. Although the FZ-PID is slow in the transient response, it provides a smaller overshoot and faster error correction in the process of the load disturbance. This simulation comparison confirms that FZ-PID gives rise to more damping and, hence, less oscillations than the conventional PID.

The integral effect of α is simulated, as shown in Figure VI.8.20 and Table VI.8.10, in which K_P and β are kept unchanged and α varies from 0.2 to 0.4. The simulation confirms that decreasing α will increase the integral action, which will speed up the response and disturbance compensation and reduce the steady-state error. However, too small a value of α may cause large oscillations and may even destabilize the system.

The derivative effect of β is demonstrated by simulation, as shown in Figure VI.8.21 and Table VI.8.11, in which K_P and α are kept unchanged and β varies from 0.1 to 0.3. The simulation confirms that increasing β will increase the derivative action, which will improve the system stability but will slow down the response and disturbance compensation. Larger values of β will cause larger tracking errors.

TABLE VI.8.12   Conventional PID for the Nonlinear Process

K_P	T_i	T_d	IAE	ITAE
0.9	0.4	0.1	0.72	0.72

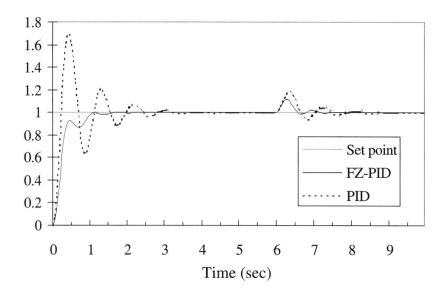

FIGURE VI.8.19   Performance comparison between conventional PID and fuzzy PID.

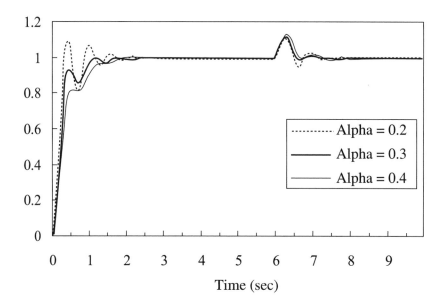

FIGURE VI.8.20   Integral effects of α to the performance of FZ-PID for the linear process when $\beta = 0.1$.

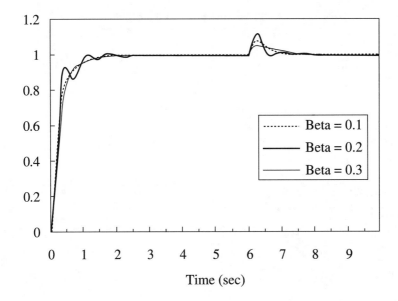

FIGURE VI.8.21   Derivative effects of β to the performance of FZ-PID for the linear process when $\alpha = 0.3$.

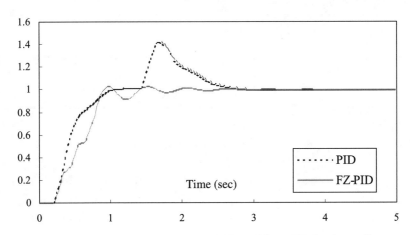

FIGURE VI.8.22   Comparison of conventional PID and fuzzy PID for the nonlinear process.

TABLE VI.8.13   FZ-PID for the Nonlinear Process

K_P	α	β	IAE	ITAE
0.9	0.4	0.1	0.63	0.34

TABLE VI.8.14   Integral Effect of α for Nonlinear Process

K_P	α	β	IAE	ITAE
0.9	0.3	0.1	0.70	0.43
0.9	0.4	0.1	0.62	0.27
0.9	0.5	0.1	0.74	0.47

TABLE VI.8.15   Derivative Effect of β for Nonlinear Process

K_P	α	β	IAE	ITAE
0.9	0.4	0.06	0.90	0.69
0.9	0.4	0.08	0.68	0.32
0.9	0.4	0.1	0.62	0.27

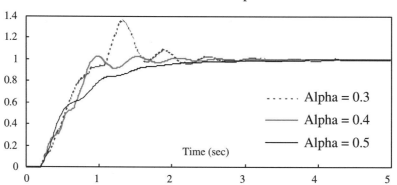

FIGURE VI.8.23   Integral effects of α to the performance of FZ-PID for the nonlinear process when $\beta = 0.1$.

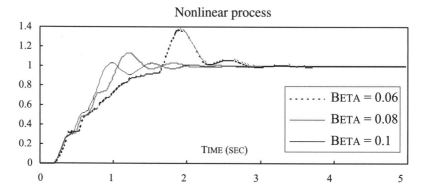

FIGURE VI.8.24   Derivative effects of β to the performance of FZ-PID for the nonlinear process when $\alpha = 0.4$.

8.6.2.2   Performance Comparison for Nonlinear Process

A nonlinear process with time-delay $\tau = 0.2$ second is chosen as in Equation VI.8.34. The process is time-varying with strong nonlinearity. The time-delay will cause the high-order features.

$$0.4\dot{x} + (1 + 0.2\sin(0.1t))x = 2u(t - \tau) - \text{sgn}\,(\dot{x}) \qquad \text{(VI.8.34)}$$

Similar to Section 8.6.1.1, by using the frequency response method, the ultimate gain and period, $\mathbf{K_c} \approx 1.5$ and $\mathbf{t_c} \approx 0.8$, are obtained. Parameters of the PID control are calculated according to Table VI.8.1 and are shown together with the performance indices obtained in Table VI.8.12. The corresponding FZ-PID parameters and performance are shown in Figure VI.8.22 and Table VI.8.13.

FZ-PID can still obtain reasonably good performance using the traditional Ziegler–Nichols method for this nonlinear process.

The gain ratio α demonstrates its integral effect for this nonlinear process, as shown in Figure VI.8.23 and Table VI.8.14. Decreasing α will speed up the response, but too much will cause large oscillation and tend to destabilize the system. The other gain ratio β also has some derivative effect, as shown in Figure VI.8.24 and Table VI.8.15. The simulation confirms that increasing β will increase the derivative action and, thus, improve the system stability.

8.6.2.3 Performance Analysis

Both theoretical analysis and simulation results show that the FZ-PID designed as described above tends to produce more damping and, therefore, less oscillation than the conventional PID. It is well-known that a system with less oscillation should be more stable than a system with more oscillation. Thus, FZ-PID could be more robust against disturbances than its conventional counterpart. The simulation results show that the FZ-PID has fairly similar characteristics as its conventional counterpart, but provides better performance in both the transient and steady states, with smaller overshoot and better disturbance rejection capability.

Although some approximations around the equilibrium state are assumed, practically, FZ-PID can be applied to a global region, which has been confirmed by simulations. The proposed gain design and tuning approach is very simple but very effective. This actually provides an important step toward optimal autotuning of FLC.

8.7 Conclusions

A hybrid design methodology, by means of a combination of nominal design and fine tuning, has been proposed for fuzzy control with emphasis on the nominal design in this chapter. Based on different features of parameters, methods from different disciplines are integrated. The rule base is designed via the qualitative approach, while the scaling gains are determined via quantitative methods. Under the linear rule base and standard triangle MFs, a mathematical model of fuzzy control can be derived, which makes quantitative analysis, designs, and tuning quite feasible. Quantitative analysis has shown that fuzzy-PID is actually a quasi-sliding control with dual features from its linear counterpart and sliding of VSC. These features are adjustable by tuning the scaling gains and membership functions. Thus, nominal gain designs of FZ-PID can be based either on the well-tuned linear PID control or VSC theory. The former design may lead to a possible autotuning. The mathematical model of FLC discloses the relationship between the lower-level scaling gains and the three higher-level proportional, integral, and derivative control action gains. Thus, the conventional Ziegler–Nichols method for PID control can be applied to the nominal scaling gain design and, possibly, the fine-tuning thereafter. Simulations have confirmed the effectiveness of this nominal design approach.

Acknowledgment

This work was partially supported by a grant from RGC of Hong Kong (CERG 9040507) and a grant from the City University of Hong Kong (SRG—7001055).

References

1. Arzen, K.E., Johansson, M., and Babuska, R., Fuzzy control versus conventional control, in *Fuzzy Algorithm for Control*, Verbruggen, H.B., Zimmermann, H.J., and Babuska, R., Eds., 1999, 59.
2. Astrom, K.J. and Hagglund, T., *Automatic Tuning of PID Controllers*, Instrument Society of America, 1988.

3. Bare, W.H., Mulholland, R.J., and Sofer, S.S., Design of a self-tuning rule based controller for a gasoline refinery catalytic reformer, *IEEE Trans. Autom. Control*, 35(2), 156, 1990.

4. Chen, G., Conventional and fuzzy PID controllers: an overview, *Int. J. Intelligent Control Syst.*, 1, 235, 1996.

5. Driankov, D., Hellendoorn, H., and Palm, R., Some research directions in fuzzy control, in *Theoretical Aspects of Fuzzy Control*, Nguyen, H.T. et al., Eds., John Wiley & Sons, New York, 1995, 281.

6. Filev, D.F. and Yager, R.R., On the analysis of fuzzy logic controllers, *Fuzzy Sets Syst.*, 68, 39, 1994.

7. Fulloy, L. and Galichet, S., Fuzzy and linear controllers, in *Fuzzy Systems*, Nguyen, H.T. and Sugeno, M., Eds., Kluwer Academic Publishers, Dordrecht, 1998, 197.

8. Galichet, S. and Foulloy, L., Fuzzy controllers: synthesis and equivalencies, *IEEE Trans. Fuzzy Syst.*, 3, 140, 1995.

9. Hung, J.Y., Gao, W., and Hung, J.C., Variable structure control: a survey, *IEEE Trans. Ind. Electron.*, 40(1), 2, 1993.

10. Lee, C.C., Fuzzy logic in control systems: fuzzy logic controller, *IEEE Trans. Syst. Man Cybernetics*, 20(2), 404, 1990.

11. Li, H.X. and Gatland, H.B., A new methodology for designing a fuzzy logic controller, *IEEE Trans. Syst. Man Cybernetics*, 25(3), 505, 1995.

12. Li, H.X., Gatland, H.B., and Green, A.W., Fuzzy variable structure control, *IEEE Trans. Syst. Man Cybernetics*, 27B(2), 306, 1997.

13. Li, H.X., A comparative design and tuning for conventional fuzzy control, *IEEE Trans. Syst. Man Cybernetics*, 27B(5), 884, 1997.

14. Li, H.X. and Tso, S.K., A hybrid design methodology for conventional fuzzy control, Proceedings of the IEEE World Conference on Computational Intelligence (WCCI '98), Alaska, May 1998, 394.

15. Mamdani, E.H., Applications of fuzzy algorithms for simple dynamic plant, *Proc. IEEE*, 121(12), 1585, 1974.

16. Mamdani, E.H., Twenty years of fuzzy control: experiences gained and lessons learnt, Proceedings of 2nd IEEE International Conference on Fuzzy Systems, San Francisco, CA, 1993, 339.

17. Mann, G.K.I., Hu, B.G. and Gosine, R.G., Analysis of direct action fuzzy PID controllers structures, *IEEE Trans. Syst. Man Cybernetics*, 29B(3), 371, 1999.

18. Mann, G.K.I., Hu, B.G., and Gosine, R.G., Two-level tuning of fuzzy PID controllers, *IEEE Trans. Syst. Man Cybernetics*, 31B(2), 263, 2001.

19. Matia, F., Jimenez, A., and Galan, R., Fuzzy controllers: lifting the linear/nonlinear frontiers, *Fuzzy Sets Syst.*, 52(2), 113, 1992.

20. Misir, D., Malki, H.A., and Chen, G., Design and analysis of fuzzy PID controller, *Fuzzy Sets Syst.*, 79, 297, 1996.

21. Mizumoto, M., Realization of PID controls by fuzzy control methods, Proceedings of the IEEE Conference on Fuzzy Systems, San Diego, CA, 1992, 709.

22. Mudi, R.K. and Pal, N.R., A robust self-tuning scheme for PI- and PD-type fuzzy controllers, *IEEE Trans. Fuzzy Syst.*, 7(1), 2, 1999.

23. Palm, R., Sliding mode fuzzy control, Proceedings of 1st IEEE International Conference on Fuzzy Systems, San Diego, CA, 1992, 519.

24. Palm, R., Tuning of scaling factors in fuzzy controllers using correlation functions, Proceedings of 2nd IEEE International Conference on Fuzzy Systems, San Francisco, CA, 1993, 691.

25. Pedrycz, W., Fuzzy control engineering: reality and challenges, Fourth International Conference on Fuzzy Systems, Japan, March 1995, 437.

26. Procyk, T.J. and Mamdani, E.H., A linguistic self-organising process controller, *Automatica*, 15, 15, 1979.

27. Qiao, W.Z. and Mizumoto, M., PID type fuzzy controller and parameters adaptive method, *Fuzzy Sets Syst.*, 78, 23, 1996.

28. Qin, S.J. and Borders, G., A multiregion fuzzy logic controller for nonlinear process control, *IEEE Trans. Fuzzy Syst.*, 2(1), 74, 1994.

29. Slotine, J.J.E. and Li, W., *Applied Nonlinear Control*, Prentice Hall, Upper Saddle River, NJ, 1991.

30. Wang, P.P. et al., Fuzzy dynamic systems and fuzzy linguistic controller classification, *Automatica*, 30(11), 1769, 1994.

31. Ying, H. et al., Fuzzy control theory: a nonlinear case, *Automatica*, 26, 513, 1990.

32. Ying, H., A nonlinear fuzzy controller with linear control rules is the sum of a global two-dimensional multilevel relay and a local nonlinear PI controller, *Automatica*, 29(2), 499, 1993.

33. Ying, H., Practical design of nonlinear fuzzy controllers with stability analysis for regulating processes with unknown mathematical models, *Automatica*, 30(7), 1185, 1994.

34. Zadeh, L.A., Outline of a new approach to the analysis of complex systems and decision processes, *IEEE Trans. Syst. Man Cybernetics*, 3(1), 28, 1973.

35. Zimmermann, H.J., *Fuzzy Set Theory and Its Applications*, Kluwer Academic Pulishers, Boston, MA, 1991.

9

Intelligent Controllers for Flexible Pole Cart Balancing Problem

9.1 Introduction VI-286
9.2 A Model of a Flexible Pole-Cart Balancing System
under its First Mode of Vibration VI-286
Mechanics
9.3 Software Simulation VI-297
The Program • The Rule-Based Controller • Analysis of
Results • Summary
9.4 Neural Network Controller for the Flexible Pole-
Cart Balancing Problem VI-301
Training and Learning in a Neural Network • Processes
Involved in the Formulation of the Flexible Pole-Cart Bal-
ancing Control: Neural Network Perspective • Discussions:
The Backpropagation Neural Network Simulator • Discus-
sions and Analysis of Results • The Physical Architecture of
the Flexible Pole-Cart Balancing System • Online Applica-
tion of the Neural Network Model to the Flexible Pole-Cart
Balancing System • Results of the Neural Network Physical
Experiments • Summary
9.5 Fuzzy Logic System Controller for the Flexible
Pole-Cart Balancing Problem VI-314
Fuzzy Logic System • Application of the Fuzzy Logic
Controller to the Real Physical Flexible Pole-Cart Balanc-
ing System • Fuzzy Associative Memory (FAM) Matrix •
Membership Functions (MFs) • Defuzzifier • Results of the
Physical Experiments • Summary
9.6 Fuzzy-Logic Based Neural Network Implemented
Controller (FLBNNIC) for the Flexible Pole-Cart
Balancing Problem VI-323
Processes Involved in Formulating the Fuzzy-Logic Based
Neural Networks Implemented Controller (FLBNNIC) •
Results of the FLBNNIC Experiments • Discussions and
Analysis of Results • Summary
9.7 Conclusions and Recommendations VI-333
References VI-334

Elmer P. Dadios
De La Salle University

9.1　Introduction

The inverted pendulum (pole-cart balancing) problem has received a great deal of attention as a model problem for the establishment of learning control systems.[1–7] That researchers are successful in this field can be seen in their published results. However, using a rigid pole as the pendulum, analysis shows that this system has only two degrees of freedom and little nonlinearity. As a result of these limitations, the learning controllers developed using such a demonstrator problem have limited power and are unlikely to have broad applications to manufacturing industries. Because of the limitations mentioned, this author modifies the pole-cart balancing problem to give a more exact test bed for learning controllers by replacing the rigid pole with an elastic pole. The dynamics of this new system are more complex and highly nonlinear when compared to the traditional rigid pole-cart balancing system as a result of the additional degree of freedom within the system, e.g., the transverse displacement of the elastic pole.[20,24,28,37,42,43]

Modeling and control of flexible robot systems has attracted much interest in recent years.[8–14] This has arisen, in particular, in the area of space and industrial robots that require lightweight and flexible links.[15] Flexible robot manipulators have many advantages compared to robot manipulators constructed from rigid links. If the advantages associated with the lightweight machine elements are not to be sacrificed, then advanced control systems for such flexible robot manipulators have to be developed.[9] The flexible pole test bed explored in this research allows the examination of some of the control issues within flexible linked robots.

This research began by investigating the limitations of the pole-cart balancing (inverted pendulum) problem as a benchmark for learning controllers. Here, the objective was to investigate the inverted pendulum problem and analyze its usefulness as a benchmark for developing learning control systems and their application in manufacturing industries. It shows particularly that the pole-cart problem may not be sufficiently testing, hence, the author extended the problem using a flexible pole as a replacement for the rigid pole.

To verify the feasibility of solving the flexible pole-cart balancing problem, the author has generated a computer simulation of this system. Equations of the dynamics of this system have been derived, and a rule-based control system has been developed. A graphic representation of the system behavior has been developed to show the cart balancing the pole along a track in real time. Having shown by computer simulation that it is possible to control the flexible pole-cart balancing problem, the next stage of the research addressed its real physical control application.

This research, therefore, focuses on developing and testing online and offline learning controllers that balance a flexible pole hinged by its root on top of a cart moving along a limited track. The capabilities of neural network algorithms, fuzzy logic systems, and neuro-fuzzy algorithms have been investigated and tested in control of the system.

9.2　A Model of a Flexible Pole-Cart Balancing System under its First Mode of Vibration

Flexible beams have been a topic of research in the field of robotics since the early 1970s.[12] Beams of this type have been used to model flexibility in robotic members, a phenomenon that has gained importance as a result of widespread attempts to lighten robotic assemblies for increased speed and efficiency.[13] The present-day industrial robot is easy to control because it is designed to be very heavy, rigid, and slow. This, however, gives high weight-to-payload ratio, which increases cost and decreases the speed of the robot. To improve this ratio, several researchers have proposed the use of lightweight robots with links that are allowed to flex during operation.[16–19]

When compared with the traditional robot manipulators constructed from rigid links, flexible robot manipulators have many advantages; among them are: the moving of larger payloads without increasing the mass of the linkages, requirements for less material and smaller actuators, less link weight, less power consumption, and the machines are more maneuverable and transportable.[10] Flexible robot manipulators are not presently used in production industries because robot manipulators are required to have a reasonable accuracy in the response of the manipulator end-effector to the input command from its control system. The experiments described in the relevant literature[10–13] were directed toward developing a controller for flexible robot manipulators. Building this type of controller is a very difficult and challenging task. One major step in making this controller is to analyze the dynamic behavior of the system. Computer simulation is necessary to evaluate whether the derived dynamics of the system are correct. It is, therefore, most appropriate to study analogues of such systems. The flexible pole-cart system provides such an analogue.

This section presents a rule-based control system for the flexible pole-cart balancing problem (the inverted pendulum using an elastic pole) that operates on a simulation of the system. The task of this system is to balance an elastic pole that is hinged on a movable cart. It is assumed that the hinge is frictionless. The cart is allowed to move along a track with limited length that has friction. Forces of different magnitude are applied to the cart in either a left or right direction. The initial angle of the pole can be varied up to 30 degrees. This is more difficult than the conventional rigid pole-cart system because of the complexity in its dynamics. The deflection of the elastic pole gives additional degrees of freedom to this system. The dynamic equations of the system were derived using Newton's laws, Bernoulli–Euler analysis, and beam theories. The system was analyzed with the presence of friction. Numerical integration using fourth order Runge–Kutta was conducted. Computer graphics of the cart balancing the pole along the track in real time have been made and are shown. Results on the analysis of the behavior of the system under various conditions have also been obtained in order to explore the practicality of attempting to control such a system.

The section begins by presenting the mechanics of the system. It then continues by describing a simulation of these mechanics. The section closes by describing the operation of the rule-based controller on the simulated dynamics.

9.2.1 Mechanics

This section discusses the dynamics of the flexible pole-cart balancing system. The analysis of the system is based on the dynamics of the rigid pole-cart, together with beam theories.

9.2.1.1 Diagram of the Flexible Pole-Cart Balancing System

Figure VI.9.1 shows the dynamics of the system.[20] The free body diagrams of the system are shown in Figures VI.9.2 and VI.9.3.

9.2.1.2 Derivation of the Equations

9.2.1.2.1 *Solution for Rigid Pole Angle, Velocity, Acceleration, and Cart Velocity*

This section presents the analysis that relates the motion of the cart to the motion of the pole. It begins by considering the dynamics of the rigid pole.

Let:

m_p = mass of the pole
m_c = mass of the cart
a_c = acceleration of the cart
L = total length of the pole
θ = the angle of the pole from the y axis
$\dot{\theta}$ = angular velocity of the pole
$\ddot{\theta}$ = angular acceleration of the pole
g = acceleration due to gravity

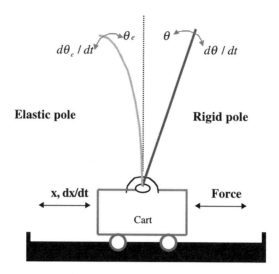

FIGURE VI.9.1   The flexible pole-cart balancing system. (From Dadios, E.P. and Williams, D.J., *IEEE Trans. Systems Man Cybernetics*, 28(6), 895, 1998. With permission from IEEE.)

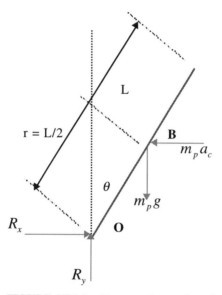

FIGURE VI.9.2   Forces acting on the pole.

Applying Newton's law to the rigid pole:

Summation of forces at the center of the pole $=$ (mass)(acceleration of pole)

$$\sum \widehat{F} = m_p a^B \qquad\qquad \text{(VI.9.1)}$$

Using definitions from Figures VI.9.2 and VI.9.3,

$e_t =$ a unit vector for a tangential component
$e_n =$ a unit vector for a normal component
$a_t =$ tangential acceleration
$a_n =$ normal acceleration

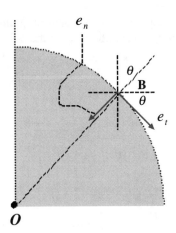

FIGURE VI.9.3 Tangential and normal forces acting on the center of the pole.

Then,

$$e_t = \cos(\theta) * e_x - \sin(\theta) * e_y \tag{VI.9.2}$$

$$e_n = -\sin(\theta) * e_x - \cos(\theta) * e_y \tag{VI.9.3}$$

$$a_t = r\alpha = \frac{L}{2} * \ddot{\theta} \tag{VI.9.4}$$

$$a_n = rw^2 = \frac{L}{2} * \dot{\theta}^2 \tag{VI.9.5}$$

and, using Equations VI.9.2–VI.9.5:

$$a^B = a_t * e_t + a_n * e_n$$

$$a^B = \frac{L}{2}\ddot{\theta}(\cos(\theta)\, e_x - \sin(\theta)e_y) + \frac{L}{2}\dot{\theta}^2(-\sin(\theta)e_x - \cos(\theta)e_y) \tag{VI.9.6}$$

Then, from Figure VI.9.2:

$$\sum \widehat{F} = (-m_p a_c + R_x)e_x + (R_y - m_p g)e_y = m_p a^B$$

$$(-m_p a_c + R_x)e_x + (R_y - m_p g)e_y = m_p \left\{ \frac{L}{2}\ddot{\theta}(\cos(\theta)e_x - \sin(\theta)e_y) \right.$$

$$\left. + \frac{L}{2}\dot{\theta}^2(-\sin(\theta)e_x - \cos(\theta)e_y) \right\} \tag{VI.9.7}$$

Equating terms of the e_x and e_y components:

$$-m_p a_c + R_x = m_p \left\{ \frac{L}{2}\ddot{\theta}\cos(\theta) + \frac{L}{2}\dot{\theta}^2(-\sin(\theta)) \right\}$$

$$-m_p a_c + R_x = m_p \left\{ \frac{L}{2}\ddot{\theta}\cos(\theta) - \frac{L}{2}\dot{\theta}^2\sin(\theta) \right\}$$

$$-m_p a_c + R_x = m_p \left\{ \frac{L}{2}\ddot{\theta}\cos(\theta) - \dot{\theta}^2\sin(\theta) \right\} \tag{VI.9.8}$$

$$R_y - m_p g = m_p \left\{ \frac{L}{2}\ddot{\theta}(-\sin(\theta)) - \frac{L}{2}\dot{\theta}^2\cos(\theta) \right\} \tag{VI.9.9}$$

From the Euler equations, the summation of the moments at point **O** is equal to the product of the moment of inertia (**I**) and the angular acceleration (α) of the pole:

$$\sum M^o = I^o \alpha \tag{VI.9.10}$$

where

$$I^o = \frac{1}{3}m_p L; \qquad \alpha = \ddot{\theta}$$

Therefore, using Figure VI.9.2, apply Equation VI.9.10:

$$m_p g \frac{L}{2}\sin(\theta) - m_p a_c \frac{L}{2}\cos(\theta) = I^o\ddot{\theta} = \frac{1}{3}m_p L^2\ddot{\theta}$$

$$a_c \frac{L}{2}\cos(\theta) = \frac{1}{3}L^2\ddot{\theta} - g\frac{L}{2}\sin(\theta); \qquad \text{since} \qquad L/2 = r$$

$$a_c r \cos(\theta) = \frac{4}{3}r^2\ddot{\theta} - gr\sin(\theta) \tag{VI.9.11}$$

Equation VI.9.11 shows the relationship of the cart acceleration to the angular acceleration of the pole. To establish the forces acting on the cart, it is assumed that the mass of the wheels is very small compared to the mass of the cart and the pole. Figure VI.9.4 shows the free body diagram of the cart.

Applying Newton's law to Figure VI.9.4:

$$\sum F = m_c a_c; \qquad F_c - R_x - f = m_c a_c; \qquad R_y = N; \qquad f = \mu N$$

$$F_c = m_c a_c + R_x + \mu R_y \tag{VI.9.12}$$

This is the force needed to move the cart. Then, substituting Equations VI.9.8 and VI.9.9 into Equation VI.9.12:

$$F_c = m_c a_c + m_p a_c + m_p \frac{L}{2}(\ddot{\theta}\cos\theta - \dot{\theta}^2\sin\theta)$$

$$+ \mu(m_p g + m_p\frac{L}{2}(-\ddot{\theta}\sin\theta - \dot{\theta}^2\cos\theta))$$

$$F_c = (m_c + m_p)a_c + \mu m_p g + \ddot{\theta}m_p\frac{L}{2}(\cos\theta - \mu\sin\theta)$$

$$+ \dot{\theta}^2 m_p\frac{L}{2}(-\sin\theta - \mu\cos\theta)$$

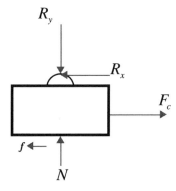

FIGURE VI.9.4   Forces acting on the cart.

$$a_c = \frac{F_c - \{\mu m_p g + \ddot{\theta} m_p \frac{L}{2}(\cos\theta - \mu\sin\theta) + \dot{\theta}^2 m_p \frac{L}{2}(-\sin\theta - \mu\cos\theta)}{(m_p + m_c)}$$

Let

$$m = m_p + m_c; \quad r = L/2$$

Then

$$a_c = \frac{F_c - \{\mu m_p g + \ddot{\theta} m_p r(\cos\theta - \mu\sin\theta) + \dot{\theta}^2 m_p r(-\sin\theta - \mu\cos\theta)}{m} \tag{VI.9.13}$$

and, substituting Equation VI.9.13 into VI.9.11:

$$(r\cos\theta)\frac{F_c - \{\mu m_p g + \ddot{\theta} m_p r(\cos\theta - \mu\sin\theta) + \dot{\theta}^2 m_p r(-\sin\theta - \mu\cos\theta)}{m}$$

$$= \frac{4}{3}r^2\ddot{\theta} - gr\sin\theta$$

$$\frac{4}{3}r\ddot{\theta} - \ddot{\theta}\frac{m_p r\cos\theta(\cos\theta - \mu\sin\theta)}{m} = g\sin\theta$$

$$- \frac{\cos\theta}{m}\{F_c - [\mu m_p g - \dot{\theta}^2 m_p r(\sin\theta + \mu\cos\theta)]\}$$

$$\ddot{\theta}\left[\frac{4}{3}mr - m_p r\cos^2\theta + \mu m_p r\cos\theta\sin\theta\right] = mg\sin\theta$$

$$- \cos\theta\{F_c - [\mu m_p g - \dot{\theta}^2 m_p r(\sin\theta + \mu\cos\theta)]\}$$

$$\ddot{\theta} = \frac{mg\sin\theta - \cos\theta\{F_c - [\mu m_p g - \dot{\theta}^2 m_p r(\sin\theta + \mu\cos\theta)]\}}{\frac{4}{3}mr - m_p r\cos^2\theta + \mu m_p r\cos\theta\sin\theta} \tag{VI.9.14}$$

This represents the angular acceleration of the rigid pole hinge root on top of the cart that moves dependent on the magnitude and direction of the applied force F_c.

9.2.1.2.2 *Solution for Elastic Pole Angle, Velocity, and Acceleration*
It is now necessary to extend this analysis to include the elastic pole.

Let

θ_t = total elastic pole's angle from the vertical axis
$\dot{\theta}_t$ = elastic pole's velocity
$\ddot{\theta}_t$ = elastic pole's acceleration

For the elastic pole, it is assumed that the total angle θ_t (the pole's actual position with respect to vertical axis) is equal to the actual angle of the rigid pole θ plus the pole's angle due to its elastic deflection θ_e.

$$\theta_t = \theta + \theta_e \tag{VI.9.15}$$

To find θ_e it is assumed that the pole behaves as a cantilever with uniform distributed load, as shown in Figure VI.9.5. From Case and Chilver:[21]

E = Young's modulus of elasticity
I = Second moment of area $BD^3/12$

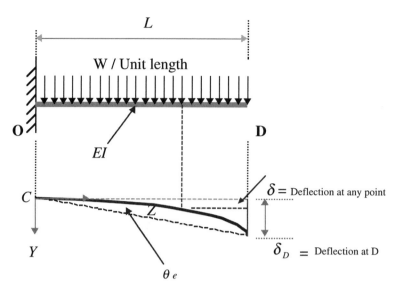

FIGURE VI.9.5   Cantilever carrying a uniformly distributed load.

B = Breadth of the beam
D = Depth of the beam
W = Weight per unit length

The bending moment at a distance Z from C is:

$$M = \frac{-1}{2}W(L - Z)^2; \text{ but} - M = EI\frac{d^2\delta}{dZ^2}$$

hence,

$$EI\frac{d^2\delta}{dZ^2} = \frac{1}{2}(L - Z)^2 = \frac{1}{2}W(L^2 - 2LZ + Z^2)$$

Integrating this equation gives

$$EI\frac{d\delta}{dZ} = \frac{1}{2}W(L^2Z - LZ^2 + \frac{1}{3}Z^3 + A)$$

Further integrating this equation gives

$$EI\delta = \frac{1}{2}W\left(\frac{1}{2}L^2Z^2 - \frac{1}{3}LZ^3 + \frac{1}{12}Z^4 + AZ + B\right)$$

At the built-in end, $Z = 0$, and we have $\frac{d\delta}{dZ} = 0$ and $\delta = 0$, Thus, $A = B = 0$. Then,

$$EI\delta = \frac{1}{24}W\left(6L^2Z^2 - 4LZ^3 + Z^4\right)$$

$$\delta = \frac{1}{24}W(6L^2Z^2 - 4LZ^3 + Z^4)/(EI) \qquad (VI.9.16)$$

At the free end, D, $Z = L$. Hence

$$\delta_D = \frac{WL^4}{8EI} \qquad (VI.9.17)$$

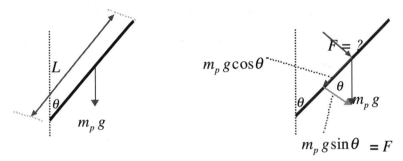

FIGURE VI.9.6 Concentrated load of the pole at any position.

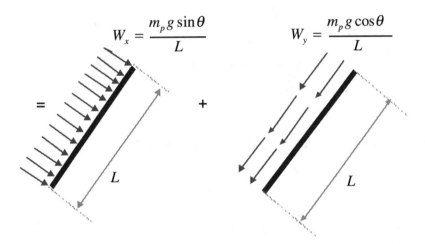

(this component is used only if there is buckling
or the hinge has friction; hence it will not affect
the value of *W*)

FIGURE VI.9.7 Uniform load of the pole at any position.

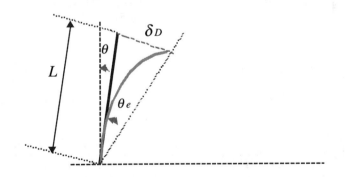

FIGURE VI.9.8 Position of the rigid and elastic pole at any time *t*.

It is now necessary to fit the cantilever analysis to the flexible pole. To find the value of *W* within the cantilever analysis, the pole can be analyzed by considering both the concentrated and uniform load, as shown in Figures VI.9.6 and VI.9.7.

For the pole at uniform load using F at Figure VI.9.6:

$$W = m_p g (\sin \theta)/L \tag{VI.9.18}$$

To determine the total elastic pole angle θ_t at any time, consider Figure VI.9.8.

$$\tan \theta_e = \frac{\delta_D}{L} \qquad \text{from Equation VI.9.17}$$

$$\tan \theta_e = \frac{WL^4}{8EIL} = \frac{WL^3}{8EI} \qquad \text{from Equation VI.9.18}$$

$$\tan \theta_e = \frac{L^2 m_p g \sin \theta}{8EI} \qquad \text{but} \quad I = BD^3/12$$

$$\tan \theta_e = \frac{12 L^2 m_p g \sin \theta}{8EBD^3}; \qquad \text{let} \quad K = \frac{12 L^2 m_p g}{8EBD^3}$$

$$\tan \theta_e = K \sin \theta; \quad \text{hence,} \quad \theta_e = \arctan (K \sin \theta) \tag{VI.9.19}$$

Substituting Equation VI.9.19 into Equation VI.9.15,

$$\theta_t = \theta + \arctan (K \sin \theta) \tag{VI.9.20}$$

To find the elastic pole's angular velocity, differentiate Equation VI.9.20:

$$\dot{\theta}_t = \dot{\theta} + \left(\frac{K \cos \theta}{1 + (K \sin \theta)^2} \right) \dot{\theta}$$

$$\dot{\theta}_t = \dot{\theta} \left\{ 1 + \left(\frac{K \cos \theta}{1 + (K \sin \theta)^2} \right) \right\} \tag{VI.9.21}$$

To find the elastic pole's angular acceleration, differentiate Equation VI.9.21:

$$\ddot{\theta}_t = \ddot{\theta} \left[1 + \frac{K \cos \theta}{1 + (K \sin \theta)^2} \right]$$

$$+ \dot{\theta} \left[\frac{(-K \sin \theta)(\dot{\theta})(1 + (K \sin \theta)^2) - (K \sin \theta)(2K \sin \theta)(K \cos \theta)(\dot{\theta})}{(1 + (K \sin \theta)^2)^2} \right]$$

$$\ddot{\theta}_t = \ddot{\theta} \left[1 + \frac{K \cos \theta}{1 + (K \sin \theta)^2} \right]$$

$$+ \dot{\theta}^2 \left[\frac{(-K \sin \theta)(1 + (K \sin \theta)^2) - (K \cos \theta)(2K \sin \theta)(K \cos \theta)}{(1 + (K \sin \theta)^2)^2} \right]$$

$$\ddot{\theta}_t = \ddot{\theta} \left[1 + \frac{K \cos \theta}{1 + (K \sin \theta)^2} \right] + \dot{\theta}^2 \left[\frac{K \sin \theta (-1 - K^2 (\sin \theta)^2 - 2K^2 (\cos \theta)^2)}{(1 + (K \sin \theta)^2)^2} \right]$$

$$\ddot{\theta}_t = \ddot{\theta} \left[1 + \frac{K \cos \theta}{1 + (K \sin \theta)^2} \right] + \dot{\theta}^2 \left[\frac{-K \sin \theta \{ 1 + K^2 ((\sin \theta)^2 + 2(\cos \theta)^2) \}}{(1 + (K \sin \theta)^2)^2} \right]$$

$$\ddot{\theta}_t = \ddot{\theta} \left[1 + \frac{K \cos \theta}{1 + (K \sin \theta)^2} \right] + \dot{\theta}^2 \left[\frac{-K \sin \theta (1 + K^2 (1 + (\cos \theta)^2))}{(1 + (K \sin \theta)^2)^2} \right] \tag{VI.9.22}$$

This presents the angular acceleration of the flexible pole hinge on top of the moving cart.

9.2.1.2.3　*Solution for Cart Acceleration and Displacement Due to Balance of the Elastic Pole*

This section discusses the mechanics to find the displacement of the cart due to the applied force in order to balance the elastic pole. The acceleration of the cart in order to balance the pole is derived from Equation VI.9.13 with $\ddot{\theta}, \dot{\theta}, \theta$ replaced by $\ddot{\theta}_t, \dot{\theta}_t, \theta_t$, respectively. Hence,

$$a_c = \frac{F_c - \{\mu m_p g + \ddot{\theta}_t m_p r (\cos\theta_t - \mu \sin\theta_t) + \dot{\theta}_t^2 m_p r(-\sin\theta_r - \mu\cos\theta_t)}{m} \tag{VI.9.23}$$

From the computer simulation using numerical integration (fourth-order Runge–Kutta), the value of the acceleration a_{ce} is found to be a cosine function. The reason for this is that the acceleration of the cart is dependent upon the force applied to it. This force is being controlled in order to balance the pole, and it is experimentally observed to be periodic. At time t equal to zero, initial force is already applied to the cart. Because of this, at this point in time, the cart is already accelerating at a magnitude equivalent to force/mass.

Thus, the acceleration of the cart for balancing the flexible pole at any time t is:

$$a_{ce} = k \cos \omega t \tag{VI.9.24}$$

The velocity of the cart at any time t for balancing the flexible pole is obtained by integrating Equation VI.9.24:

$$v_{ce} = \int_0^t a_{ce} = \int_0^t k \cos wt = \frac{k}{w} \sin wt \tag{VI.9.25}$$

The displacement of the cart at any time t for balancing the flexible pole is obtained by integrating Equation VI.9.25.

$$x_{ce} = \int_0^t v_{ce} = \int_0^t \frac{k}{w} \sin wt = \frac{-k \cos wt}{w^2}$$

but, from Equation VI.9.24, $a_{ce} = k \cos \omega t$, hence,

$$x_{ce} = \frac{-a_{ce}}{\omega^2} \quad \text{and} \quad \omega = 2\pi f; \quad f = average\ frequency$$

$$x_{ce} = \frac{-a_{ce}}{4\pi^2 f^2} \tag{VI.9.26}$$

To find the frequency f, it is necessary to obtain the total number of cycles during the total time of pole balancing. Below is the algorithm to determine this.

1. Determine the highest value of the acceleration for the entire time of balancing excluding the first one (a_{ch}).
2. Starting from time $t > 0.0$, record the value of time for the first a_{ch} $(Time_{ach1})$.
3. Record the time it takes to have another acceleration approximately equal to a_{ch} $(Time_{ach2})$.
4. Record the frequency from $(Time_{ach1})$ to $(Time_{ach2})$ = one cycle.
5. Repeat steps 2 and 3 by substituting $Time_{ach2}$ to $Time_{ach1}$ for I to N. I and N can be any value of time from $t > 0.0$ to the final time of balancing the pole. Record the total number of cycles for this process (tot_cycles).
6. Get the sum of the time recorded from $Time_{achI}$ to $Time_{achN}$ (tot_time):

$$tot\_time = \sum_{I=1}^{N}(Time_{achI} + Time_{achI+1})$$

7. Average frequency is:

$$f = tot\_cycles/tot\_time \tag{VI.9.27}$$

9.2.1.2.4 *Solution for the Location of the Pole at Any Time in the XY Plane*

This section discusses the mechanics to find the coordinates of the flexible pole on the XY plane at any angle (see Figure VI.9.9). The equations derived from this analysis are very important in displaying the pole graphically. Every point of the pole is plotted. This analysis uses Equation VI.9.16 as its starting point.

Let:

$(x1, y1)$ = the coordinate at any point of the pole without elastic deflection (say, **p1**)
$(x2, y2)$ = the new coordinate of **p1** due to elastic deflection

The deflection of the elastic pole at any point is derived from Equation VI.9.16.

$$\delta = \frac{1}{24} W (6L^2 Z^2 - 4LZ^3 + Z^4)/(EI)$$

From Figure VI.9.9, the value of $L1$ is from 0.0 to L.

$$x1 = (\sin \theta)\,(L1) \tag{VI.9.28}$$

$$y1 = (\cos \theta)\,(L1) \tag{VI.9.29}$$

$$L2 = \sqrt{(L1)^2 + \delta^2} \tag{VI.9.30}$$

$$\theta_e = arctan\,(\delta/L1) \tag{VI.9.31}$$

$$x2 = \sin\,(\theta + \theta_e)(L2) \tag{VI.9.32}$$

$$y2 = \cos\,(\theta + \theta_e)(L2) \tag{VI.9.33}$$

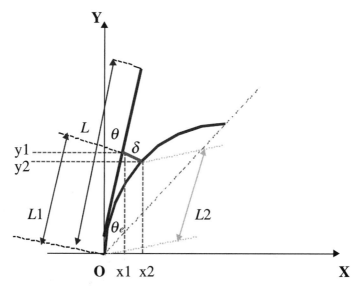

FIGURE VI.9.9   The coordinates of the elastic pole at any point in the xy plane.

9.3   Software Simulation

9.3.1   The Program

A computer program was developed simulating the derived dynamics of the pole-cart balancing system. The program can simulate both rigid and elastic pole-cart balancing with or without friction. The data input to the simulation represents the characteristics of the pole, the system initial conditions, and the simulation time. Once the data are entered into the computer, the program, using numerical integration (fourth-order Runge–Kutta), calculates the values of the dynamic equations presented previously. The outputs of this process are the values at any given time of the angle of the pole from the vertical axis, the angular velocity and acceleration of the pole, the force applied to the cart, the velocity of the cart, the acceleration of the cart, and the displacement of the cart. All of these data are stored in an external file for future use and can be displayed subsequently.

9.3.2   The Rule-Based Controller

The task of the controller is to balance the flexible pole on top of the cart moving along a limited track for a given time. The algorithm for this process is shown in Figure VI.9.10. This process uses numerical integration. Subprocess RUNGE( ) will calculate the values of the pole's angle, velocity, and acceleration for every time step set by the user. These values are taken from the parameters V and F of RUNGE( ). For every increment of the time step, the controller will check if it has exceeded the total simulation time. If the total simulation time has been attained, process *numerical_integration* will end; otherwise, subprocess RUNGE( ) will be executed.

The motion of the cart is dependent on the force applied to its body. Hence, it is necessary to control the magnitude and direction of this applied force. However, this force is directly proportional to the angle of the pole from the vertical axis and the total mass of the cart and pole. The angle of the pole is obtained from Equation VI.9.14 by applying numerical integration using fourth-order Runge–Kutta. This is the subprocess RUNGE( ) shown in Figure VI.9.10.

To determine the actual magnitude and direction of the force, the controller will first check the magnitude and direction of the pole's angle. If the angle of the pole exceeds the prescribed limit, then it will report a failure and go back to the main menu; otherwise, the process will continue. If the inclination of the angle of the pole is going left (negative), then the direction of the force applied to the cart is going left (also negative); otherwise, it is in the opposite direction. The magnitude of the force is chosen using a simple rule-based system in a manner of a look-up table, as shown below:

> *If the pole angle reached limit then*
>> *failure, exit program*
> *else if (pole angle* $>= -0.0009$ *and pole angle* $<= 0.0009$) *then*
>> *force applied* $= 0.0$ *Newton*
> *else if (pole angle* > -0.001 *and pole angle* $<= -0.0009$) *then*
>> *force applied* $= -0.1$ *Newton*
>>
>> *...*
> *else if (pole angle* > -0.427 *and pole angle* $<= -0.424$) *then*
>> *force applied* $= -14.9$ *Newton*
> *else if (pole angle* > -0.424 *and pole angle* $<= 0.427$) *then*
>> *force applied* $= 14.9$ *Newton*

For 0.0009 to 0.001 degrees inclination, this corresponds to 0.1 Newton of force applied to the cart. Above this value, an increment of 0.003 degrees angle will correspond to an increase of 0.1 Newton

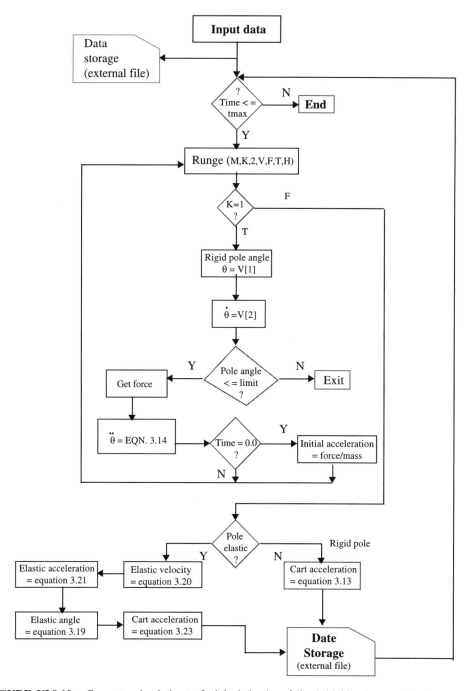

FIGURE VI.9.10   Computer simulation to find the behavior of elastic/rigid pole-cart balancing system.

in the applied force. The controller can apply a maximum force of 15 Newtons, although the user may enter an initial force greater than this. The author conducted a number of experiments in simulating the values relating the applied forces on the cart and the angles of the pole to determine satisfactory values of the parameter. The control algorithm can be modified to accommodate changes in mass. Examples of these parameters are shown below. The results and graphical representations of the

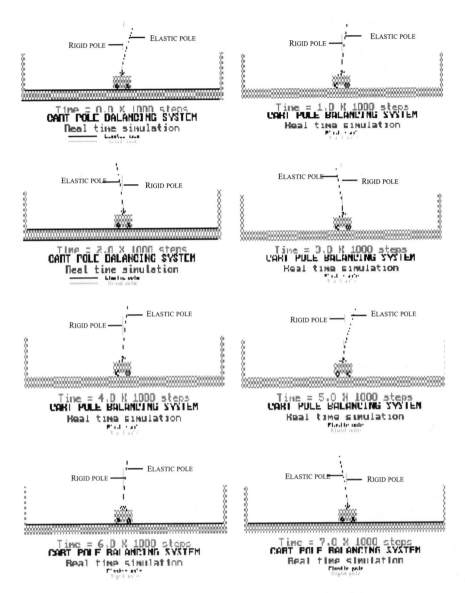

FIGURE VI.9.11   Real-time movement of the cart and the whip of the pole.

motion of the entire system are shown in Figures VI.9.11 and VI.9.12. In this simulation, the values of the size of the cart, the length of the pole, and the initial force can be changed. Below are the parameters necessary to run the program.

Mass of the pole	= 0.0500 kg
Total mass of the pole and the cart	= 0.5050 kg
Total length of the pole	= 0.5000 meters
Breadth of the pole	= 0.0150 meters
Depth of the pole	= 0.0025 meters
Elasticity of the pole	= 0.1800 Pascal
Initial force applied	= 1.0000 Newton

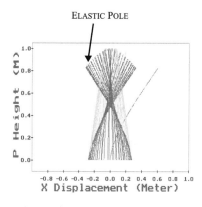

a) Initial angle = 10 degrees
Initial force = 10 Newtons

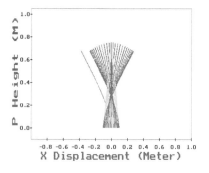

c) Initial angle = -10 degrees
Initial force = 7 Newtons

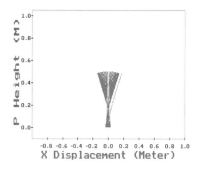

b) Initial angle = 5 degrees
Initial force = 1 Newton

FIGURE VI.9.12   Animation of the flexible pole-cart balancing system at different parameter values.

Coefficient of friction	= 0.0800
Step size (H)	= 0.0010
Upper limit of integration (t max)	= 2.0000
Freq. intermediate printouts (I freq)	= 20
Initial time (t sec)	= 0.0
Initial pole angle (theta in deg)	= 5.0000 degrees
Limitations of pole angle	= 50.000 degrees
Initial pole velocity (theta/dt)	= 0.0
Acceleration due to gravity	= 9.81 m/sq sec.

9.3.3   Analysis of Results

A computer program has been used to simulate the mathematical equations derived in Section 9.2.1. The output of this program indicates that this analysis is qualitatively correct. This needs to be verified by experiment. In order to balance the pole on top of the cart, it is necessary to control the force that is applied to the cart. The magnitude of this force is directly proportional to the angle of the pole from the vertical axis and the total weight of the cart and pole. If the angle of the pole increases or the total weight of pole and cart increases, then the magnitude of the force also increases. The direction of the force is dependent on the direction of the angle. If the pole is inclined to the left (angle is negative), then the direction of the force is also to the left; otherwise, it is the opposite. The value of the force used is selected based upon experiment within the simulation. Figure VI.9.11 shows the algorithm of the controller for this.

The initial condition of the program is, at time *t* equal to zero, the velocity and displacement of the cart are zero. The force and the pole angle can be initialized to any value (positive or negative). However, since there is a limitation on the length of the track, the value of the pole angle is limited to plus and minus forty degrees. The movement of the cart is dependent on the rate of change of the magnitude of the force and direction. The faster the applied force changes in one direction, the more oscillation there is in the movement of the cart. Furthermore, the larger magnitude of force applied to the cart, the further it travels.

In order to center the cart on the track, the distance it travels should be controlled. The average frequency of the cart's acceleration is very important here. The frequency should not be small in order to prevent the cart from hitting the track limit. The frequency should also not be too large because in that case the cart will not move sufficiently far (see Equation VI.9.26).

The strength, efficiency, and capability of the developed program have been tested by running it under various conditions. Results indicated that the flexible pole cart-balancing problem can be controlled.

9.3.4 Summary

A computer simulation of the simple rule-based control of a cart balancing a flexible pole under its first mode of vibration was presented. The appropriate dynamic equations of the system have been derived using Newton's laws, Euler analysis, and beam theories. The system can be assumed to be with or without friction. Numerical integration using fourth-order Runge–Kutta was implemented. A real time graphic representation of the cart balancing the flexible pole on a limited track can be displayed. The behavior of the system can be analyzed and observed by viewing the graphs of the pole's angle vs. time (either with or without friction), acceleration of the cart vs. time, velocity of the cart vs. time, and displacement of the cart vs. time. The Turbo Pascal language has been used to implement the computer program on an IBM PC machine.

The simulation program indicated that the possibility of balancing a flexible pole-cart system was likely. It was, therefore, decided to proceed to demonstrate on a real system without the necessary simplifications made to the model system and to explore the applicability of non-conventional control techniques to the system.

The next sections of this chapter are focused, therefore, on the development and testing of online and offline intelligent controllers using neural network algorithms and fuzzy logic systems on a real system, as well as the simulation of the application of fuzzy-neuro algorithms as an extension of the intelligent control approaches.

9.4 Neural Network Controller for the Flexible Pole-Cart Balancing Problem

For the past decade, neural networks have received a great deal of attention and are being proposed for use as powerful computational tools.[22,23] The structures of neural networks are roughly based on our present understanding of the biological nervous system. This section presents a simulation of the flexible pole-cart balancing problem as a test bed for neural network applications. As has been discussed earlier, this type of problem is more complex and highly nonlinear when compared to the classical rigid pole-cart balancing problem because it gives an additional degree of freedom to the classical system, e.g., its transverse displacement. The author has derived (see Section 9.2) the mathematical equations of the dynamics of this system and has used computer simulation to test the validity of the mathematical model. The results of this computer simulation have been used as the training data for the neural network.

The objective of the work presented in this section is to develop and test neural-network based software that learns to predict the value of the force applied to the cart at any given time in order to balance the flexible pole hinged at its root on the top of the cart. The backpropagation neural network architecture has been used to test the capability of neural networks to control the flexible pole-cart balancing problem in simulation. A backpropagation neural network has been trained by supervised learning. The network was presented with a training data set made up of pairs of patterns, i.e., an input pattern paired with a target output. Upon the presentation of this data, weights within the network were adjusted to decrease the difference between the network's output and the target output. The inputs are the elastic pole angle, rigid pole angle, velocity of the cart, and the displacement of the cart, while the output is the force applied to the cart.

9.4.1   Training and Learning in a Neural Network

Training and learning are fundamental to nearly all neural networks. A network in which learning is employed must be trained. Training is an external process or regimen. It is the procedure by which the network learns. Learning is the result that takes place internal to the network. It is the process by which a neural network modifies its weights in response to external inputs. Weights are changed when the output(s) are not what is expected.

Training is done using examples, and it can take place in three distinct ways,[23] namely, supervised, reinforcement, and unsupervised. In supervised training, the network is provided with an input stimulus pattern along with the corresponding desired output pattern. The learning law for such a network typically computes an error, that is, how far from the desired output the network's actual output really is. This error is then used to modify the weights on the interconnections between the PEs. Initial weights can be set randomly. Using this technique, a network can do things like make decisions, map associations, memorize information, or generalize.

Reinforcement training is similar to supervised training except that the exact desired output is not provided, only a grade of how well the network is working. In this type of training, the neural network only receives feedback indicating the value of the system's action. The weights are reinforced for properly performed actions and are punished for inappropriate ones. This technique is useful in those cases where supervisory information is not available.

Unsupervised training is sometimes called self-organization training. In this type of training, the network is presented only with a series of input patterns and is given no information or feedback at all about its performance levels. The network uses no external influences to adjust its weights. It looks for regularities or trends in the input signals, and it makes adaptations according to the function of the network. Even without being told whether it's right or wrong, the network still must have some information about how to organize itself. Competition between PEs can also form the basis for learning. Training of competitive clusters can amplify the responses of specific groups to specific stimuli and associate those groups with each other. For example, processing elements could be organized to discriminate between various pattern features such as vertical edges or left-hand and right-hand edges.

9.4.2 Processes Involved in the Formulation of the Flexible Pole-Cart Balancing Control: Neural Network Perspective

This section describes the various processes involved in formulating the problem from a neural network perspective and provides an effective specification of the application of a neural network to the flexible pole-cart balancing system. These processes are briefly described below:[24]

1. The decision on how the information for presentation to the neural network should be represented is very important. Since neural networks are pattern matchers, the representation of the data contained in the training sets is critical to a successful neural network solution. Clear understanding of the problem is necessary. Writing a brief narrative description of what the neural network will do is known to support this. For this work, the goal is to develop a neural network that learns to predict the amount of force exerted on the cart to balance the pole given the position of the pole, displacement of the cart, and the velocity of the cart.

2. It is important to have enough data to yield sufficient training and test sets to train and evaluate the performance of the neural network effectively. The architecture of the network, the training method, and the problem being addressed are dependent on the amount of data required for training a network. In this research, the data used to train the network are elastic pole angle, rigid pole angle, velocity of the cart, displacement of the cart, and the force applied to the cart.

3. The data sets in the input training set, as well as the desired output, should be as orthogonal as possible; that is, the variables contained in the data sets should be independent with no correlation.

4. Generally, the majority of effort in developing a neural network goes into collecting data examples and preprocessing them appropriately. The standard process is to normalize the data. Here, the requirement is that the input to each input processing element should be in the interval between -1.0 to 1.0, and the output to each output processing element should be between 0.0 to 1.0. The following approaches have been adopted for normalizing the raw data to the pole balancing problem before using it in the neural network.

   - For input values:

     - el_ang_n = el_ang_r/max_el_ang
       where:
       el_ang_n = normalized value of the flexible pole's angle
       el_ang_r = raw value of the flexible pole's angle
       max_el_ang = largest absolute value of the flexible pole's angle
     - ri_ang_n = ri_ang_r/max_ri_ang
       where:
       ri_ang_n = normalized value of the rigid pole's angle
       ri_ang_r = raw value of the rigid pole's angle
       max_ri_ang = largest absolute value of the rigid pole's angle
     - cart_vel_n = cart_vel_r/max_car_vel
       where:
       cart_vel_n = normalized value of the cart velocity
       cart_vel_r = raw value of cart velocity
       max_cart_vel = largest absolute value of cart velocity
     - cart_dis_n = cart_dis_r/max_car_dis
       where:
       cart_dis_n = normalized value of the cart displacement
       cart_dis_r = raw value of cart displacement
       max_cart_dis = largest absolute value of cart displacement

- For output values:

  force_n = force_r/max_force

  where:

  force_n = normalized value of the force exerted to the cart

  max_force = maximum value of the force exerted to the cart

  = 15 Newton

  Since the output range is from 0.0 to 1.0, the author has used two output vectors in order for the network to identify the direction of the force (negative and positive values) as well as its magnitude.

  For example, for a normalized force of −0.5, the corresponding output data are 0.5 and 0.0. 0.0 indicates that the force is going left (−). For a normalized force of 0.5, the corresponding output data are 0.5 and 1.0. 1.0 indicates that the force is going right (+).

5. Experiments must be carried out to train and test the neural network. The architecture is a specification of the neural network topology, with other attributes of the neural network such as the learning rule, activation function, update function, and learning and momentum factors. It should be kept in mind that the number of hidden layers and number of nodes in each layer are problem dependent and are empirically selected. It is necessary to vary the parameters used in the neural network such as the learning rate, error tolerance, momentum, etc. in order to get the fastest convergence.

9.4.3   Discussions: The Backpropagation Neural Network Simulator

The architecture of backpropagation neural network has been discussed by Dadios and Williams.[24] For the problem of interest, the input layer consists of four PEs because there are four input variables to the network. The output layer has two PEs since the neural network needs two outputs in order to identify the direction and magnitude of the force. In this program, the best result was obtained by using two hidden layers, each layer consisting of eight PEs (see Figure VI.9.16 for the complete structure). The equations used in this neural network program are explained in detail by Dadios and Williams.[24]

The program needs the following information from the user:

- Error tolerance — this is the difference between desired output and networks' computed output. If this is attained, the program simulation will stop.
- Learning parameter — used in scaling the adjustment to weights.
- Maximum number of cycles — a cycle is one pass for the whole training data. This will ensure the program stops even if the error tolerance is not attained.
- The total number of layers.
- The total number of processing elements for every layer.
- Momentum parameter — used in scaling the adjustments from the previous iteration and adding the adjustments in the current iteration.
- Noise — a random number added to each input component of the input vector as it is applied to the network. This will avoid getting stuck to local minima.

There are two major processes to be undertaken to construct the backpropagation network. The first is the training process and the second is the testing process. All of these processes use external files for data storage. The training process uses files input.dat, weights.dat, and results.dat. File input.dat contains exemplar pairs or patterns. Each pattern has four input variables and two output variables. Once the training process reaches the error tolerance or the maximum number of cycles, the program maintains the state of the network by saving all its weights in file weights.dat. Results of the last pattern are stored in file results.dat. In the testing process, the user will enter only the

TABLE VI.9.1 Samples of the Training Data for a Backpropagation Model with 2 Outputs

Input Data				Output Data	
Rigid Pole Angle	Elastic Pole Angle	Cart Displacement	Cart Velocity	Force Magnitude	Force Direction
0.838203	0.846334	0.482706	−0.866357	0.847458	1.0
0.647774	0.660336	0.685901	−0.694884	0.661017	1.0
0.402434	0.413959	0.845998	−0.457932	0.423729	1.0
0.121891	0.126050	0.952937	−0.165491	0.152542	1.0
−0.459220	−0.471551	0.368853	0.521115	0.474576	0.0
−0.186079	−0.192290	0.125581	0.230196	0.203390	0.0
0.105567	0.109182	−0.125581	−0.146711	0.135593	1.0
0.387285	0.398548	−0.368852	−0.440091	0.406780	1.0

Note: This is the training data used for Examples 1, 2, 3, and 4.

number of layers of the network and the processing elements for each layer. The program has assumed that the network has already been trained. External files testing.dat, weights.dat, and results.dat are used in this process. File testing.dat contains only the input patterns. When this file is presented to the network, it then uses the weights from file weights.dat to evaluate the output. The outputs from the network for all input patterns are then generated and stored in the file results.dat.

9.4.4 Discussions and Analysis of Results

The author conducted a number of different sets of experiments in this program. The training data consisted of 40 patterns (see Table VI.9.1). Different methods were used to normalize the data, the best method being the one described in Section 9.4.3. In the application of the backpropagation algorithm, a number of different layers and processing elements were tried. For a three layer architecture, the simulation did not converge. Good results were obtained for four layers. The time of convergence depended on the number of processing elements in each hidden layer. The addition of momentum parameter and noise factor also helped the simulation to converge. In this program, the best result was obtained using the following input parameters.

- Error tolerance = 0.007039.
- Learning parameter = 0.01.
- Maximum number of cycles = 3050.
- Total number of layers = 4.
- Total number of processing elements for every layer (input, hidden, hidden, output) = 4, 8, 8, 2.
- Momentum parameter = 0.01.
- Noise factor = 0.05.

Samples of the results of the backpropagation simulation are shown in Table VI.9.1.

Example 1:
The inputs are:

1. Error tolerance = 0.007039.
2. Learning parameter = 0.01.
3. Maximum number of cycles = 3050.

4. Total number of layers = 4.
5. Total number of processing elements for every layer:
   (input, hidden, hidden, output = 4, 8, 8, 2).
6. Momentum parameter = 0.01.
7. Noise factor = 0.05.

Outputs of Example 1

Input Vectors				Output Values	
Rigid Pole Angle	Elastic Pole Angle	Cart Displacement	Cart Velocity	Force Magnitude	Force Direction
0.838203	0.846334	0.482706	−0.866357	0.841673	0.999996
0.647774	0.660336	0.685901	−0.694884	0.673766	0.999993
0.402434	0.413959	0.845998	−0.457932	0.384845	0.999978
0.121891	0.126050	0.952937	−0.165491	0.162709	0.999701
−0.459220	−0.471551	0.368853	0.521115	0.487629	0.000157
−0.186079	−0.192290	0.125581	0.230196	0.177145	0.002171
0.105567	0.109182	−0.125581	−0.146711	0.143045	0.995773
0.387285	0.398548	−0.368852	−0.440091	0.399558	0.999948

Example 2:

The inputs are:

1. Error tolerance = 0.0075.
2. Learning parameter = 0.01.
3. Maximum number of cycles = 2525.
4. Total number of layers = 4.
5. Total number of processing elements for every layer:
   (input, hidden, hidden, output = 4, 12, 12, 2).
6. Momentum parameter = 0.01.
7. Noise factor = 0.05.

Outputs of Example 2

Input Vectors				Output Values	
Rigid Pole Angle	Elastic Pole Angle	Cart Displacement	Cart Velocity	Force Magnitude	Force Direction
0.838203	0.846334	0.482706	−0.866357	0.853668	0.999984
0.647774	0.660336	0.685901	−0.694884	0.693899	0.999962
0.402434	0.413959	0.845998	−0.457932	0.392296	0.999860
0.121891	0.126050	0.952937	−0.165491	0.167283	0.998675
−0.459220	−0.471551	0.368853	0.521115	0.483844	0.000078
−0.186079	−0.192290	0.125581	0.230196	0.176161	0.002150
0.105567	0.109182	−0.125581	−0.146711	0.131581	0.991830
0.387285	0.398548	−0.368852	−0.440091	0.382929	0.999813

Example 3:
The inputs are:

1. Error tolerance = 0.0075.
2. Learning parameter = 0.01.
3. Maximum number of cycles = 1360.
4. Total number of layers = 4.
5. Total number of processing elements for every layer:
   (input, hidden, hidden, output = 4, 16, 16, 2).
6. Momentum parameter = 0.01.
7. Noise factor = 0.05.

Outputs of Example 3

Input Vectors				Output Values	
Rigid Pole Angle	Elastic Pole Angle	Cart Displacement	Cart Velocity	Force Magnitude	Force Direction
0.838203	0.846334	0.482706	−0.866357	0.861767	0.999988
0.647774	0.660336	0.685901	−0.694884	0.726439	0.999981
0.402434	0.413959	0.845998	−0.457932	0.411485	0.999946
0.121891	0.126050	0.952937	−0.165491	0.142722	0.999201
−0.459220	−0.471551	0.368853	0.521115	0.474354	0.000141
−0.186079	−0.192290	0.125581	0.230196	0.186945	0.003093
0.105567	0.109182	−0.125581	−0.146711	0.134045	0.990485
0.387285	0.398548	−0.368852	−0.440091	0.390488	0.999867

9.4.5   The Physical Architecture of the Flexible Pole-Cart Balancing System

A photograph of the real system and the hardware architecture are shown in Figures VI.9.13 and VI.9.14, respectively. The specifications of the physical system are given below.

- Track length = 91.4 cm.
- Pole length = 41.0 cm.
- Mass of the cart and camera sensor = 0.755 kg.
- Additional load on the tip of the pole = 0.35 kg.
- Period of the elastic pole = 2 seconds.
- Camera system = coupled at the base of the pole and a light bulb is attached to the tip of the pole.

In the real physical system, a CCD camera is used to detect the deflection of the pole. This is coupled at the base of the flexible pole and will detect the light coming from the bulb attached to the tip of the pole. The deflection of this light corresponds to the deflection of the pole. A potentiometer is attached to the base of the elastic pole in order to obtain its angular position. To determine the distance travelled by the cart, another potentiometer is attached to the wheel that rolls on the track. The values of these sensors are then fed to the computer via an analog to digital/digital to analog converter (AD/DA converter); see Figure VI.9.14. In order to make the problem more complex, an additional load was attached on the tip of the pole equivalent to 0.35 kilograms. This has the effect of increasing the period of the elastic pole to 2 seconds.

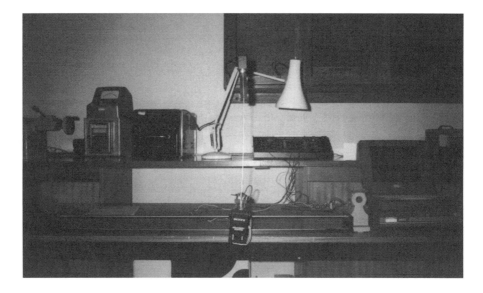

FIGURE VI.9.13   A photograph of the real flexible pole-cart balancing system (controllers developed balanced the flexible pole for an infinite length of time).

9.4.6   Online Application of the Neural Network Model to the Flexible Pole-Cart Balancing System

Figure VI.9.15 shows the online hybrid controller block diagram for the flexible pole-cart balancing problem. The backpropagation neural network controller described in Section 9.4.2 was applied to the real system with the outputs directly mapped as voltages to the actuator (see Figure VI.9.16). The training data was taken from observations of the inputs and outputs of the real system with its existing controller. The backpropagation neural network controller successfully balanced the pole for a limited period. However, this control system frequently failed due to the cart running out of track. In order to solve this problem, a hybrid control system (see Figure VI.9.15) was then applied to the physical system, the backpropagation system being overridden in extreme cases by a small rule-based supervisory system that periodically corrected extreme angles of the pole that caused the cart to decentralize on the track.

It can be seen from Figure VI.9.16 that the feedforward neural network (backpropagation algorithm) used to control the system has 2 hidden layers. Each hidden layer has 8 processing elements (neurons). The input layer has 4 processing elements and the output layer has 2 processing elements. The actual value of the weights connecting each processing element is shown in Table VI.9.2. Here, the leftmost value corresponds to the number of the layer, and the number of lines having the same leftmost value corresponds to the number of processing elements for that layer. For example, the first four lines have a leftmost value of 1. The number one corresponds to the first layer, and the four lines correspond to the four processing elements of this layer. The next 8 lines have a leftmost value of 2, indicating the second layer, and the last 8 lines have a leftmost value of 3 for the third layer.

The values next to the leftmost number on each line correspond to the weights that connect a processing element of that layer to all of the processing elements of the next layer. For example, each processing element of the first layer is connected to 8 processing elements of the second layer; hence, there are 8 values next to the leftmost number 1. The same is true for the second layer; each processing element is connected to 8 processing elements of the third layer. Hence, there are 8 values after the leftmost number 2. Finally, the third layer has only two values after number 3 because each

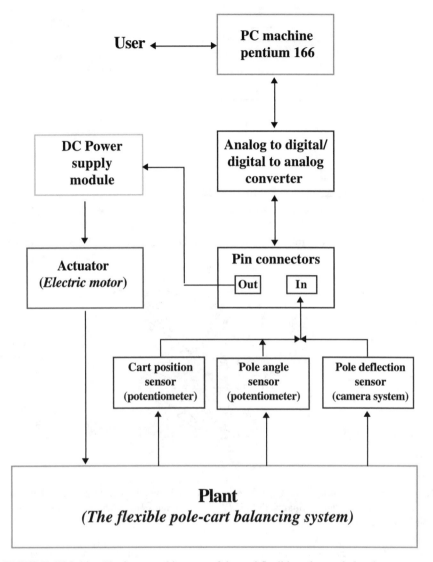

FIGURE VI.9.14   Hardware architecture of the real flexible pole-cart balancing system.

processing element is connected only to the two processing elements of the output layer. It is worth noting that the arrangement of the weight values are sequential, meaning that a processing element of a layer is connected to the first processing element, second processing element, third processing element, and so on, of the next layer. The first line of the same leftmost value corresponds to the first processing element of this layer, the second line corresponds to the second processing element of this layer, etc.

Table VI.9.3 describes the rule base of the evaluator used to correct extreme behaviors of the neural network.

It can be seen from these rules that Rules 1 to 4 take care of balancing the pole under extreme conditions, and Rules 5 to 8 bring the cart to the center of the track. Rule 1 is the condition when the pole angle inclines more to the right, while Rule 2 inclines more to the left. Rules 3 and 4 are the condition when the pole moves fast toward the inclination. Rule 5 is the condition when the cart stays to the right, the pole angle inclines to the right, and the cart moves to the right. Rule 6 is the

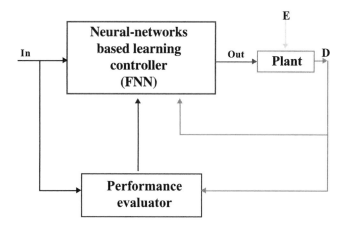

FIGURE VI.9.15 Hybrid controller block diagram for the flexible pole-cart balancing problem (online).

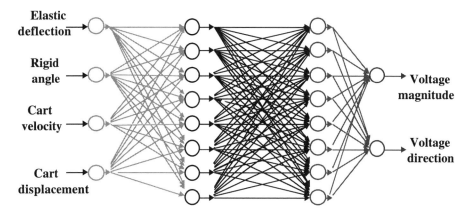

FIGURE VI.9.16 Back propagation neural network model for the flexible pole-cart balancing problem (online). (From Dadios, E.P. and Williams, D.J., *IEEE Trans. Systems Man Cybernetics*, 28(6), 895, 1998. With permission from IEEE.)

same as Rule 5 but the cart moves to the left. Rule 7 is the condition when the cart stays to the left, the pole angle inclines to the left, and the cart moves to the left. Rule 8 is the same as Rule 7 but the cart moves to the right.

9.4.7 Results of the Neural Network Physical Experiments

Experiments on the real physical system were conducted to test the neural network controller developed. The results are shown in Figures VI.9.17–VI.9.20, depicting the actual behavior of the flexible pole-cart balancing system under different conditions. Each graph shows the motion and position of the system at any time, the *X*-coordinate. The *Y*-coordinate corresponds to the measurements of the angle of the pole in degrees, the deflection of the pole in centimeters, and the location of the cart on the track in centimeters.

The motion of the cart can be analyzed by reviewing the graph of the cart displacement. The graph of the pole angle and the pole deflection shows the motion and position of the flexible pole on top of the cart. For example, Figure VI.9.17 shows the behavior of the system when it was initialized at 15.4 degrees. Here, in order to balance the flexible pole, the cart moves quickly

TABLE VI.9.2 The Values of the Weights Connecting each Processing Element of the Online Feedforward Neural Network Controller

1 −0.188662 −14.036287 −3.352839 −2.473303 1.985948 19.490234 11.239498 0.481953
1 4.956312 −2.398801 −4.827594 −7.715945 0.969971 0.219576 0.509925 −12.291900
1 1.993420 −2.410884 −16.832338 9.828149 0.335660 1.427019 −0.122610 −14.757426
1 7.953069 −4.432342 7.229471 0.493617 1.000431 −0.147754 5.767792 −7.396082
2 −2.260085 2.721589 −0.617301 0.991330 −7.796785 1.129860 5.016824 −3.787388
2 −0.694484 −13.195033 −4.532439 4.279198 1.135099 3.270727 −7.059445 −2.270613
2 −7.269268 −14.795992 −4.052489 5.050180 1.482505 1.696126 −2.924294 2.483483
2 5.767020 1.752683 −0.856498 1.665344 −7.655233 −3.964407 −0.737276 −3.415481
2 0.008674 0.932936 1.678631 0.618250 −3.230417 0.894155 −1.421489 −1.976010
2 0.980527 13.268960 5.889343 −1.131561 0.221078 −7.700460 0.936903 3.213932
2 0.087941 10.657167 4.582888 −0.344960 −2.330751 1.065558 −0.837586 −1.921857
2 2.037073 −14.794135 −5.376902 3.622092 7.990932 1.187436 1.665400 3.076303
3 −0.107427 −8.428615
3 1.923418 26.984537
3 −7.034989 −0.344512
3 0.184818 −5.594577
3 3.655073 10.139201
3 −1.026488 −8.348601
3 8.367657 1.654985
3 −0.067262 6.702623

TABLE VI.9.3 The Rule-Based Evaluator

Rule 1   If (pole angle > 2.31 degrees) then applied voltage = 5.0.
Rule 2   If (pole angle < −2.31 degrees) then applied voltage = −5.0.
Rule 3   If (pole angle > 1.01 degrees) and (pole angular velocity > 0.01 deg/s)
         then applied voltage = 3.0.
Rule 4   If (pole angle < −1.01 degrees) and (pole angular velocity < −0.01 deg/s)
         then applied voltage = −3.0.
Rule 5   If (displacement of the cart > 14 cm.) and (velocity of the cart
         > 0.01 cm/sec) and (pole angle > 1.01 degrees)
         then applied voltage = 2.0.
Rule 6   If (displacement of the cart > 14 cm.) and (velocity of the cart < −0.01
         cm/sec) and (pole angle > 1.01 degrees)
         then applied voltage = 1.1.
Rule 7   If (displacement of the cart < −14 cm.) and (velocity of the cart < −0.01
         cm/sec) and (pole angle < −1.01 degrees)
         then applied voltage = −0.5.
Rule 8   If (displacement of the cart < −14 cm.) and (velocity of the cart > 0.01
         cm/sec) and (pole angle < −1.01 degrees)
         then applied voltage = −0.2.

to the right and, after 0.35 seconds, the pole angle reached −1 degrees. To bring back the pole angle to the center, the cart then moved back to the left. The flexible pole then stabilized after 0.5 seconds, and the controller tried to bring the cart to the center of the track. Figure VI.9.18 shows the behavior of the system when it was initialized nearly at the end of the track. The controller effectively balances the flexible pole and gradually brings the cart to the center of the track. The

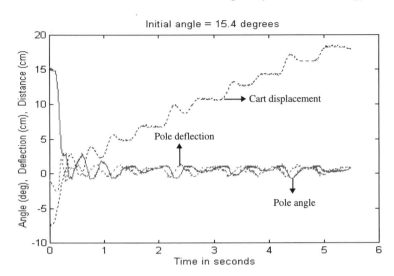

FIGURE VI.9.17   Flexible pole-cart behavior when the pole initially inclined to the right.

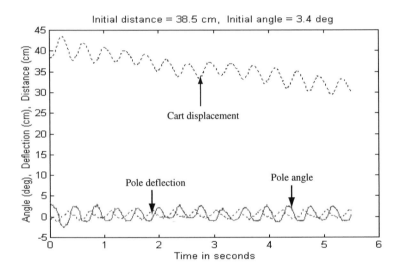

FIGURE VI.9.18   Flexible pole-cart behavior when the cart started from the right side of the track.

controller developed was tested to establish how it reacts to external disturbances applied to the system. Figure VI.9.19 shows the graphical results of the behavior of the system when the left ends of the track were elevated. Here, the controller balances the pole easily and the cart oscillates toward the center of the track. Figure VI.9.20 shows the behavior of the system when an external force is applied to the pole. Here, at 4.15 seconds, the pole was pushed to the right. Immediately, the controller reaction was to move the cart quickly to the right. The controller easily stabilizes the system and brings back the cart to the center of the track. It should be emphasized that for all of the test cases presented, the controller developed was able to control the system for an infinite amount of time.

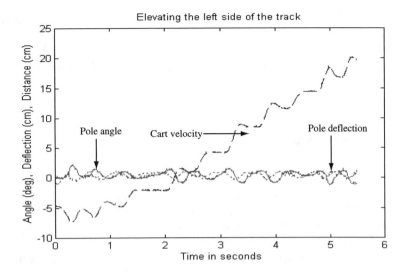

FIGURE VI.9.19 Flexible pole-cart behavior when external disturbance is applied to the system.

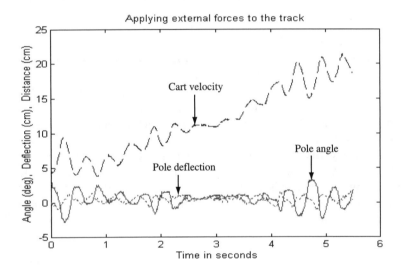

FIGURE VI.9.20 Flexible pole-cart behavior when external disturbance is applied to the system.

9.4.8 Summary

This section has demonstrated the use of neural networks in the control of a highly nonlinear system. A computer simulation of a neural network controlling a model of a cart balancing a flexible pole has been presented. The backpropagation algorithm has been used for neural network examples. The networks learned from a set of training data taken from the results of a computer simulation of the derived dynamics of the flexible pole-cart balancing system.

An online reinforcement learning hybrid neural network controller was developed to balance a flexible pole hinged root on top of the cart moving along a limited track. The physical experiments show that the controller not only balances the flexible pole for an infinite amount of time but also brings the cart to the center of the track. The learning controller developed is sufficiently robust to control the system at different initial pole angles and different initial cart positions on the track.

The stability, flexibility, and adaptability of this learning controller was tested by applying external disturbances to the plant.

The next section of this work discusses the development and test of an online intelligent controller that controls the flexible pole-cart balancing system without knowing the mathematical descriptions of the dynamics of the system, using a fuzzy logic control system.

9.5   Fuzzy Logic System Controller for the Flexible Pole-Cart Balancing Problem

This section presents a fuzzy logic system controller that controls the flexible pole-cart balancing problem. Here, the objective is to develop and test an online fuzzy logic controller that predicts the value of the force applied to the cart at any given time in order to balance the flexible pole hinged at its root on top of the cart. In this work, multiple fuzzy logic systems have been used to fuzzify the input data from the environment. There are six input data to the system (the elastic pole deflection, deflection velocity, angular position, angular velocity, cart displacement, and cart velocity).

This section begins with the discussion of the concepts and architecture of a fuzzy logic system and it continues with the development of a fuzzy logic controller for the flexible pole-cart balancing problem. The results of online experiments conducted on this controller are presented.

9.5.1   Fuzzy Logic System

A fuzzy logic system has a system design that is based on how the human brain thinks.[28] The idea arose from the desire to describe complex systems linguistically.[25] Fuzzy logic looks at the world in imprecise terms in much the same way that our own brain takes in information. The information is described in terms of fuzzy linguistic terms. These fuzzy linguistic terms are called fuzzy sets and can be regarded as sets of singletons, the grades of which are not only 1 but also range from 0 to 1. Each singleton is an element of fuzzy sets.

The concept of fuzzy sets is made precise through the definition of an associated membership function. This membership function indicates a grade of membership of each element (physical value) in a fuzzy linguistic term of interest. Fuzzy membership functions are the mechanism through which the fuzzy system interfaces with the outside world.[26] The domain of the membership function is the set of possible values for a given variable. The possible output value of the membership functions

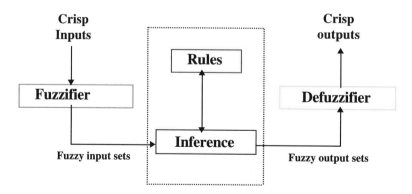

FIGURE VI.9.21   The fuzzy logic system (FLS).

is the set of all real numbers from 0 to 1. A typical choice of the shape of the fuzzy membership function is a triangular, trapezoidal, or a gaussian function.

Fuzzy sets can be combined through fuzzy rules to define specific actions. The fuzzy system can provide insight into its own operation because the fuzzy rules provide a common-sense description of the system's own action. The technique used to store and represent fuzzy rules is the fuzzy associative memory matrix (see Section 9.5.3). This matrix may have dimensions higher than two. Usually, the number of inputs, or antecedents, to the fuzzy rules determines the dimension of the matrix.

Figure VI.9.21 depicts a fuzzy logic system that is widely used in fuzzy logic controllers and signal processing applications.[27] It contains four components: fuzzifier, rules, inference engine, and defuzzifier. Once the rules have been established, a fuzzy logic system can be viewed as a mapping of inputs to outputs.

9.5.2 Application of the Fuzzy Logic Controller to the Real Physical Flexible Pole-Cart Balancing System

Figure VI.9.22 shows the fuzzy logic controller developed for the flexible pole-cart balancing problem.[24] Five fuzzy logic systems (FLS) and a rule-based evaluator are used to control the flexible pole-cart balancing system. The importance of using multiple FLSs is to minimize the memory consumption of the computer, and each FLS serves as a good filter to the noise on the input data. FLS1 is a fuzzy logic system that maps the cart displacement and cart velocity to crisp output1. Crisp output1 corresponds to the crisp numerical value that will compensate for the effect of the movement of the cart on the overall system. FLS2 is a fuzzy logic system that maps the pole angle and angular velocity to crisp output2. Crisp output2 corresponds to the crisp numerical value that will compensate for the effect of the movement of the flexible pole on the overall system. FLS3 is a fuzzy logic system that maps the pole's deflection and deflection velocity to crisp output3. Crisp output3 is the crisp numerical value that will compensate for the effect of the movement of the flexible pole, due to its deflection and deflection velocity, on the overall system. Since the contribution of the effects of crisp output2 and crisp output3 to the plant are similar, those two can be fuzzified further using FLS4. This maps crisp output2 and crisp output3 to crisp output4. Crisp output4 is the crisp numerical value that will compensate the fuzzified effect of the movement of the pole due to its angular position, angular

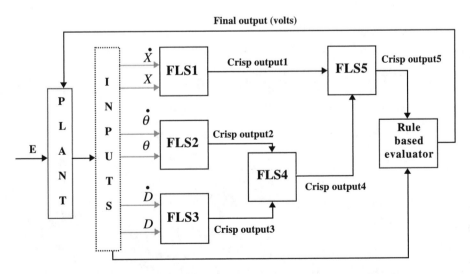

FIGURE VI.9.22 Multiple fuzzy logic controller block diagram. (From Dadios, E.P. and Williams, D.J., *IEEE Trans. Systems Man Cybernetics*, 28(6), 895, 1998. With permission from IEEE.)

velocity, deflection, and deflection velocity on the overall system. In order to obtain the overall crisp value that will compensate for the effect of the total movement of the system, crisp output1 and crisp output4 map to crisp output5 through FLS5, which will further fuzzify the fuzzified effect in FLS1 and FLS4. The FLS5 process filters the noise on the final data required to control the system. Finally, to ensure that the cart stays at the center of the track, a rule-based evaluator is used to evaluate the condition of the plant. The evaluator adds additional constant forces to those supplied by the fuzzy system when the cart exceeds particular displacements. The output of the rule-based evaluator is then fed to the plant for appropriate action. External disturbances can be applied to the plant at any time without affecting the performance of the controller.

9.5.3 Fuzzy Associative Memory (FAM) Matrix

The FAM matrix is a method of storing and representing fuzzy rules. In this controller, each fuzzy logic system (FLS) has two inputs. Each input variable has 5 fuzzy sets associated with it, which are labeled NL (negatively large), NS (negatively small), ZE (zero), PS (positively small), and PL (positively large). Note that here, ZE is a fuzzy set that would typically represent a range of values near 0, not just a single numerical value 0. The output variable has 7 fuzzy sets associated with it: *NL* (negatively large), *NM* (negatively medium), *NS* (negatively small), *ZE* (zero), *PS* (positively small), *PM* (positively medium), and *PL* (positively large). The number of inputs, or antecedents, to the fuzzy rules determines the dimension of the FAM matrix. Thus, in this controller, we are using a two-dimensional FAM matrix. Note that we are particularly concerned with the pole position if the pole angle is too large and increasing. The same is true for its deflection. These need to be corrected, regardless of the location of the cart in the track, by applying maximum force to the cart with the same direction as the inclination of the pole. Similarly, if the cart is too near the end of the track, this should be corrected regardless of the state of the flexible pole (angle and deflection) by applying maximum force to the cart toward the end of the track, making the flexible pole incline more in the other direction, thus, in turn, allowing the controller action to balance the flexible pole by applying more force to the opposite side of the cart and, at the same time, bringing it to the center of the track.

The FAM matrix for FLS1 is shown in Table VI.9.4.[28] This is used to fuzzify the cart's displacement and velocity to obtain a crisp result (output) that can compensate for the effect of the dynamic movement of the cart (due to its displacement and velocity) on the system. These rules can be interpreted as:[43]

Rule 1: IF X is NL and $\dot{X}$ is NL THEN the output result is *NL*.
Rule 2: IF X is NL and $\dot{X}$ is ZE THEN the output result is *NM*.
Rule 3: IF X is NL and $\dot{X}$ is PL THEN the output result is *PS*.

Note that X = the displacement of the cart and $\dot{X}$ = the velocity of the cart.

The FAMs of the other FLSs are shown in Tables VI.9.5–VI.9.8. The operations of these are similar to that in Table VI.9.4.

TABLE VI.9.4 Fuzzy Associative Memory Matrix for FLS1

		$\dot{X}$			
		NL	NS	ZE	PS	PL
	NL	*NL*		*NM*		*PS*
	NS		*NM*		*NS*	
X	ZE	*NS*		*ZE*		*PS*
	PS		*PS*		*PM*	
	PL	*NS*		*PM*		*PL*

TABLE VI.9.5 Fuzzy Associative Memory Matrix for FLS2

			$\dot\theta$		
	NL	NS	ZE	PS	PL
NL	*NL*		*ZE*		*ZE*
NS		*NM*		*NS*	
θ ZE	*NS*		*ZE*		*PS*
PS		*PS*		*PM*	
PL	*ZE*		*ZE*		*PL*

Note: This is used to fuzzify the pole's angle and angular velocity to obtain a crisp result (output) that will compensate for the effect of the movement of the pole (due to its angular position and angular velocity) to the entire system.

TABLE VI.9.6 Fuzzy Associative Memory Matrix for FLS3

			$\dot D$		
	NL	NS	ZE	PS	PL
NL	*NL*		*ZE*		*ZE*
NS		*NM*		*NS*	
D ZE	*NS*		*ZE*		*PS*
PS		*PS*		*PM*	
PL	*ZE*		*ZE*		*PL*

Note: This is used to fuzzify the pole's deflection and deflection velocity to obtain a crisp result (output) that will compensate the effect of the movement of the pole (due to its deflection and deflection velocity) to the entire system.

TABLE VI.9.7 Fuzzy Associative Memory Matrix for FLS4

			$\theta fs\,\dot\theta$		
	NL	NS	ZE	PS	PL
NL	*NL*		*NS*		*PS*
NS		*NM*		*PS*	
$Dfs\,\dot D$ ZE	*ZE*		*ZE*		*ZE*
PS		*NS*		*PM*	
PL	*NS*		*PS*		*PL*

Note: This is used to further fuzzify the fuzzified effect of the movement of the pole's angle and angular velocity with the fuzzified effect of the movement of the pole's deflection and deflection velocity to obtain a crisp result (output) that will compensate for the effect of the movement of the pole (due to its angular position, angular velocity, deflection, and deflection velocity) to the entire system.

TABLE VI.9.8 Fuzzy Associative Memory Matrix for FLS5

		$(\theta fs\, \dot{\theta}) fs (D fs\, \dot{D})$				
		NL	NS	ZE	PS	PL
	NL	*NL*		*NM*		*NS*
	NS		*NM*		*NS*	
$X fs \dot{X}$	ZE	*NS*		*ZE*		*PS*
	PS		*PS*		*PM*	
	PL	*PS*		*PM*		*PL*

Note: This is used to further fuzzify the fuzzified effect of the movement of the pole's angle, angular velocity, deflection, and deflection velocity with the fuzzified effect of the movement of the cart's displacement and velocity to obtain a crisp result (output) that will compensate for the effect of the total movement of the entire system.

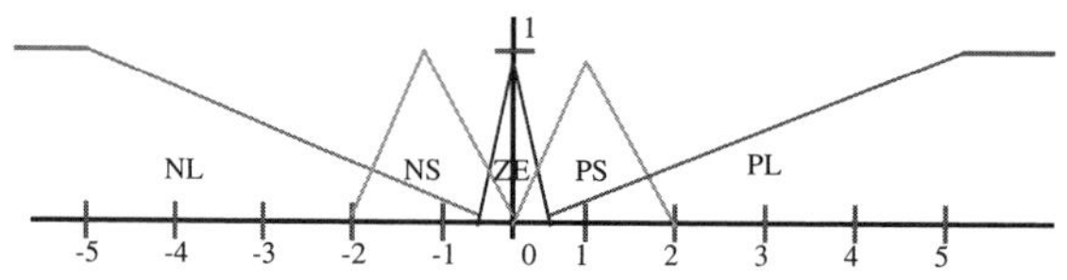

FIGURE VI.9.23 Membership functions for the cart's displacement. (From Dadios, E.P. and Williams, D.J., *IEEE Trans. Systems Man Cybernetics*, 28(6), 895, 1998. With permission from IEEE.)

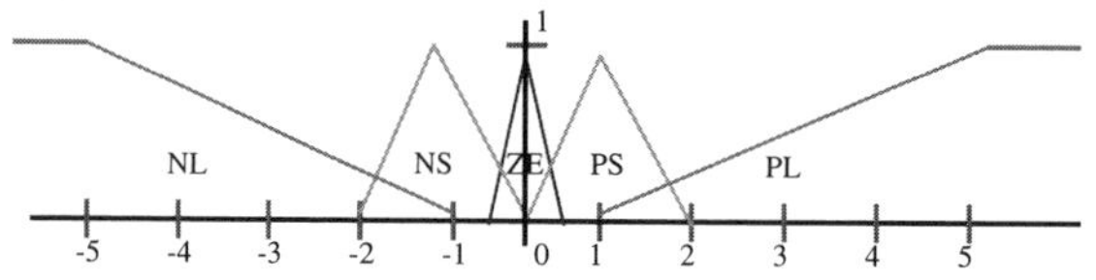

FIGURE VI.9.24 Membership functions for the cart's velocity, pole's angular position, and pole's angular velocity.

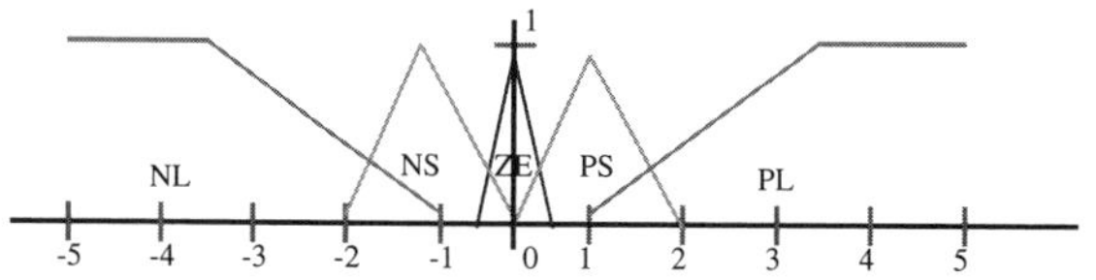

FIGURE VI.9.25 Membership functions for pole's deflection and pole's deflection velocity.

9.5.4 Membership Functions (MFs)

Membership functions map each element of a universe of discourse to a continuous membership value (or membership grade) between 0 and 1. The universe of discourse may contain either discrete objects or continuous values. In this controller, each membership function is sampled to discrete grades, whose representation depends on the type of input variables (e.g., −5.0 to 5.0 centimeters for the cart displacement, −5.0 to 5.0 degrees for the pole angle, etc.). The shape and the regions of the membership function can be changed by reassigning its grade distribution, as shown in Figures VI.9.23–VI.9.25. Determination of the shapes of each membership function usually requires some trial and error.[26] The exact shape of the functions, as well as where they intersect the horizontal axis and how much overlap exists between adjacent functions, is open to experimentation.

There are two shapes of membership functions used in this controller.

- Trapezoidal MFs, specified by four parameters $\{a, b, c, d\}$ which determine the x coordinate as follows:

$$trapezoid\ (x; a, b, c, d) = \max\left(\min\left(\frac{x - a}{b - a}, 1, \frac{d - x}{d - c}\right), 0\right) \qquad (VI.9.34)$$

- Triangular MFs, specified by three parameters $\{a, b, c\}$ which determine the x coordinate as follows:

$$triangle\ (x; a, b, c) = \max\left(\min\left(\frac{x - a}{b - a}, \frac{c - x}{c - b}\right), 0\right) \qquad (VI.9.35)$$

The leftmost and rightmost regions of the MF of Figures VI.9.23–VI.9.25 are an open trapezoid whose values for d and c are equal to 0. Other shapes of MF are triangles. Obviously, a triangular function is a special case of a trapezoidal function.

9.5.5 Defuzzifier

A defuzzifier is a way of obtaining a deterministic value, in the universe of discourse, from a fuzzy value (membership function). The most popular method of defuzzification is a center-of-gravity method.[29,30] In this research, our universe of discourse contains discrete objects, thus, our membership function is represented by sampled data (a set of elements). The center-of-gravity (CG) for discrete membership functions can be calculated using Equation VI.9.36.

$$\text{Crisp output} = \text{CG} = \frac{\sum_{i=0}^{n} O_i \cdot \mu_i}{\sum_{i=1}^{n} \mu_i} \qquad (VI.9.36)$$

where n represents the number of elements of the sampled membership function, μ_i the grade of the ith element, and o_i the output variable of the ith fuzzy set. The maximum value of n is equal to the total number of fuzzy rules in the FLS. The value of μ_i can be calculated using Equation VI.9.34 or VI.9.35. For this particular system, the output variable o_i has 7 fuzzy sets associated with it. (e.g., NL, NM, NS, ZE, PS, PM, PL). The value of the output variable is based on the voltage capacity of the actuator and, hence, has specific values in volts:

$$
\begin{aligned}
NL &= -4.75 \\
NM &= -2.65 \\
NS &= -1.35 \\
ZE &= 0.0 \\
S &= 1.35 \\
PM &= 2.65 \\
PL &= 4.75
\end{aligned}
$$

These values were derived through experimental observation of the flexible pole-cart balancing process.

9.5.6 Results of the Physical Experiments

9.5.6.1 Discussion and Analysis

The author conducted a number of different sets of experiments in this work. The first experiment was to attempt to develop a single fuzzy logic system (FLS) to control the plant (the flexible pole-cart balancing system). The application of this controller was not encouraging. The controller was very sensitive to noise from the pole deflection as the values of the deflection of the pole and the deflection velocity can vary abruptly. In order to further eliminate this noise, a controller was built with multiple FLS. This technique is effective because each FLS acts as a noise filter.

In this work, different total numbers of fuzzy rules were also applied. Attempts were made using 27, 75, and 135 rules, etc. Unfortunately, the results of the application of these controllers are not appropriate because increasing the number of rules increased the memory consumption of the computer program. Attempts to change the number of input regions (number of membership functions) to 3 regions did not give encouraging results. Best results were obtained using 5 regions (see Figures VI.9.23–VI.9.25). The size of these regions plays an important role. The more regions overlap each other, the better the result because of the design aim to use a minimum number of rules. Choosing the exact position where regions overlap is critical and depends on knowledge of the physical structure of the plant and the capability of the sensors (e.g., knowing the exact size of the track, the minimum and maximum deflection of the pole, as well as its angle for the system to operate, etc.). The shape of the regions is also important. Figures VI.9.23–VI.9.25 show that open trapezoids were used in the leftmost and rightmost regions. Whenever the plant reaches these positions (the beginning of the horizontal line and beyond), the controller gives a maximum output value to the system.

Since the output of the FLS is based on generalized results, the pairing of input variables (antecedents of the fuzzy rules) for fuzzification is particularly important. Good results occur when input data that have similar characteristics (e.g., $[\theta, \dot{\theta}]$, $[X, \dot{X}]$, $[D, \dot{D}]$, etc.) are combined. This technique is effective in building multiple FLSs. As discussed earlier, this eliminates the excessive noise on the flexible pole's deflection sensor. After fuzzifying $[\theta, \dot{\theta}]$ and $[D, \dot{D}]$, we further fuzzify the two results together.

Selecting the fuzzy associative memory (FAM) matrix plays a vital role in the process. FAM matrices of smaller dimensions are easier to deal with. Although we have 6 inputs from our plant, we have been able to reduce the size of our FAM matrix by considering two inputs at a time. It can be seen from the membership functions that the input of each FAM matrix has 5 fuzzy sets. This means that we have $5 \times 5 = 25$ possible fuzzy rules generated. However, there is no need to assign all of these rules because membership functions (see Figures VI.9.23–VI.9.25) are assigned in such a way that neighboring membership functions penetrate each other. This means that a defect in one rule can be compensated (interpolated) by the surrounding four rules. Thus, this technique enables minimization of the total number of fuzzy rules in our FLS.

In this work, the accuracy of the sensor initial values (offsets) are important. The fuzzy controller design assumes that there are zero sensor values when the system is balanced. Unfortunately, in the real physical system, it is extremely difficult to achieve this (i.e., all the values of the sensors equal zero when the pole is perfectly balanced on the center of the track). This initialization difficulty causes a slight offset in the data to the controller that leads to the cart to traverse off the track. This problem was resolved using the rule-based evaluator (see Figure VI.9.22), thus correcting for the initial transducer offset errors.

9.5.6.2 Graphical Results

The graphs in Figures VI.9.26–VI.9.29 present the results of online application of the fuzzy logic controller. These figures show the complete status of the system with respect to time (i.e., pole angle, pole deflection, and cart displacement). The movement of the cart is shown by the graph of the cart displacement, and the movement of the pole is shown by the graphs of pole deflection and angle.

Figure VI.9.26 shows the result of operating the system at initial conditions of pole angle $= -20.5$ degrees, pole deflection $= 1$ cm, and cart displacement $= 5$ cm. Here, the cart initially moved quickly to the left in order to balance the pole. After 0.5 seconds, the pole position changed to 10 degrees, causing the cart to move back to the right. Because of this movement, the pole moved back toward the

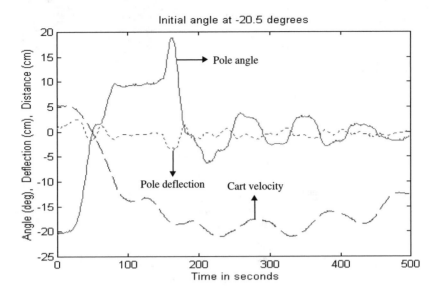

FIGURE VI.9.26   Flexible pole-cart behavior when the pole initially inclined to the left. (From Dadios, E.P. and Williams, D.J., *IEEE Trans. Systems Man Cybernetics*, 28(6), 895, 1998. With permission from IEEE.)

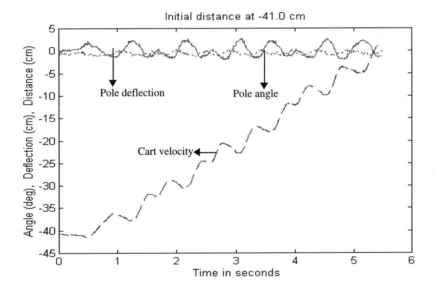

FIGURE VI.9.27   Flexible pole-cart behavior when the cart started from the left side of the track.

left, even though it reached 18 degrees at 0.8 second. The movement of the pole going left is best seen on the graph of the pole deflection. It can be seen that at 0.75 seconds, the pole deflection is −3.0 cm. This means that the pole moved quickly towards the left. Finally, at 1.0 second, the system stabilized.

Figure VI.9.27 shows the result of operating the system initially on the left end of the track. It can be seen that the controller brings the cart to the center of the track after 4.7 seconds without any difficulties in balancing the pole. Figure VI.9.28 shows the result of applying external forces to the pole. Here, at 1.3 seconds, the pole was pushed toward the left and stabilized at 2.3 seconds.

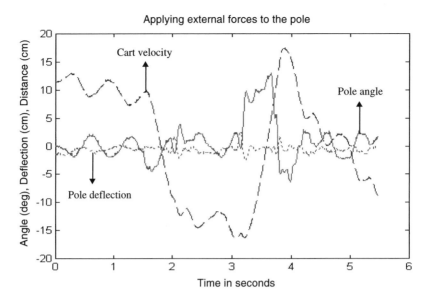

FIGURE VI.9.28   Flexible pole-cart behavior when external disturbance is applied to the system. (From Dadios, E.P. and Williams, D.J., *IEEE Trans. Systems Man Cybernetics*, 28(6), 895, 1998. With permission from IEEE.)

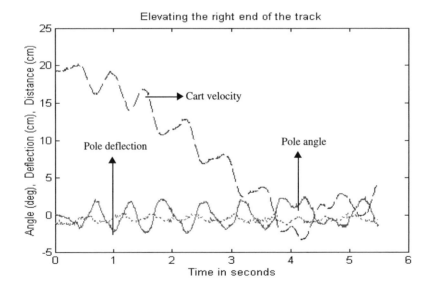

FIGURE VI.9.29   Flexible pole-cart behavior when external disturbance is applied to the system.

At 3.1 seconds, the pole was again pushed toward the right direction and stabilized at 4.1 seconds. Figure VI.9.29 shows the result of elevating the right end of the track. Here, the graph shows that the cart moved toward the center of the track, keeping the pole balanced. It can be seen from the graphs that the deflection of the pole stays below 1.0 cm as soon as the system stabilizes. The figures also show a superposed vibration at the natural frequency of the pole/mass system.

9.5.7 Summary

This section presented the applicability of fuzzy-logic based algorithms to control a cart balancing a flexible pole. The controller design includes 5 fuzzy logic systems (FLS) and a single rule-based evaluator that centers the pole on the track. Results of physical experiments show that the controller not only balances the flexible pole indefinitely but also brings the cart to the centers of the track. The controller can also easily adapt to disturbances from the external environment (e.g., moving or shaking the track randomly, elevating the height of the track on either side, pushing the pole in any direction, preventing the pole from moving further by putting an obstacle in its path). The operation of the system can also be initialized anywhere in the track. The controller is sufficiently fast to balance the system from an initial angle of 20 degrees.

The next section presents the application of a combination of neural networks and fuzzy logic system techniques in the control of the flexible pole-cart balancing problem.

9.6 Fuzzy-Logic Based Neural Network Implemented Controller (FLBNNIC) for the Flexible Pole-Cart Balancing Problem

This section presents a learning controller based on the fuzzy logic reasoning concepts that are implemented by neural networks to control flexible pole-cart balancing problem. The neural networks (feedforward and Kohonen self-organizing map) are used to determine the appropriate fuzzy rules (antecedent and consequent), to generate the membership functions, and to implement the fuzzy-logic based controller.[42] The use of this method in building the controller eliminates heuristic knowledge needed from a human expert.

The application of fuzzy logic to control problems was introduced by Mamdani in 1975[31] and was used to solve the flexible inverted pendulum problem by Dadios and Williams.[28] There are problems, however, in designing conventional fuzzy logic controllers. When a rule base representing the experience and intuition of a human expert is not available, efficient control may not be possible.[32] This problem has been addressed by several researchers and is still undergoing investigation.[33–36] Design of the input/output membership functions is another drawback as it requires the adjustment of many parameters simultaneously and is difficult to do manually. The traditional way is to assume the membership functions to be triangular or trapezoidal at the beginning and, if found to be unfit, heuristic tuning has to be tried, which shows that there is no straightforward method for designing membership functions.[33] An example of genetic approach to solve this problem can be seen in Dadios and Williams.[37]

9.6.1 Processes Involved in Formulating the Fuzzy-Logic Based Neural Networks Implemented Controller (FLBNNIC)

The objective of this research is to develop a learning controller based on fuzzy logic reasoning technique implemented by neural networks to solve the flexible inverted pendulum problem. There are 4 major processes involved in the development of this controller, namely, (1) normalization or preprocessing of the data; (2) implementation of the Kohonen self-organizing map to cluster or group the data (the total number of clusters is equal to the total number of fuzzy rules used by the controller); (3) implementation of the feedforward neural networks to calculate the fuzzy input degree of membership and the fuzzy output values; and finally, (4) the integration of the neural networks developed to execute the task of solving the problem.

The architecture of the control rules that are modeled by the NNs in this research is based on that of Sugeno et al.[38,39] wherein the rule consequents are expressed as functions of the input variables and the rule combination is done by forming a weighted sum of the output function values. The formulation of the fuzzy logic rules in Sugeno et al.'s research is in the form of:

$$\text{if } x_1 \text{ is } A_1 \text{ AND } x_2 \text{ is } A_2 \ldots, \text{AND } x_n \text{ is } A_n \text{ then } y = f(x_1, \ldots, x_n)$$

Details pertaining to this work can be seen in Sugeno et al.[38,39] In this FLBNNIC research, the consequent function is substituted by feedforward neural network. For example, as shown in Table VI.9.8, a formulated rule is of the form:

$$\text{if } (x_1, x_2, \ldots,) \text{ is } A^s, \text{ then } y^s = NN_s(x_1, x_2, \ldots)$$

where $x = (x_1, x_2, \ldots)$ is the vector of inputs, $y^s = NN_s(x_1, x_2, \ldots)$ is a neural network that determines the output y^s of the sth rule, and A^s is the membership value of the antecedent of the sth rule. All A^s or the degree of membership values in particular are generated by NN_s where $s \in (1, 2, \ldots, r)$ and r is the total number of rules.

With the foregoing form of rules, the following equation derives the final control value y^* for a given input vector x_i:

$$y^* = \frac{\sum_{s=1}^{r} \mu_A^S(x_i) \bullet \mu_s(x_i)}{\sum_{s=1}^{r} \mu_A^s(x_i)} \tag{VI.9.37}$$

where

$\mu_A^s(x_i) = $ membership function of the IF part of the sth rule.
$\mu_s(x_i) = $ output function representing the consequent of the sth rule.
$r = $ number of rules.

Figure VI.9.30 shows the overall architecture of the FLBNNIC.[42] The Nor block will preprocess or normalize the input data to be used by the neural networks. The 8 clusters formulated by the Kohonen self-organizing map were used as groups of feedforward neural networks represented by NN1 to NN8 (see Table VI.9.9). The outputs of these neural networks are equivalent to the consequent y^i of the fuzzy inference rules, as shown in Table VI.9.10. Each of these neural networks was trained using the data elements present in each cluster. For example, NN3 was trained using data elements 5, 6, 30, 31, 56, and 81 (see Table VI.9.10). The NNmem is another feedforward neural network that will calculate the value of the antecedent w^i or degree of membership of the fuzzy logic inference rule (see Table VI.9.8). This neural network was trained using all the data elements present in all clusters.

The overall crisp output of this controller is calculated based on Equation VI.9.2. This is similar to the center-of-area defuzzification formula.[40] Finally, the crisp output is de-normalized using the Denor block to acquire the required force applied to the cart to balance the flexible pole.

$$y^* = \frac{\sum_{i=1}^{8} (W^i) \bullet (y^i)}{\sum_{i=1}^{8} (W^i)} \tag{VI.9.38}$$

9.6.2 Results of the FLBNNIC Experiments

There are 250 sets of raw data used in this experiment. These data were taken from the results of a real-time computer simulation of the derived dynamics of the flexible inverted pendulum problem.[20] Each data set has 5 inputs (rigid angle, elastic angle, cart acceleration, cart velocity, and cart displacement) and one output (force applied to the cart). These data were labeled 1 to 250 and are normalized as described by Dodios and Williams.[24] Items 1 to 126 are used as the training data, and items 127 to 250 are used as the checking or testing data. The normalized training data were clustered using the Kohonen self-organizing map neural networks. Refer to Rao and Rao[41] for more discussions about

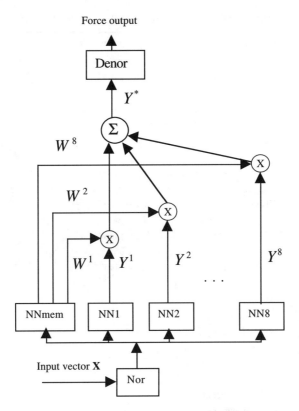

FIGURE VI.9.30 Block diagram of the fuzzy-logic based neural network implemented controller.[42] (From Dadios, E.P. and Williams, D.J., *IEEE Trans. Systems Man Cybernetics*, 28(6), 895, 1998. With permission from IEEE.)

this type of neural network. In this paper, there were 8 clusters formulated, as shown in Table VI.9.11. These 8 clusters represent 8 fuzzy inference rules of the developed FLBNNIC. Also, the total number of data sets in each cluster corresponds to the size of each membership function.

The architecture and the training parameters of each NN that determine the control value of the consequent part of the fuzzy inference rule are shown in Table VI.9.10. Here, the NN's rows of entries represent the training steps. For example, NNmem was trained in 4 steps, and the initial weights of the first step (first row) was calculated in random. However, for the succeeding steps (succeeding row), the initial weights were based on the last weights of the previous step. Note that the NN rows marked with a * in their architecture entry were used in the final developed FLBNNIC (refer to Figure VI.9.30).

Figure VI.9.31 shows the comparison of the outputs of the developed controller against the desired output of the plant at any given time. It can be seen from this figure that the developed controllers accurately obtain and control the required behavior of the plant. Figures VI.9.32–VI.9.36 present the results of several tests made on the controller when started at different initial conditions. These results show the robustness of the developed controller because it can still achieve the desired behavior even when the system experiences different and extreme conditions.

In this research, each NN that represents a consequent of a rule is considered a module of the system. To test the adaptability and modularity of the developed controller, an experiment was conducted when modules are removed. Figures VI.9.37–VI.9.40 show the results of this experiment. Figure VI.9.41 shows the prediction error when the elastic angle input variable is eliminated from all rule consequent networks. It is evident from these results that the developed controller exhibits its modularity and adaptability.

TABLE VI.9.9   Fuzzy Inference Rule of the FLBNNIC

RULE 1:	IF $x = (x_1, x_2, x_3, x_4, x_5)$ is W^1, THEN
	$y^1 = NN_1(x_1, x_2, x_3, x_4, x_5)$.
RULE 2:	IF $x = (x_1, x_2, x_3, x_4, x_5)$ is W^2, THEN
	$y^2 = NN_2(x_1, x_2, x_3, x_4, x_5)$.
RULE 3:	IF $x = (x_1, x_2, x_3, x_4, x_5)$ is W^3, THEN
	$y^3 = NN_3(x_1, x_2, x_3, x_4, x_5)$.
RULE 4:	IF $x = (x_1, x_2, x_3, x_4, x_5)$ is W^4, THEN
	$y^4 = NN_4(x_1, x_2, x_3, x_4, x_5)$.
RULE 5:	IF $x = (x_1, x_2, x_3, x_4, x_5)$ is W^5, THEN
	$y^5 = NN_5(x_1, x_2, x_3, x_4, x_5)$.
RULE 6:	IF $x = (x_1, x_2, x_3, x_4, x_5)$ is W^6, THEN
	$y^6 = NN_6(x_1, x_2, x_3, x_4, x_5)$.
RULE 7:	IF $x = (x_1, x_2, x_3, x_4, x_5)$ is W^7, THEN
	$y^7 = NN_7(x_1, x_2, x_3, x_4, x_5)$.
RULE 8:	IF $x = (x_1, x_2, x_3, x_4, x_5)$ is W^8, THEN
	$y^8 = NN_8(x_1, x_2, x_3, x_4, x_5)$.

where

x_1 = rigid angle

x_2 = elastic angle

x_3 = cart acceleration

x_4 = cart velocity

x_5 = cart displacement

W^I = membership values, $I = 1, 2, \ldots, 8$

y^I = output of NN_i, $I = 1, 2, \ldots, 8$

9.6.3   Discussions and Analysis of Results

The Kohonen self-organizing map neural network (KSOMNN) was used to cluster the data necessary to formulate the total number of fuzzy rules needed by the controller. Note also that the data element in each cluster of information is considered to be the range of values of the fuzzy membership functions. In this experiment, there are 126 sets of training data to be clustered. Each training data set has 5 inputs and one output, hence, the KSOMNN input layer consists of 5 neurons. Initial experiment results gave 18 clusters and was further enhanced to 8 clusters by adjusting the KSOMNN parameters and the distance criterion. In this experiment, the computations stopped when it was observed that the neighborhood size reached zero, no further changes could be observed in the average winning distance, and the distance criterion was satisfied. Below are the parameters used in obtaining the final 8 clusters.

Initial gain, $\alpha = 0.9$.

Initial neighborhood size, $N = 30$.

Period of α and N updates $= 50$.

Maximum cycles set to 1000.

Input size $= 5$.

Kohonen layer size $= 50$.

Average winning distance $= 0.18^+$.

Total number of cycles $= 210$.

TABLE VI.9.10   Feedforward Neural Network Architecture and Training Parameters used for the FLBNNIC to Solve the Flexible Pole-Cart Balancing Problem

Neural Networks	Architecture Input/Hidden/Output Layers	Training Parameters				
		Learning Rate	Momentum	Noise Factor	Total Cycles	Error
NNmem	5-20-20-8	0.001	0.001	0.00	187	0.1353
	5-20-20-8	0.005	0.005	0.00	790	0.0057
	5-20-20-8	0.005	0.005	0.00	1000	0.0036
	5-20-20-8*	0.010	0.010	0.00	97	0.0010
NN1	5-12-1*	0.025	0.025	0.00	4088	0.0040
NN2	5-12-1	0.025	0.020	0.05	2277	0.0083
	5-12-1*	0.030	0.050	0.00	5000	0.0037
NN3	5-12-1	0.025	0.020	0.05	5000	0.0061
	5-12-1*	0.200	0.150	0.00	84	0.0010
NN4	5-12-1*	0.025	0.050	0.00	1532	0.0058
NN5	5-12-1*	0.025	0.020	0.05	5000	0.0034
NN6	5-12-1*	0.020	0.020	0.05	289	0.0050
NN7	5-12-1*	0.020	0.020	0.05	1123	0.0050
NN8	5-12-1	0.025	0.020	0.05	65	0.0155
	5-12-1*	0.030	0.025	0.00	1185	0.0039

Source: Dadios, E.P. and Williams, D.J., *IEEE Trans. Systems Man Cybernetics*, 28(6), 895, 1998. With permission from IEEE.

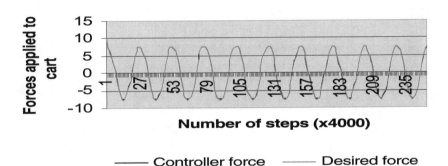

FIGURE VI.9.31   Controller behavior vs. desired behavior.

The KSOMNN clustered data (see Table VI.9.11) were investigated and analyzed, and each cluster was found to show distinct behavior as presented below.

For cluster 1:

Applied force is to the right and is decreasing.
Cart is moving toward the center from left.
Pole is going vertical from right.
Cart acceleration is decreasing (from +).
Cart velocity changes direction (from − to +).

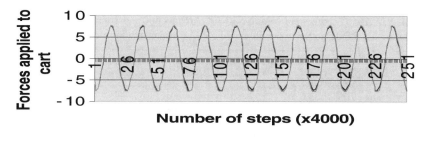

FIGURE VI.9.32   Controller behavior vs. desired behavior (initial force = −5 Newton: initial angle = −12 deg).

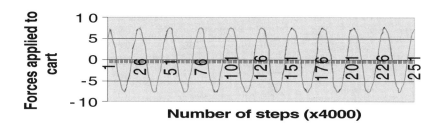

FIGURE VI.9.33   Controller behavior vs. desired behavior (initial force = 5 Newton; initial angle = 12 deg).

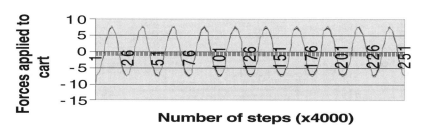

FIGURE VI.9.34   Controller behavior vs. desired behavior (initial force = −10 Newton; initial angle = −12 deg).

For cluster 2:

Applied force is to the right and is decreasing.
Cart is moving toward the center from left.
Pole is going vertical from right.
Cart acceleration is decreasing (from +).
Cart velocity is increasing (from +).

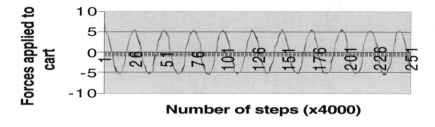

FIGURE VI.9.35 Controller behavior vs. desired behavior (initial force $= -8$ Newton; initial angle $= 8$ deg).

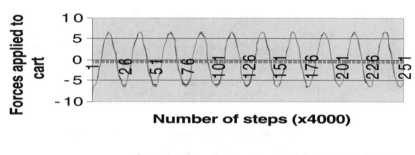

FIGURE VI.9.36 Controller behavior vs. desired behavior (initial force $= -8$ Newton; initial angle $= -10$ deg).

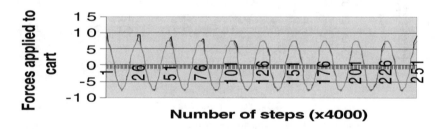

FIGURE VI.9.37 Controller behavior vs. desired behavior (Module 1 was removed).

For cluster 3:

Applied force is to the right and is decreasing.
Cart is moving toward the center from left (the cart is almost at the center of the track).
Pole is going vertical from right (the pole is almost vertical).
Cart acceleration is decreasing (from $+$).
Cart velocity is increasing (from $+$).

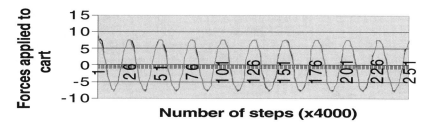

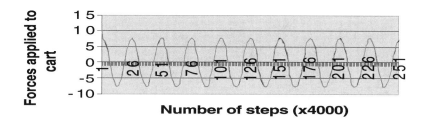

FIGURE VI.9.38   Controller behavior vs desired behavior (Module 2 was removed).

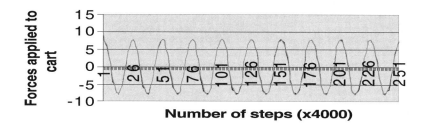

FIGURE VI.9.39   Controller behavior vs. desired behavior (Module 3 was removed).

FIGURE VI.9.40   Controller behavior vs. desired behavior (Module 4 was removed).

For cluster 4:

Applied force is to the left and is increasing.
Cart is moving away from the center to the right.
Pole is increasingly inclining and deflecting to the left.
Cart acceleration is increasing (from −).
Cart velocity is increasing (from −).

For cluster 5:

Applied force is to the left, increases, then decreases.
Cart changes direction from right to left.
Pole is increasingly inclining and deflecting to the left then goes vertical (from left).
Cart acceleration is increasing (from −) then decreases (to −).
Cart velocity is decreasing (from +).

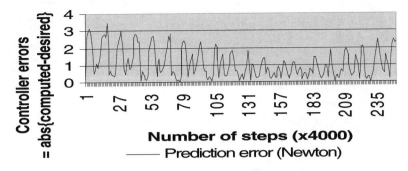

FIGURE VI.9.41 The FLBNNIC prediction error when elastic angle input variable is eliminated from all rule consequent networks.

TABLE VI.9.11 Data Clustered using Kohonen Self-Organizing Map Neural Network for the Flexible Pole-Cart Balancing Problem

Clusters/ Control Rule	Input-Output Set Number of Training Data	Total Number of Data Sets in a Cluster
R1	1, 25–27, 51–53, 77–79, 103–105	13
R2	2–4, 28, 29, 54, 55, 80, 106	9
R3	5, 6, 30, 31, 56, 81	6
R4	7, 8, 32–34, 57–59, 82–85, 107–110	16
R5	9–12, 35–38, 60–65, 86–91, 111–117	27
R6	13–16, 39–42, 66, 67, 92, 93, 118	13
R7	17–19, 43, 44, 68, 69, 94, 119	9
R8	20–24, 45–50, 70–76, 95–102, 120–126	33

Source: Dadios and Gunay.[42]

For cluster 6:

Applied force is to the left and is decreasing.
Cart is moving toward the center from right.
Pole is going vertical from left.
Cart acceleration is decreasing (from −).
Cart velocity changes direction (from + to −).

For cluster 7:

Force changes direction from left to right.
Cart crosses the track center from right to left.
Pole crosses the vertical axis from left to right.

Cart acceleration changes (from − to +).
Cart velocity is increasing (from −).

For cluster 8:

Force is to the right and is increasing.
Cart is moving away from the center to the left.
Pole is increasingly inclining and deflecting to the right.
Cart acceleration is increasing (from +).
Cart velocity is decreasing (from −).

These results demonstrate that the KSOMNN is able to accurately classify closely interrelated data into a cluster that are considered to be the values of the fuzzy membership functions. As has been shown in Figure VI.9.30, the overall architecture of the developed controller is similar to the implementation of a Sugeno-type fuzzy logic system. This architecture is composed of 9 independent feedforward neural networks trained using a supervised learning. The first neural network (NNmem) uses 126 training data sets similar to the one used by the KSOMNN. The outputs of this neural network are the fuzzy logic degree of memberships whose values are between 0 to 1. The remaining neural networks (NN1 to NN8) were trained individually using the data found for each cluster; for example, NN1 uses 13 training data and NN2 uses 9 training data (see Table VI.9.11 for the data of each cluster). The output of NN1, NN2 , . . . , NN8 is the fuzzy output value or consequent of each fuzzy logic rule whose values are between −1 and +1. The final output of the system is calculated similar to the center-of-gravity formula shown in Equation VI.9.38.

The performance of the developed controller was tested when a neural network module was removed from the system. This experiment is similar to observing the performance of a person when part of his body is disabled (e.g., losing a hand or a foot). It was found from the result of this experiment, as shown in Figures VI.9.37–VI.9.40, that the developed controller is still able to control the system with high accuracy. However, a particular observation has been made that minute errors occurred on data sets that belonged to the removed module. This observation shows explicit reason similar to a person's ability and difficulty in holding multiple objects when one of his hands is not available. Hence, the result of this experiment evidently confirmed the adaptability and modularity of the developed system.

The robustness of the developed controller was tested by operating the system at different conditions, for example, by changing the initial pole angle and the initial forces applied to the cart. The results shown in Figures VI.9.32–VI.9.36 confirmed that this controller is robust. The performance of the controller is also tested when the data of the elastic angle are eliminated from all of the rule networks. This idea evaluates the significance of the elastic angle in generating the desired behavior of the system. It was found from this experiment that significant errors occurred, as shown in Figure VI.9.41. Therefore, this result established the importance of the information about the elastic angle for the controller to solve the problem. Furthermore, this also supports existing claims that the dynamics of the flexible inverted pendulum problem are sufficiently testing and more complex compared to the rigid inverted pendulum problem.

9.6.4 Summary

This section showed the applicability of developing a controller based on fuzzy logic reasoning implemented by neural networks to control a highly nonlinear system. The controller developed was tested off-line to solve the flexible inverted pendulum problem and gave excellent results. The controller was found to be robust, modular, and adaptive. As compared to the traditional fuzzy logic controller, this developed controller does not need heuristic knowledge from the human expert because the neural networks are what formulate the fuzzy rules, membership functions, fuzzy output values, and degrees of membership.

The use of the Kohonen self-organizing map neural network in partitioning the training information led to an acceptable and minimal number of fuzzy rules needed by the controller and accurately allocated the ranges of each membership function. The elastic angle input variable showed significant correlation to the desired control behavior, as exhibited and demonstrated in Figure VI.9.41. Thus, it can be inferred that the flexible pole-cart balancing system is a sufficient testing benchmark, more complex, and more highly nonlinear than the traditional rigid inverted pendulum system for developing advanced learning controllers.

Work to be performed in the future is the verification of possible modularization of the neural network that will generate the membership functions.

9.7   Conclusions and Recommendations

Table VI.9.12 shows the comparison of functions of the intelligent controllers developed for the flexible pole-cart balancing system and the route of their evolution. The following should be noted.

The neural network experiment determined that the pole deflection was the variable most prone to error. This was, therefore, measured directly in the fuzzy controller.

The fuzzy controller uses six inputs, $x, \dot{x}, \theta_r, \dot{\theta_r}, d_e, \dot{d_e}$, while the neural network and evaluator only uses $x, \dot{x}, \theta_r, \theta_e$ (noting that $\theta_e = d_e/L$, where d_e is deflection and L is the length of the pole). The neural network structure does not use $\dot{d_e}$ and $\dot{\theta_r}$ as input variables. A fuzzy controller using $x, \dot{x}, \theta_r$, and d_e as inputs was not able to balance the pole for an infinite length of time. The 4-8-8-2 neural network structure was built upon the understanding of the mechanics and control of the system determined during the construction of the simulation. It would appear that the 4-8-8-2 structure includes some implicit knowledge of the nonlinear dynamics of the system and the relationships between $\theta_e, \dot{\theta_r}$, and $\dot{d_e}$. The weight structures of the network used to control the simulation and the test bed are different.

The fuzzy control system was the only system to use the input $\dot{d_e}$. This is a particularly noisy input. The cascaded approach was therefore implemented to remove the effect of the noise. An off-line fuzzy-neuro controller showed very interesting results. The neural network unsupervised learning method was able to determine the appropriate number of fuzzy rules to control the system.

There are two areas that can be identified for future work:

- The construction of an online fuzzy neural network controller to control the flexible pole-cart balancing system. Critical here is the determination of whether there is an improvement in the performance of the overall system by identifying new sets of parameters for fuzzy logic system decision-making using gradient-descent optimization techniques based on a neural network formulation of the problem.
- Perhaps the most demanding task is to develop an intelligent controller that can control the position of the tip of the flexible pole irrespective of the position and movement of the cart. This problem is much more representative of the true needs of the application of flexible robots. Determining the appropriate input-output variables for the task would make a good starting point for this work. It may be possible to apply the multiple fuzzy logic systems technique described earlier. In this case, knowledge of the dynamic equations of the system may not be needed.

TABLE VI.9.12   Comparison of Functions of Controllers Generated

	Simulation Inputs	Real Time/Real System Inputs		Uniqueness
Problem				Unique problem Extending the pole balance test case. Flexible robot link application.
Simulation	Rule base/Force base			First mode approximate dynamics.
Neural Network	**Kohonen**			Neural network simulates rule base.
	Feedforward 4-8-8-2 $x, \dot{x}, \theta_r, \theta_e$ Force output	**Feedforward** $x, \dot{x}, \theta_r, d_e$ 4-8-8-2	Volt output. Limited time. Runs out of track	Neural network control of flexible pole-cart balancing.
		Feedforward + $x, \dot{x}, \theta_r, d_e$ 4-8-8-2	Volt output. Evaluator. Balance infinite time. Cart brought to center of the track. With external disturbance.	Robust to external disturbances.
Multiple Fuzzy Logic		$x, \dot{x}, \theta_r, d_e$	Balance for limited time.	Cascade system with filtering. Robust to external disturbances. Compact for memory. Real time. No need for plant dynamic equation.
		$x, \dot{x}, \theta_r, \dot{\theta}_r,$ $d_e, \dot{d}_e$	Volt output. Balance infinite time. Cart brought to center of the track. With external disturbance. Needs *a priori*	
Neuro Fuzzy Algorithm	INPUTS: Rigid angle. Elastic angle. Cart acceleration. Cart velocity. Cart displacement.			Neural network archetecture is derived from fuzzy logic concept. No *a priori* knowledge required to build the fuzzy logic rules and neural network architecture.

References

1. Nguyen, D.H. and Widrow, B., Neural networks for self learning control system, *IEEE Control Syst. Mag.*, 18, 1990.
2. Anderson, C.W., Learning to control an inverted pendulum using neural networks, *IEEE Control Syst. Mag.*, 15, 31, 1989.
3. Barto, A.G., Sutton, R.S., and Anderson, C.W., Neuronlike adaptive elements that can solve difficult learning control problems, *IEEE Syst. Man Cybernetics*, 13, 834, 1983.
4. Widrow, B. and Smith, F.W., Pattern recognition control systems, Computer Information Science (COINS) Symposium Proceedings, Washington, DC., 1963, 288.

5. Tolat, V.V., and Widrow, B., An adaptive broom balancer with visual inputs, Proceedings of the International Conference on Neural Networks, San Diego, CA, July 1988, 2–641.

6. Geva, S. and Sitte, J., A cartpole experiment benchmark for trainable controllers, *IEEE Control Syst. Mag.*, 40, 1993.

7. Zhang, B., Experiments in learning control using neural networks, Ph.D. thesis, University of Strathclyde, 1991.

8. De Luca, A. and Ulivi, G., Iterative learning control of robots with elastic joints, Proceedings of the IEEE International Conference on Robotics and Automation, Nice, France, May 1992, p. 1920.

9. Li, C.J. and Sankar, T.S., Systematic methods for efficient modelling and dynamics computation of flexible robot manipulators, *IEEE Trans. Syst. Man Cybernetics*, 23(1), 77, 1993.

10. Meng, C.H. and Chen, J.S., Dynamic modelling and payload-adaptive control of a flexible manipulator, Proceedings of thee IEEE International Conference on Robotics and Automation, 1988, 488.

11. Wang, D., and Vidyasagar, M., Control of a flexible beam for optimum step response, Proceedings of the IEEE International Conference on Robotics and Automation, 1987, 1567.

12. Gervater, W.B., Basic relations for control of flexible vehicles, *AIAA J.*, 8(4), 666, 1970.

13. Chapnik, B.V., Heppler, G.R., and Aplevich, J.D., Modelling impact of a one-link flexible robotic arm, *IEEE Trans. Robotics Automat.*, 7(4), 479, 1991.

14. Chapnik, B.V., Heppler, G.R., and Aplevich, J.D., Controlling the impact response of a one-link flexible robotic arm, *IEEE Trans. Robotics Automat.*, 9(3), 346, 1993.

15. Matsuno, F., Asano, T., and Sakawa, Y., Modelling and quasi-static hybrid position/force control of constrained planar two-link flexible manipulators, *IEEE Trans. Robotics Automat.*, 10(3), 287, 1994.

16. Cannon, R.H. and Schmitz, E., Initial experiments on the end-point control of a flexible one-link robot, *Int. J. Robotic Res.*, 3(3), 62, 1984.

17. Fukuda, T., Flexibility control of elastic robotic arm, *J. Robotic Syst.*, 2(1), 73, 1985.

18. Hastings, G.G. and Book, W.J., A linear dynamic model for flexible robotic manipulators, *IEEE Control Syst. Mag.*, 7, 61, 1987.

19. Hastings, G.G. and Book, W.J., Experiments in the optimal control of a flexible arm, Proceedings of the IEEE International Conference on Robotics and Automation, 1986, 1024.

20. Dadios, E.P. and Williams, D.J., The flexible pole-cart balancing system, Proceedings of the 27th International Symposium on Industrial Robots, Milan, Italy, October 6–8, 1996, 515.

21. Case, J. and Chilver, A.H., *Strength of Materials and Structures*, 2nd ed., 1971.

22. Asar, A.U. and McDonald, J.R., A specification of neural network applications in the load forecasting problem, *IEEE Trans. Control Syst. Technol.*, 2, 135, 1994.

23. Butler, C. and Caudill, E., *Understanding Neural Networks*, MIT Press, Cambridge, MA, 1992.

24. Dadios, E.P. and Williams, D.J., Application of neural networks to the flexible pole-cart balancing problem, Proceedings of the IEEE Systems, Man, and Cybernetics International Conference, Vancouver, Canada, October 22–25, 1995, 2506.

25. Yamakawa, T., A fuzzy inference engine in nonlinear analog mode and its application to a fuzzy logic control, *IEEE Trans. Neural Networks*, 4(3), 496, 1993.

26. Welstead, T., *Neural Networks and Fuzzy Logic Applications*, John Wiley & Sons, New York, 1994.

27. Mendel, J.M., Fuzzy logic systems for engineering: a tutorial, *Proc. IEEE*, 83(3), 354, 1995.

28. Dadios, E.P. and Williams, D.J., Multiple fuzzy logic systems: a controller for the flexible pole-cart balancing problem, Proceedings of the 1996 IEEE International Conference on Robotics and Automation, Vol. 3, Minneapolis, MN, April 24–26, 1996, 2276.

29. Pedrycz, W., *Fuzzy Control and Fuzzy Systems*, John Wiley & Sons, New York, 1989.

30. Kosko, B., *Neural Networks and Fuzzy Systems*, Prentice Hall, Englewood Cliffs, NJ, 1992.

31. Mamdani, E.H. and Assilian, S., An experiment in linguistic synthesis with a fuzzy logic controller, *Int. J. Man Mach. Studies*, 7(1), 1, 1975.

32. Parl, Y.M., Moon, U., and Lee, K.Y., A self-organizing fuzzy logic controller for dynamic systems using a fuzzy auto-regressive moving average (FARMA) model, *IEEE Trans. Fuzzy Syst.*, 3(1), 75, 1995.

33. Takagi, H. and Hayashi, I., NN-driven fuzzy reasoning, *Int. J. Aprroximate Reasoning*, 5(3), 191, 1991.

34. Hayashi, I. et al., Construction of Fuzzy Inference Rules by NDF and NDFL, *Int. J. Approximate Reasoning*, 6, 241, 1992.

35. Keller, J.M., Hayashi, Y., and Chen, Z., Additive hybrid networks for fuzzy logic, *Fuzzy Sets Syst.*, 66(3), 307, 1994.

36. Jang, J.S. and Sun, C.T., Neuro-fuzzy modelling and Control, *Proc. IEEE*, 83(3), 378, 1995.

37. Dadios, E.P. and Williams, D.J., A fuzzy-genetic controller for the flexible pole-cart balancing problem, Proceedings of the IEEE 3rd International Conference on Evolutionary Computation (ICEC '96), Nagoya, Japan, May 20–2, 1996, 223.

38. Sugeno, M. and Nishida, M., Fuzzy conrol of a model car, *Fuzzy Sets Syst.*, 16, 103, 1985.

39. Takagi, T. and Sugeno, M., Fuzzy identification of systems and its application to modelling and control, *IEEE Trans. Syst. Man Cybernetics*, SMC-15(1), 116, 1985.

40. Tsoukalas, L.H. and Uhrig, R.E., *Fuzzy and Neural Approaches in Engineering*, John Wiley & Sons, New York, 1997.

41. Rao, B. and Rao, H.V., *C++ Neural Networks and Fuzzy Logic*, 1st ed., 1993.

42. Dadios, E.P. and Gunay, N.S., A fuzzy logic based neural network controller for highly nonlinear systems, *Int. J. Knowledge Based Intelligent Eng. Syst.*, 4(4), 254, 2000.

43. Dadios, E.P. and Williams, D.J., Nonconventional control of the flexible pole-cart balancing problem: experimental results, *IEEE Trans. Systems Man Cybernetics*, 28(6), 895, 1998.

10

A Control Architecture for Mobile Robotics Based on Specialists

C.V. Regueiro
University of A Coruña

M. Rodriguez
University of Santiago de Compostela

J. Correa
University of Santiago de Compostela

D.L. Moreno
University of Santiago de Compostela

R. Iglesias
University of Santiago de Compostela

S. Barro
University of Santiago de Compostela

10.1 Introduction VI-337
10.2 General Description of Specialist-Based
     Architecture VI-339
     Components of a Specialist • Communication Between
     Specialists • Control Agent of a Specialist
10.3 Application of Specialist-Based Architecture
     to Mobile Robotics VI-343
     Specialist Structure • Reactivity
10.4 Operation of SBA-Robotics VI-346
     Execution of a Task Order in Pilot • Reactivity
10.5 Comparison with Other Architectures VI-354
     Comparison Criteria • Related Work
10.6 Discussion VI-358
     Acknowledgments VI-359
     References VI-359

In this chapter, we present a new control architecture for mobile robots that emphasizes the encapsulation of data, tasks, and control into independent modules that specialize in specific functions of the system and which we call specialists. The resulting architecture is modular and flexible and facilitates the distributed implementation and heterogeneity of its components. The experiments carried out on a Nomad 200 robot executing various navigational tasks demonstrate that this specialist-based architecture is suitable for the design and synthesis of autonomous intelligent control systems that need to tackle multiples tasks such as avoiding static and moving obstacles in real time and in real, dynamic, and complex environments, as occurs in the majority of mobile robot applications.

10.1 Introduction

The design of a control architecture for a mobile robot gives rise to a number of important challenges.[1–3] Robots operate in and interact with complex, dynamic, and unpredictable environments by means of imperfect sensors and effectors. The environment in which a robot moves cannot be modelled completely and has its own, often unknown, dynamics. Sensorial information is always incomplete since sensors can only measure a very small number of environmental variables. Measurements are usually imprecise and even ambiguous. All this implies that the sensorial data have to be constantly monitored and interpreted in order to obtain sufficient knowledge of the environment.

0-8493-1121-7/03/$0.00+$1.50

Conversely, a robot needs to be capable of planning the realization of multiple objectives, which may even be self-contradictory. Robots must be autonomous, i.e., they must act independently, without outside assistance, due to which they must be capable of resolving any difficulty that may arise during interaction with the environment. And, of course, they must operate in real time, which means that time for decision making is always limited.

In general, these requirements demand high processing power, which comes into conflict with the very limitations of the mobile robot with regard to load capacity, size, and available energy. Due to this, the different tasks that a mobile robot carries out internally must be continually prioritized in accordance with the requirements and resources available at each instant. All these facts make control architecture a critical element for mobile robots, which totally conditions its applicability to real problems.

The first control architectures used in mobile robotics were highly hierarchical and operated along the sense–model–think–act cycle.[4,5] These were quickly shown to be too inefficient and fragile for real-time operation. Reactive architectures[6,7] solved this problem by carrying out each task in a single module (behavior) that receives and processes only those data that it needs. Nevertheless, these architectures stored little information about the environment, and it is difficult to coordinate the different behaviors and carry out complex, high-level tasks.

Hybrid deliberative/reactive control architectures[8,9] resolve the dilemma by breaking down the functions of the system into levels with decreasing amounts of abstraction. The lower levels are of a reactive nature, while the higher ones tend toward deliberation and even to the planning of tasks to be realized, often by means of a symbolic-type processing. All this facilitates the design, coordination, and execution of the tasks without losing the capability to react when faced with unexpected events. Nevertheless, one drawback is that there is no specialization in each layer, and, generally, these architectures cannot by easily implemented in a distributed manner.

There are also distributed control architectures,[1,10] which facilitate the exploitation of the capacity for the joint calculation of multiple processing elements, thus permitting acceptable response times. The main drawbacks of this approach are that task coordination at a global level becomes more difficult, and communications between modules form a bottleneck in the operation of the system.

Besides considerations that are characteristic of the domain, in the implementation of a control architecture, the properties required for any software system should be taken into account: modularity, flexibility (easily modified), scalability (easily expanded), robustness, efficiency, etc. It is also important to reuse the highest number of components or elements possible.

In this work, a new software architecture model is presented that emphasizes the encapsulation of data, tasks, and control in modules that specialize in specific system functions, which we call specialists. Specialization can be continually refined according to need, giving rise to an architecture with various functional and abstraction levels.

In the same manner as with hybrid architectures, the upper levels deal with high-level tasks, mainly by using abstract information of a symbolic nature, while the lower-level specialists, which are closer to the sensors and effectors, operate in a more reactive manner, with simpler and more data-specific processing.

The relative independence between the specialists, which is a product of encapsulation, allows the heterogeneous implementation of each specialist and gives a distributed implementation with the consequent improvement in the use of computational resources. Unlike traditional distributed architectures, successive specialization facilitates the implementation of complex tasks and considerably reduces communication costs, albeit at the expense of a certain degree of rigidity in the architecture as a whole.

The rest of this work is structured in the following manner. In the following sections, we describe the specialist-based architecture (SBA) and its application in mobile robotics. We then go on to illustrate how the proposed architecture runs a task and how it responds to unforeseen situations, and we compare our proposal with other more well-known and important control architectures in mobile robotics.

10.2   General Description of Specialist-Based Architecture

The idea behind specialist-based architecture is to encapsulate those tasks that are closely linked among themselves, together with the control necessary to realize these tasks and the data needed in order to do so, into blocks called specialists. Each specialist is responsible for carrying out and controlling the tasks that have been assigned to it, for processing the data that are necessary for carrying out these tasks and, generally, for resolving any problems that may arise while they are being carried out.

Each specialist can be made up of a number of agents (Figure VI.10.1), which carry out and control some of the sub-tasks and process part of the data. Likewise, an agent may be configured in turn as a specialist with its own agents on an internal level. This means that the architectural scheme can be replicated in each agent and, in this sense, we can also call it specialist fractal architecture (SFA).

Specialists and their agents form a structure which is based on a hierarchical task organization, where complex tasks are progressively broken down in a natural manner the deeper one gets (Figure VI.10.2), until those tasks are simple enough to be carried out by the lower-level agents. Reciprocally, the data that are associated with the tasks become more and more abstract in the opposite direction, starting with the simplest (reactive) tasks, which are carried out in the lower specialists and which principally use the raw data supplied by the sensors.

The hierarchical organization of the architecture imposes a certain degree of rigidity to the system, but, in exchange, it facilitates the design and synthesis of complex tasks and reduces communications. Nevertheless, it may also excessively limit the exchange of information between

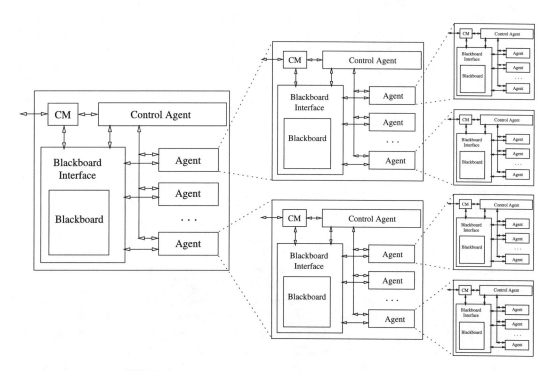

FIGURE VI.10.1   Overall scheme of specialist-based architecture.

Tasks Descomposition

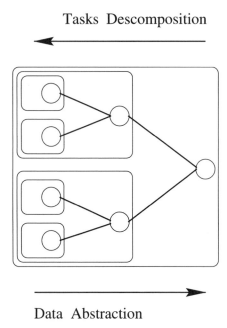

Data Abstraction

FIGURE VI.10.2   Illustration of the hierarchical breakdown of tasks and data abstraction in specialist-based architecture.

specialists, due to which there is a need for mechanisms that speed up these interchanges as far as possible (Section 10.2.1).

The encapsulation that is proposed in the specialist-based architecture has a number of advantages. On one hand, it implies a greater degree of modularity for the system, greater design organization, and more simplicity in the design of the control. The organization of the architecture is adapted directly to the existence of the hierarchy at both task and data levels.

On the other hand, the proposed encapsulation facilitates an independent implementation for each specialist, as well as their distributed and efficient realization. Distributed execution makes it possible to obtain greater calculation power and to process various tasks in parallel, even in a redundant manner, which translates into greater robustness for the system as well as a lower response time.[10] It also improves the scalability of the system and leads to a more efficient use of computational resources.

Although distributed systems usually have a communications bottleneck, the proposed encapsulation minimizes the effects of this problem since, in general, the microworld that it affects, and which is affected by a specialist, is included in it, which considerably reduces the need for the exchange of information.

10.2.1   Components of a Specialist

Each specialist can be implemented in a different manner. Nevertheless, for the sake of greater uniformity and simplicity, we have opted to use one single software architecture for all of them. Of course, this architecture must be relatively simple to implement and sufficiently flexible in order to be able to adapt to the needs and requirements of each specialist, which may differ greatly.

Figure VI.10.1 shows that each specialist is basically implemented using a blackboard-based architecture,[2,11] i.e., it is made up of a set of agents that share data by means of a shared memory (blackboard), and which are coordinated by a single control agent. The exchange of information with the rest of the specialists is established by means of a communications module (CM in Figure VI.10.1).

The advantage of using a blackboard-based architecture over other software architectures is that it facilitates the heterogeneous implementation of the agents, the inclusion of new data and agents, and the modification of the existing ones. This flexibility is of prime importance in domains where new methods and algorithms are continually being developed and where information is still not clearly structured, nor are dependencies completely established.

Conversely, the carrying out of tasks in a specialist is delegated to the agents, whose operation is coordinated and supervised with the control agent. If one of these agents stops functioning or functions incorrectly, the specialist does not stop working, it simply loses the function that is associated with that agent.

Nevertheless, there are certain differences between a specialist and a classical blackboard architecture. The first is that the specialist blackboard should not be merely a store for data, which may be structured to a greater or a lesser degree, rather it should be an active system that is capable of generating events each time the information it contains is updated. These events are fundamental for improving the reactivity of the specialist and of the system as a whole, as shall be seen in the following sections.

The second difference is that the agents of a specialist can, in turn, be implemented as specialists, giving rise to the possibility that the architecture is replicated at any level of depth, even at the lowest level, where the specialists are simple independent processes which we call terminal agents. There is no limit as to how these terminal agents may be implemented.

Lastly, the control agent of a specialist responds to the task orders coming from its immediately higher specialist and to the events, both internal (generated in the blackboard itself) and external (generated by other specialists) that affect the behavior of the specialist. Therefore, in the same manner as classical distributed architectures, activity may be initiated in any agent/specialist of the system. In this way, we succeed in permitting the specialist and the architecture to react in time to any change in the environment. Section 10.4 illustrates how this works by way of examples.

10.2.2 Communication Between Specialists

Internally, in a specialist (Figure VI.10.1) the control agent deals with the exchange of control information with the agents, and the blackboard interface is responsible for the interchange of data between the agents, including the control agent, and the blackboard. At an external level, all communications are centralized in the communications module (CM in Figure VI.10.1). On one hand, it deals with exchanging the control information with the control agent of its immediately higher specialist. On the other hand, it is responsible for the interchange of data with any specialist in the architecture, due to which the data may be shared independently of the hierarchy of the specialists.

The communication system must fulfill the following requirements. First, it must be independent of the place of execution of each specialist. On the other hand, it must be orientated toward an exchange of information that is fundamentally asynchronous (task orders and events), and, except in specific cases, not very intense (low bandwidth). Lastly, and perhaps most importantly, it must be sufficiently flexible and must converse on the basis of specifications with a high level of abstraction. In this way, communications are easily interpreted, and modifications, which, conversely, are inevitable, are simple to implement.

The KQML communications language[12] fulfills all of these requirements. With this, it is possible to define a complex, abstract, and modular task order syntax, data can be exchanged independently of their nature, and events can be defined and sent. Furthermore, since the messages are in ASCII code, it is extremely portable and general. For all of these reasons, KQML was chosen for our implementation.

Another type of communication system, which recently has been greatly used, is that based on the CORBA standard.[13] With this protocol, two objects may exchange information in a manner that is transparent for the programmer and independently of the language in which they are defined and of the platform and operating system in which they are carried out. Even though the use of CORBA would be very simple given that all dependencies regarding communications are concentrated in the CMs, it was rejected due to its lower efficiency, higher consumption of computational resources, and the lesser degree of portability with regard to KQML.

10.2.3   Control Agent of a Specialist

The objective of the control agent is to coordinate and supervise the operation of the rest of the elements of the specialist. Its function is two-fold: on the one hand it waits for the task orders (TOs) remitted by its immediately higher specialist, carries them out, and communicates the responses; on the other hand, it must be alert and act suitably in the case of an event for which it is responsible.

The control agent divides each TO into a set of tasks that constitute a plan P. In this plan, the resources needed by each task, to which agent it is delegated, and the priority of each task are specified; all this determines at what instant and in which manner it will be run. A plan also sets (e.g., inhibits) the conditions under which events in the specialist blackboard are generated (*Reactivity Schemes* in Figure VI.10.3).

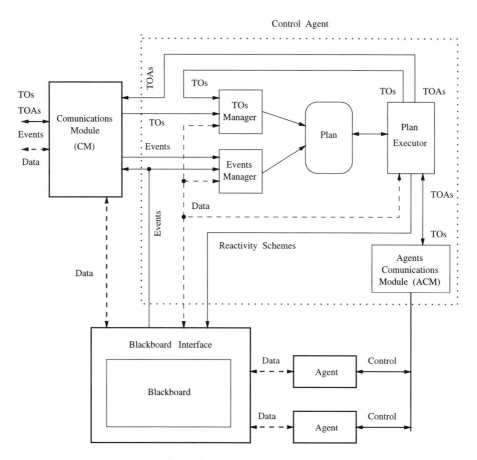

FIGURE VI.10.3   Internal scheme of the control agent of a specialist.

In order to fulfill its dual functions, the control agent requires three concurrent modules (Figure VI.10.3): the TOs_Manager, the Events_Manager and the Plan_Executor. The first holds the TOs from its immediately higher specialist, and, in order to execute them, associates to them a plan that depends on the state of the system and on the knowledge available about the environment, both of which are stored in the blackboard. The second also associates a plan, but to the events received, either for the specialist blackboard itself or from another specialist.

The Plan_Executor module interprets the plan: it translates each task into a TO, sends it to the appropriate agent, and awaits its reply (TOA, task order answer). On the basis of each reply (e.g., whether the execution has been carried out correctly or not), it continues interpreting the rest of the plan. When the plan is terminated or when there is an unsolvable error, the Plan_Executor sends the corresponding TOA to its immediately higher specialist (Figure VI.10.3). The dialog between the control agent and its agents is channelled through the Agents_Communications_Module (Figure VI.10.3).

10.3 Application of Specialist-Based Architecture to Mobile Robotics

The properties of SBA make it interesting in a number of domains.[14,15] The mobile robotics domain is, nevertheless, one of the most demanding since the environment is dynamic, complex, unstructured, and unpredictable; sensors and effectors are imperfect and one has to operate in real time, combining reactive behaviors with others of a deliberative and planned nature.

The basic objective of a mobile robot consists of carrying out the tasks that are assigned to it, moving from one point to another in the environment as efficiently as possible and without colliding with any other object, stationary or mobile. In order to do so, certain aspects regarding the priority of tasks, resources and the time available for realizing them, and the knowledge available about the environment have to be considered.

10.3.1 Specialist Structure

The specialist structure of SBA in its application to mobile robotics (SBA-Robotics) is shown partially in Figure VI.10.4, where it is possible to see the breakdown of the tasks and the abstraction of the data. Thus, Master deals with planning the realization of the tasks that are assigned to the robot. However, it delegates the planning and execution of movements to Navigator and the interaction with other systems, people, or robots to Dialogue.

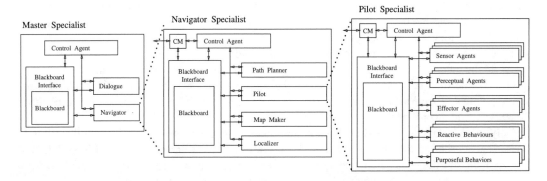

FIGURE VI.10.4   Specialist-based architecture applied to mobile robotics (SBA-Robotics).

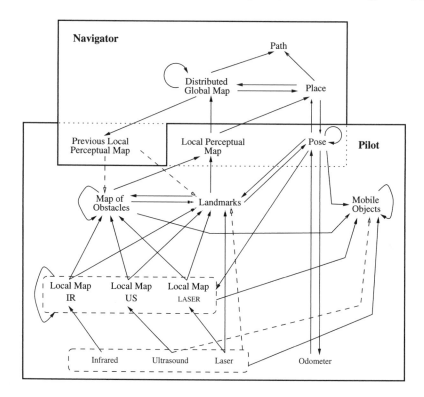

FIGURE VI.10.5   Partial structure of data in SBA-Robotics. The solid arrows show dependencies and the dotted arrows show signs, which are useful for focusing the robot's attention. Rectangles show to which specialist each datum belongs.

The specialist Navigator uses Pilot to move the robot around in its local environment and to extract information from sensorial data, with which Map_Maker draws up and maintains a map of the environment (distributed global map in Figure VI.10.5). This map is used by Localizer to determine in which region of the map (place) the robot currently is, according to the position (pose) and the objects present in the environment (local perceptual map) that are perceived by Pilot.

Lastly, Path_Planner uses the distributed global map to trace a route (path) that the robot can follow in order to reach its selected destination from its current position (place). The route obtained in the map is later translated into a series of steps or tasks that are to be carried out by the agents in Navigator. The control agent of Navigator supervises and coordinates the execution of the route.

Pilot carries out the tasks ordered by Navigator by means of purposeful behaviors and includes reactive behaviors for the robot's safety. Each behavior is implemented by an agent that calculates a movement command according to the data perceived in the environment. The effector agents integrate the commands that are generated by the different active behaviors (there may be more than one) and send the final command to each of the robot's actuators. The purposeful behaviors that are implemented are Move_To and Follow_Wall.[16–18] The reactive behaviors are Avoid_Contact, Keep_Off, and Avoid_Mobile_Object.[19]

Only local and recent information that is supplied by the robot's sensors and which the different sensor agents send to the blackboard (ultrasound, infrared, laser, and odometer in Figure VI.10.5) are available in Pilot. The perceptual agents deal with integrating these sensorial data with the data from the same region of the environment stored in Navigator (previous local perceptual map) in order to obtain pertinent information for the robot: the local perceptual map. To improve perception

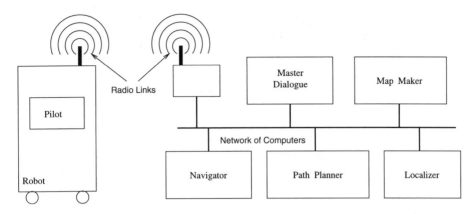

FIGURE VI.10.6   Possible distribution of the specialists and agents in specialist-based architecture for application in mobile robotics.

and obtain more or better data, the perceptual agents can also generate movement commands (*active perception*).

At present, there are four perception agents implemented in Pilot. One draws up and maintains a short-term map with the obstacles that surround the robot (obstacle map).[20] Another detects moving objects near the robot (mobile objects). A third perceives the pertinent elements in the environment (landmarks), mainly walls, corridors, and doors, which are fundamental for the relocalization of the robot. This task is carried out by the fourth agent, which constantly updates the position of the robot (Positioner).

In order to implement our architecture, we have a network of computers to which it is possible to connect a robot by means of radio links (Figure VI.10.6). The main aim is to maintain the most vital functions in the robot's computers so that, in the case of problems, the safety of the robot can be ensured and so that there is sufficient capability for reaction. The remaining specialists are distributed according to their computational needs and their requirements with regard to response time. Nevertheless, the ideal situation would be for both Navigator and Master to be run in the robot, probably with simplified versions of their agents, or under a multiprocessor architecture with sufficient computation capacity.

10.3.2   Reactivity

A mobile robot must respond in real time to any significant change in its environment. In SBA, the blackboards are active information stores capable of generating events which require a suitable and sufficiently rapid response. These events are linked to certain changes in the data available in the blackboard. When the reaction is carried out in the same specialist in which the event was produced, we talk of local reactivity. In the following section, its operation is described by way of an example.

At times, the reactive response is not adapted to the hierarchy that is followed by the breakdown of tasks and data abstraction, as the dependencies between the data and their use in a mobile robot are numerous and complex (Figure VI.10.5). This means that, on these occasions, an event that is produced in a specialist has to have a response in another specialist. An example of this is the presentation of a warning to a user when battery levels are low: the event occurs in Pilot, while the response is in Dialogue.

To avoid this difficulty, SBA sends the event from where it is generated to where it may find a response, so the system has the capacity to obtain reactive responses on a global level (*global reactivity*). The following section shows how this reactivity works.

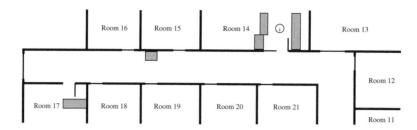

FIGURE VI.10.7   Partial view of the Electronics and Computer Science Department. The circle represents the robot and the obstacles appear in grey.

10.4   Operation of SBA-Robotics

In order to illustrate how our architecture operates, we now go on to describe how Navigator operates when it receives the task order* (TO_1): "**Move To** *Room 17*." For these tests, we use the environment partially shown in Figure VI.10.7 and a Nomad 200 mobile robot that is capable of turning with a zero radius and which has a rangefinder laser, one ring of 16 ultrasound sensors, another ring of 16 infrared sensors, two tactile sensors, and one radio Ethernet (Figure VI.10.8).

The control agent of Navigator associates a plan (P_1) to TO_1 that can be divided into the stages shown in Figure VI.10.9. The localization of the robot is carried out by the control agent of Navigator in two separate parts. First, it orders Pilot (TO_1^1) to detect and send the relevant details on the immediate environment (local perceptual map) and the robot's position (pose) to the blackboard in Navigator. Second, with the data that has been received, the control agent orders Localiser (TO_1^2) to determine in which place of the environment it is most probable that the robot is located (*Room 14*). Knowing where the robot is, Navigator control orders Path_Planner (TO_1^3) to calculate, on the basis of the stored map of the environment (global distributed map), the best route for the required displacement. The global distributed map is a topological map (a graph), but it has metric annotations, due to which a route is made up of an ordered set of nodes from this map. Figure VI.10.10 shows the route found for the trajectory in Figure VI.10.7.

Once a route has been selected, the control agent of Navigator executes it: it translates the first section into one or various task orders that its agents can carry out. If the execution is adequate, the following section is translated, and the process is repeated until the route has been completed. For example, the nodes of the route in the same passageway are followed in one single wall-following task. Figure VI.10.11 shows the path that the robot has followed after Pilot has successfully executed all the tasks received from Navigator control (Figure VI.10.12). Figure VI.10.13 shows to which agent each TO is aimed.

Once the route has been executed, the robot is localized once again (TO_1^{11}). If the robot is at the destination, P_1 finishes and Navigator sends a response of adequate execution (TOA_1) to its immediately higher specialist, Master. Otherwise, the route is replanned (TO_1^{14}) and the process is repeated. If it is concluded that the required displacement cannot be carried out (e.g., there is no feasible route or the number of attempts is too high), then the control agent aborts the execution of the plan and reports the motive of the cancellation to Master. Figure VI.10.14 shows some of the possible instantiations of P_1.

---

* Video recordings of some tests are available at http://www-gsi.dec.usc.es/areas/robotica_cas.htm.

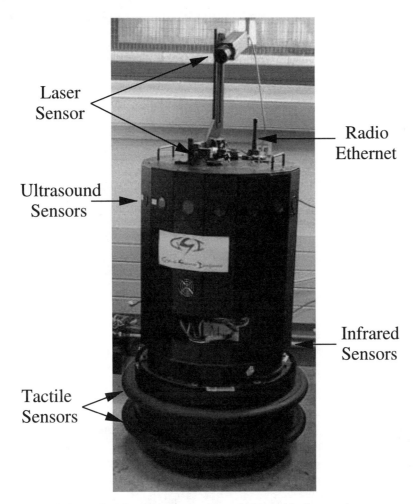

Laser
Sensor

Radio
Ethernet

Ultrasound
Sensors

Infrared
Sensors

Tactile
Sensors

FIGURE VI.10.8　　Mobile robot Nomad 200 used in the experiments.

1.　Localise robot
2.　Plan route to desired destination
3.　Translate and execute each section of route
4.　Localise robot
5.　If robot is not at destination, go back to 2

FIGURE VI.10.9　　Stages into which the **Move To** task is divided.

The plan is executed dynamically and, if a TO is not successfully concluded, the control agent must redirect the situation. For example, if the position of the robot is checked and it is not the one that was predicted (e.g., point A in Figure VI.10.15), then the route in progress is aborted, Pilot (TO_1^{12} in Figures VI.10.13 and VI.10.14) is ordered to remit its local perceptual map, and Map_Maker (TO_1^{13}) is ordered to update this information in the distributed global map. Lastly, the route is replanned (TO_1^{14}) and the process is repeated.

An example of how the Navigator control responds when confronted with an error in the execution of a section of the route can be seen in Figure VI.10.16, which shows the first part of the robot's

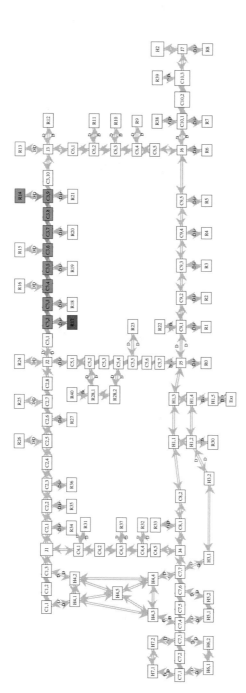

FIGURE VI.10.10 Route found between rooms 14 and 17 marked on the topological map of the entire environment (Department of Electronics and Computer Science). In this graph, nodes represent distinguishable places in the environment. These are classified into corridors (C), corridor juntions (J), rooms (R), and halls (H). Connections representing free space are not labelled, and doors are marked as D.

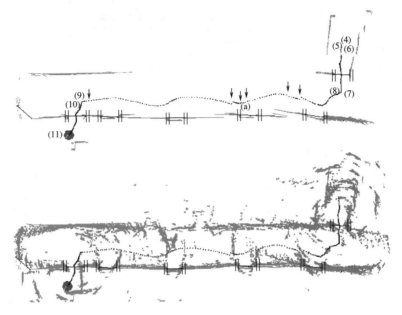

FIGURE VI.10.11   The robot's trajectory: (up) only laser data and detected doors; (down) also sonar data. Arrows show recalibrations of the odometrical system with respect to corridors detected by sonar.

TO_1^4	Turn Turret 1.1 Right
TO_1^5	Localise Robot R14
TO_1^6	Go Through Door
TO_1^7	Go to Angle 90 right
TO_1^8	Follow Wall left until Door 4 left Or Distance 1310
TO_1^9	Localise Robot C3,2
TO_1^{10}	Go Through Door

FIGURE VI.10.12   Sequence of tasks generated by Navigator in order to execute the route shown in Figure VI.10.11.

route between room 17 and room 15 (area *A* in Figure VI.10.15). In this case, Pilot has not detected the door to room 15, partly because the robot has had to go around an obstacle close to that door. The Navigator control has detected the problem, first, because Pilot has only found one door, and second, because the location calculated by Localizer as the final position of the robot was not the one that was predicted (*C3, 7*). As has already been mentioned, the solution consists of attempting the maneuver once again, and also planning and executing a new route from the place actually occupied by the robot. As can be seen in the general view given in Figure VI.10.15, this new attempt was successful. As was to be expected, the correct detection of the doors in a passageway is very important for the satisfactory functioning of the system.

Another demonstration of the robustness of the execution of a route takes place in the section labelled *B* in Figure VI.10.15. In this section, Pilot has detected only two of the three doors that are in the passageway leading to room 9. Nevertheless, as the final location of the robot is the appropriate one (*C6, 4*), and the last door to be detected is at the correct distance with respect to the beginning of the passageway, Navigator control decides to adhere to the plan and go through the last door that was detected.

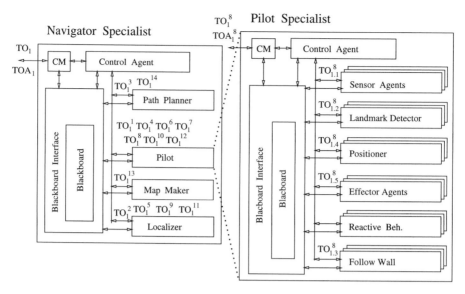

FIGURE VI.10.13   TOs sent during the execution of the displacement task shown in Figure VI.10.11.

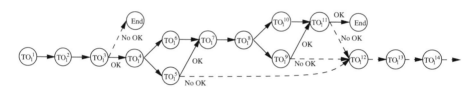

FIGURE VI.10.14   Possible instantiations of the plan associated in Navigator to "**Move to** *destination*" for the task sequence shown in Figure VI.10.12.

10.4.1   Execution of a Task Order in Pilot

Figure VI.10.13 also shows how Pilot executes the task order TO_1^8: "Follow Wall *left* until *Door 4 left* Or *Distance 1310*." Plan P_1^8 from Figure VI.10.17 is associated to this order. All TOs in Pilot are carried out by following the same plan but with a different purposeful behavior.

The first task of the plan consists of configuring the sensors through sensor agents ($TO_{1.1}^8$) as well as reactivity schemes (e.g., modifying security distance). Then, each time new sensorial data are received, a series of tasks are carried out in a cyclical manner until the termination conditions are fulfilled. These conditions are either imposed by the TO that is received, or are implicit (e.g., maximum execution time). When one of the conditions is fulfilled, the plan is concluded and Pilot tells Navigator (TOA_1^8) whether the order has been adequately carried out (TO conditions) or not (implicit conditions).

After receiving new sensorial data, the first perceptual agent and the purposeful behavior that must guide the movement of the robot are activated in parallel, in our case, Landmark_Detector ($TO_{1.2}^8$) and Follow_Wall ($TO_{1.3}^8$). A second perceptual agent, Positioner, is activated once the first has finished ($TO_{1.4}^8$). The last task of the cycle is to integrate the generated commands and order the turret and base effector agents that remit the result to the robot ($TO_{1.5}^8$).

The Landmark_Detector agent deals with the detection of landmarks of interest in the environment. It currently uses the data supplied by the laser to detect walls and doors and the data from the ultrasound sensors to detect walls and passageways. Moreover, and within the philosophy of

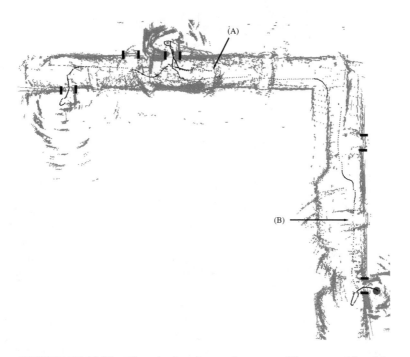

FIGURE VI.10.15   The robot's trajectory from room 17 to rooms 15 and 9.

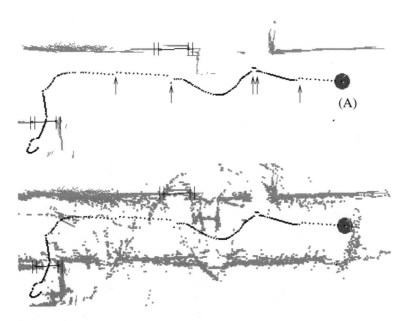

FIGURE VI.10.16   First part of the robot's route between room 17 and room 15 shown in Figure VI.10.15: (up) only laser data and detected doors; (down) also sonar data. Arrows show odometric recalibrations.

active perception, it also calculates the commands that are necessary in order to move the robot and to improve its perception. In practice, movements are concentrated on maintaining the correct orientation of the turret (at 45°) toward the contour in which the doors are to be detected. This orientation is fundamental due to the strong limitations of the laser on the Nomad 200 (it can only

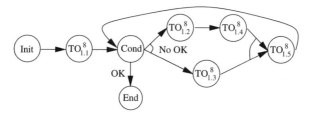

FIGURE VI.10.17   Plan P_1^8 associated to TO_1^8 in Pilot.

detect objects at a range of between 0.5 and 2.5 m and within an angle of approximately 15° with respect to the sensor axis).

As can be seen, on the basis of the same sensorial acquisition, it is possible to generate multiple commands within the execution of the same plan. In order to integrate them, we use an arbitration technique based on selecting the highest priority command, i.e., the one generated by the most important agent. There are more complex techniques for arbitration among behaviors that result in a smoother final movement for the robot (e.g., those based on fuzzy logic[21,22]) but the selection method that has been chosen is more than sufficient for our current requirements.

Lastly, we would point out that the Positioner agent deals with calculation of the positioning and orientation of the robot (pose) and determining any errors made by the odometry. Real pose is determined by comparing the landmarks that are detected (principally, passageways) with previously stored landmarks, either detected in a prior exploration phase or supplied by the user. When the pose error is excessive (arrows in Figures VI.10.11 and VI.10.16), the recalibration of the odometrical system and the repositioning of the landmarks detected are calculated. The most difficult task of the Positioner agent is to calculate the angle of the base because there is no sensor attached to it (except tactile ones), so indirect measurements must be used (nowadays, laser data).

10.4.2   Reactivity

While Pilot responds to a task order, there may be unforeseen events in the presence of which it has to react. In SBA, the blackboards are active information stores capable of generating events linked to certain changes in their data. One such situation arises when an object is too close to the robot (point (a) in Figure VI.10.11). Pilot's blackboard detects the problem because it checks all sensorial data each time they are stored, and, if any distance measurement is too low, it generates the corresponding event (E_a in Figure VI.10.18) and remits it to its control agent (*internal reactivity*).

Pilot's response depends on its current state and the information in its blackboard. Most frequently a plan, P_a, is activated and consists of two tasks: first, ordering the reactive behavior Keep_Off (TO_a^1) to calculate how to move the robot in order to maintain a minimum distance from the objects that surround it; second, activating the turret and base effector agents (TO_a^2) in order for them to realize this maneuver.

Figure VI.10.19 shows an enlargement of the point labelled (a) in Figure VI.10.11. Here, the response of the Pilot specialist to an event can be seen. In this case, the robot goes too close to the wall while following the left-hand contour, confused by false echoes from the sonar that are generated by the door. When the robot is too close to the wall, the Pilot blackboard generates the event E_a, and as this is an internal event, it is sent to the control agent (Figure VI.10.18). As has already been mentioned, the response of the latter is to activate the Keep_Off agent and send the command that is generated to the robot. The situation is repeated with each new sensorial acquisition until the robot is sufficiently distant from the wall and no event is generated.

The correct integration of behaviors is always important, but it is especially relevant in tasks such as passing through a door. This type of situation puts the reactivity of a control architecture to the test since the robot has to move very close to the door frames and events are continuously activated. Although in this work only a small number of examples are shown, in all tests that have been carried out the robot has successfully passed through all doors. The robot only collides with a frame when it is orientated in such a manner that its ultrasound sensors cannot detect it. This happens very rarely,

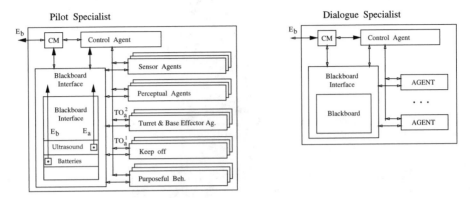

FIGURE VI.10.18   Implementation of internal and external reactivity in SBA.

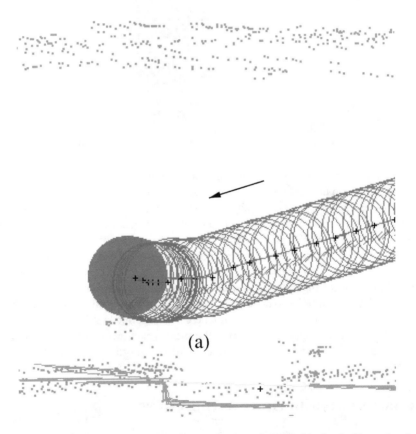

FIGURE VI.10.19   Enlargement for the region (a) shown in Figure VI.10.11, clearly illustrating the integration of the reactive responses (dark circles) with the execution of a plan (light circles). Dots and lines represent sonar and laser data, respectively.

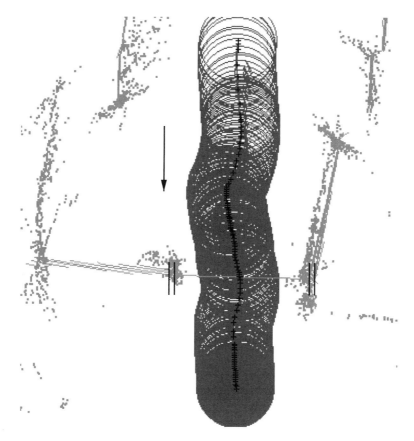

FIGURE VI.10.20   Example of how the robot passes through a door.

only when the frame is very narrow (e.g., double doors). Figure VI.10.20 shows how the robot passes through the first door on the route, i.e., the double door of room 14.

At times, the reactive response is not adapted to the hierarchy that is followed by the breakdown of tasks and data abstraction because the dependencies between the data and their use in a mobile robot are numerous and complex (Figure VI.10.5). On these occasions, an event that is produced in a specialist has to have a response in another specialist (*external reactivity*). An example of this is the presentation of a warning to a user when battery levels are low. This event is detected in the Pilot blackboard (E_b); however, given that it is an external event, it is sent to the communications module (CM in Figure VI.10.18) in order to be sent by the appropriate specialist, in this case, Dialogue. The Pilot CM determines the direction of the Dialogue CM and remits a KQML message with the data of the event. When the Dialogue CM receives the message and recognizes it as an event, it sends the message directly to its control agent (Figure VI.10.18) in order for the suitable response to be executed in the same manner as if it were an internal event.

10.5   Comparison with Other Architectures

10.5.1   Comparison Criteria

As with any software application that must control a physical device that interacts with a real environment, mobile robot control architecture must comply with certain design (modularity,

TABLE VI.10.1   Degree of Compliance of the Proposed Comparison Criteria
for Each of the Principal Paradigms for Control Architectures In Mobile Robotics

Architecture	Robustness	Reactivity	Behavior Integration	Flexibility	Scalability
Hierarchical	−−	−−	−−	−	−−
Reactive	++	++	−−	+−	+−
Hybrid	+	+	+	+−	−
Blackboard-based	−	+	++	++	−
Agent-based	+	+−	+−	−	+−
SBA	+	+	++	++	++

flexibility, reliability, efficiency, portability, etc.) and control (predictability, robustness, reactivity, behavior integration, adaptability, autonomy, etc.) requirements. In order to compare the different control architectures, we have selected a limited and manageable number of these requirements which, on one hand, stress the characteristics that we deem to be essential for the control of an autonomous mobile robot (AMR) and, on the other hand, highlight the most significant differences existing between the different proposals. We have selected the following comparison criteria:

- Flexibility — ease in carrying out modifications coming about from changes in system specification, e.g., in data representation
- Scalability — ease of the incremental implementation of the functional characteristics, modules, or tasks that it is necessary to incorporate into the system without this adversely affecting performance
- Robustness — capability of carrying on operating in anomalous situations (faulty input data, unexpected events, failures, etc.) without the system crashing
- Reactivity — capability of the system to act suitably in real time when confronted with any sudden modification in the environment
- Integration of behaviors — of a reactive, deliberative, and planned nature with the aim of obtaining a coherent and intelligent final behavior for the robot.

There are three fundamental paradigms for control architectures developed specifically for mobile robotics: hierarchical architectures, reactive architectures, and hybrid deliberative/reactive architectures. In addition, other general computational paradigms have been used to implement control architectures in robotics, notably, blackboard-based and agent-based architectures. Table VI.10.1 gives a summary of our evaluations of the characteristics of each of these architectures and their comparison with SBA.

The two main limitations of hierarchical architectures[4,5,23] are their low degree of robustness and their lack of reactivity. In the former case, given that the system functions are executed sequentially, a fault in any of them will result in the system crashing. Conversely, the activation of any reaction implies that the information must run through the entire execution cycle, giving rise to totally prohibitive response times which preclude the execution of any type of reactive behavior and, thus, its integration with other behaviors.

Finally, as hierarchical architectures are organized around an internal model of the world, flexibility and scalability are severely affected. Any change in the type of data, in the manner of their representation, or, simply, in the algorithms used to process them results in a high level of complexity. At the same time, adding new tasks implies a more complete model of the world and, thus, results in more complexity, more difficult maintenance, and less efficiency.[24]

Reactive architectures[6,7,25,26] have been designed to be extremely robust and reactive. On one hand, behaviors are executed independently and in parallel. On the other hand, they have direct access to the sensorial data and can directly generate commands for effectors with very short delay times. Nevertheless, they are incapable of developing deliberative or planned behavior since no information about the world is stored.[24]

Flexibility in reactive architectures is high within each behavior module. On a global level, the arbitration and fusion methods mean that the architectures are fairly rigid and are not suitable for modifications. Theoretically, reactive architectures should be scalable, but, in practice, the resulting behavior is more difficult to predict and control as the number of behaviors increases and their mutual interactions become more complex and unexpected.

Hybrid deliberative/reactive architectures[8,9] combine reactive and deliberative planning parts in order for the benefits of one to compensate for the shortcomings of the other, which is reflected in the criteria. On one hand, the behaviors of the reactive part supply the robustness and reactivity that are necessary for real-time operation in dynamic and unpredictable environments. The reactivity is perhaps somewhat slower than in reactive architectures because it is subjected to the supervision and control of the higher layers, which makes it possible to modulate reactive responses. On the other hand, reactive planning supplies a flexible and robust method for operating in open and not totally modelable environments, since it is specifically designed to adapt the execution of plans to the conditions of the environments, thus making the integration of reactive, deliberative, and planned behaviors simpler and more immediate.

The flexibility of hybrid architectures is greater than for hierarchical ones, but it is still limited by the strong dependence on the models of the environment. It can be said that these architectures are not designed to facilitate changes in the representation of data or their processing. The scalability of these architectures is still low. The principal reason for this is that when new tasks are added, new behaviors are required in the lower levels. This increase means that these layers become more difficult to implement and, due to possible undesirable interactions between behaviors, their coordination by the higher layers becomes more sensitive.

Blackboard-based architectures[2,11] are not very robust, e.g., if the blackboard or control module ceases to function, the system crashes. There is, however, robustness at the knowledge base (KB) level. Reactivity is possible given that the KBs are activated on the basis of the data from the blackboard (including sensorial data), but special attention has to be paid to the design of the control module in order to speed up the execution of the KBs and to avoid unnecessary waiting time, which is very common in traditional implementations of blackboard-based architectures. The integration of different types of behaviors is carried out in a natural manner in the control module, where it is decided which knowledge bases activated by data (reactive behaviors) are executed in accordance with the objectives of the system (planned and deliberative behavior).

The flexibility of blackboard-based architecture is high as the KBs are totally independent among themselves, and all interchanges of data are based in the blackboard. Nevertheless, they are not scalable given that as new data, knowledge bases, and tasks are added, the problem of bottlenecks arising at both the blackboard interface and the control module becomes more acute.

Agent-based architectures[1,10] are robust in the sense that there may be multiple agents to carry out the same function. Their real weak point is to be found in communications: if they fail, the system is rendered useless. Moreover, they do not have a specific support for implementing reactivity, which appears to be restricted to the local field of each agent. There is no support in these architectures for the integration of planned, deliberative, or reactive behaviors except collaboration between the agents of the architecture. Nevertheless, this collaboration is always decentralized and lacks global coordination.

Agent-based architectures are very flexible in the sense that agents may be modified and communication protocols are usually open. However, as new agents and protocols must be compatible with the already-existing ones, possible modifications are limited in practice. For similar reasons, the scalability of these architectures is also low, as adding new agents involves amplifying the communication protocol and making it more complicated, i.e., each agent must store more information about the other agents of the system in order to interact with them.

SBA is, on the whole, robust, since if one of the specialists or one of the communication paths fails, the system carries on functioning and only loses the functionality of the specialist that is affected. Nevertheless, at specialist level, the robustness is similar to that of a blackboard-based architecture. Regarding communications, in SBA, they are implemented in a totally distributed manner and are mainly concentrated on the interchanges between a specialist and its immediately higher specialist. This preference for point-to-point interchanges simplifies communications and strengthens the SBA's robustness.

Regarding reactivity, special attention has been paid in SBA to the design and implementation of the control module in order to avoid the inconveniences that are inherent to blackboard-based architectures. The great advantage of specialization is that it allows a greater focusing of attention in both the detection of events of interest within each specialist and in the generation of the suitable response. The dividing up of information about the world into different specialists does not affect the reactivity of the system as a whole as there are mechanisms that make it possible for any event that is detected in an SBA specialist to be redispatched to the specialist in which the response is generated.

The integration of planned, deliberative, and reactive behaviors is very similar to that which takes place in a blackboard-based architecture and, thus, is more flexible than those methods used in current hybrid deliberative/reactive architectures. Another advantage of SBA is the integration that takes place within each specialist and at all levels of the architecture. Once again, this sharing of behavior integration functions simplifies design and implementation.

At the specialist level, the flexibility of SBA is very similar to blackboard-based architectures. Nevertheless, this very flexibility is reflected in the whole of the architecture as, lest we forget, in SBA, each specialist does not cease to be yet another agent of its immediately higher specialist. Thus, the flexibility of SBA is much higher than both agent-based architectures and hybrid architectures.

Of all the properties of SBA, the most significant is its high degree of scalability, the product of various design decisions in the architecture. First, the encapsulation of specialists gives rise to a hierarchical division of the architecture, which greatly reduces the need for interchanges and mitigates, to a great degree, the communication bottleneck. Second, the recursive implementation of the agents/specialists by means of blackboard-based architectures allows a regular, independent, and heterogeneous implementation of all its components. Third, the use of message-based communications facilitates the delocalization of the specialists, independence of computational platform, and, consequently, the distributed execution of the architecture. In this manner, and thanks to the hierarchical and recursive organization of the architecture, the inclusion of new specialists does not significantly affect the performance of the system.

10.5.2   Related Work

Having carried out the general evaluation, we now go on to compare SBA with the control architectures in mobile robotics which, to date, have been the most successful:[27–29] hybrid deliberative/reactive architectures. There are many architectures which can be classified in this category, but here we concentrate solely on the most representative ones.

With Saphira,[30,31] we share the idea of integrating all sensorial information over time in a single, coherent form for ease of filtering and processing. The same is true for the special attention that is paid to sharing perceptual information (in SBA, a blackboard is employed, and in Saphira, the local perceptual space is used). Another point in common is the joint integration of the calculation of the displacement commands with the processing of sensorial information, which makes it possible to simplify the algorithms and improve performance and the robustness of perception. Nevertheless, in SBA, the information is structured, and it is shared among the specialists according to its level of abstraction and complexity, thus minimizing the need for communications and alleviating the bottleneck. The same occurs with the control, which is not centralized into one single module but rather is distributed among the different control agents in the specialists. This simplifies the control implementation and makes the incorporation of new tasks easier.

In 3T architecture[8] and, generally, in three-layer architectures,[9,32,33] the integration of reactive, deliberative, and planned behaviors is carried out in the intermediate layer or Sequencer. In SBA, integration takes place in the control agent of each specialist and is carried out in all layers of the architecture with no architectural limitation. This characteristic lends a greater degree of flexibility and facilitates the treatment and resolution of problems at different levels of abstraction.

In autonomous robot architecture (AuRA),[34,35] task division is very similar to that in SBA, but with the great difference that the AuRA Pilot does not incorporate reactive-type characteristics which are delegated out in the control module. A further aspect of AuRA which we deem to be unfavorable is the division of perception and control into independent modules, which hinders the integration of sensorial data as they cannot be easily related to the actions and tasks that the robot carries out in each moment. SBA is more flexible as it integrates part of the control closest to reactive and perceptual behaviors. On the other hand, it seems that in behavior implementation, it is difficult to employ techniques other than those of potential fields.

Task control architecture (TCA)[36] has been used to construct the control system of a number of robots, of which Xavier is the most advanced.[37,38] In this implementation, it can be seen that the division of the architecture into layers facilitates the implementation of the different elements, although one senses a strong dependence with regard to the navigation model that is used, based on a representation of the environment by means of a partially observable Markov decision process (POMDP). Once again, there is the impression that the architecture does not facilitate modifications such as the type of data representation or task division. In the same manner as in SBA, the control is centralized, but the data may travel freely throughout the architecture. Nevertheless, in spite of centralization, control in SBA is encapsulated in the specialists, thus facilitating its implementation and modification as well as the scalability of the system.

10.6 Discussion

A number of advantages are obtained with encapsulation. First, a high degree of simplicity and modulation in design, implementation, and control of the global architecture are attained. Second, the specialization reduces intermodule communications and strengthens the architecture's distributed nature. Distributed implementation gives rise to greater calculatory power, increased scalability and reliability for the system, and improves adaptation to new requirements.

In this work, we have proposed the distributed execution of specialists for mobile robotics on the basis of computational load and the relative importance of each specialist. According to these criteria, the vital tasks, those of Pilot, are carried out on board; the rest are realized in a network of computers which the robot accesses by means of an Ethernet radio link.

Specialisation entails the system being divided into levels of responsibility in the same way as in modern hybrid deliberative/reactive architectures. Nevertheless, unlike the latter, specialization also gives a hierarchical character to task distribution, even within a same level. Hence, the execution of complex tasks is facilitated and increases system scalability. At the same time, specialization also simplifies and speeds up the implementation and execution of reactive mechanisms, which are crucial in any system that operates in real time.

All the aforementioned properties make it possible to reaffirm the suitability of the SBA for the design and synthesis of intelligent, autonomous control systems that need to operate in real time in unstructured environments, carrying out complex tasks. The examples included herein illustrate and endorse the suitability of SBA for mobile robots in both task execution and response to unforeseen situations.

Acknowledgments

This work was made possible thanks to projects XUGA20608B97 and PGIDT99PXI20603A and the availability of a Nomad 200 mobile robot acquired through an infrastructure project, all funded by the Xunta de Galicia.

References

1. Elfes, A., A distributed control architecture for an autonomous mobile robot, *Artif. Intelligence*, 1, 1986.
2. Liscano, R. et al., Using a blackboard to integrate multiple activities and achieve strategic reasoning for mobile-robot navigation, *IEEE Expert*, 24, 1995.
3. Gat, E., Integration reaction and planning in a heterogeneous asynchronous architecture for mobile robot navigation, *SIGART Bull.*, 2, 70, 1991.
4. Albus, J.S., Lumia, R., and McCain, H.G., NASA/NBS standard reference model for telerobot control system architecture (NASREM), Technical Report 1235, NASA SS-GFSC-0027, National Bureau of Standards, 1986.
5. Meystel, A., Knowledge based nested hierarchical control, in *Advances in Automation and Robotics*, Saridis, G., Ed., 1990, 63.
6. Brooks, R.A., A robusted layered control system for a mobile robot, *IEEE J. Robotics Automat.*, RA-2, 14, 1986.
7. Arkin, R.C., Motor schema-based mobile robot navigation, *Int. J. Robotics Res.*, 8, 92, 1989.
8. Bonasso, R.P. et al., Experiences with an architecture for intelligent, reactive agents, *J. Exp. Theor. Artif. Intelligence*, 9, 1997.
9. Gat, E., Three-layer architectures, in *Artificial Intelligence and Mobile Robots: Case Studies of Successful Robot Systems*, Kortenkamp, D., Bonasso, R.P., and Murphy, R., Eds., AAAI Press/ MIT Press, Cambridge, MA, 1998, 195.
10. Hu, H. and Brady, M., Towards advanced mobile robots for manufacturing, in *Artificial Intelligence and Mobile Robots: Case Studies of Successful Robot Systems*, Kortenkamp, D., Bonasso, R.P., and Murphy, R., Eds., AAAI Press/MIT Press, Cambridge, MA, 1998, 297.
11. Hayes-Roth, B., An architecture for adaptive intelligent systems, *Artif. Intelligence*, 72, 329, 1995.
12. Labrou, Y. and Finin, T., A proposal for a new KQML specification, Technical Report CS-97-03, Computer Science and Electrical Dept., University of Maryland—Baltimore, 1997.
13. Siegel, J., *CORBA Fundamentals and Programming*, John Wiley & Sons, New York, 1996.
14. Lama, M. et al., Design of a therapeutic specialist for acute myocardial infarct, *Cybernetics Sys.*, 30, 227, 1999.
15. Fraga, S. et al., A proposal for a real time signal perception specialist, in *Proceedings of EIS-98*, Alpaydin, E., Ed., Tenerife, Spain, 1998, 261.
16. Iglesias, R. et al., Implementation of a basic reactive behavior in mobile robotics through artificial neural networks, in *Proceedings of the International Work-Conference on Artificial Neural Networks (IWANN '97)*, No. 1240, Lecture Notes in Computer Science, Springer-Verlag, Lanzarote, Spain, 1997, 1364.
17. Iglesias, R. et al., Supervised reinforcement learning: application to a wall following behavior in a mobile robot, in *Tasks and Methods in Applied Artificial Intelligence, Lecture Notes in Artificial Intelligence*, Vol. 2, 1998, 300.
18. Mucientes, M. et al., A fuzzy temporal rule-based velocity controller for mobile robotics, *Fuzzy Sets Syst.*, in press.

19. Mucientes, M. et al., Avoidance of mobile obstacles in real environments, in *Proceedings of the IJCAI-2001 Workshop on Reasoning with Uncertainty in Robotics*, Fox, D. and Saffiotti, A., Eds., Seattle, WA, 2001, 69.

20. Rodriguez, M. et al., Probabilistic and count methods in map building for autonomous mobile robots, in *Advances in Robot Learning, Lecture Notes in Artificial Intelligence*, vol. 1812, Springer Verlag, Lausanne, Switzerland, 2000, 120.

21. Saffiotti, A., Konolige, K., and Ruspini, E.H., A multivaluated logic approach to integrating planning and control, *Artif. Intelligence*, 76, 481, 1995.

22. Saffiotti, A., The uses of fuzzy logic in autonomous robot navigation, *Soft Computing*, 1, 180, 1997.

23. Nilsson, N.J. et al., Shakey the robot, Technical Report, 323, SRI, 1984.

24. Nehmzow, U., *Mobile Robotics: A Practical Introduction*, Springer, 2000.

25. Brooks, R.A., A robot that walks; emergent behaviors from a carefully evolved network, *Neural Comput.*, 1, 253, 1989.

26. Connell, J.H., *Minimalist Mobile Robotics, A Colony-style Architecture for an Artificial Creature*, Academic Press, New York, 1990.

27. Arkin, R.C., *Behavior-Based Robotics*, MIT Press, Cambridge, MA, 1998.

28. Murphy, R.R., *Introduction to AI Robotics*, MIT Press, Cambridge, MA, 2000.

29. Kortenkamp, D., Bonasso, R.P., and Murphy, R., Eds., *Artificial Intelligence and Mobile Robots: Case Studies of Successful Robot Systems*, AAAI Press/MIT Press, Cambridge, MA, 1998.

30. Konolige, K. et al., The Saphira architecture: a design for autonomy, *J. Exp. Theor. Artif. Intelligence*, 9, 215, 1997.

31. Konolige, K. and Myers, K., The Saphira architecture for autonomous mobile robots, in *Artificial Intelligence and Mobile Robots: Case Studies of Successful Robot Systems*, Kortenkamp, D., Bonasso, R.P., and Murphy, R., Eds., AAAI Press/MIT Press, Cambridge, MA, 1998, 211.

32. Connell, J.H., SSS: a hybrid architecture applied to robot navigation, in *Proceedings of the 1992 IEEE International Conference on Robotics and Automation*, 1992.

33. Andersson, M. et al., ISR: an intelligent service robot, in *Sensor Based Intelligent Robots, Lecture Notes on Computer Science*, vol. 1724, Springer Verlag, 1999, 287.

34. Arkin, R.C. Integrating behavioral, perceptual, and world knowledge in reactive navigation, *Robotics Auton. Syst.*, 6, 105, 1990.

35. Arkin, R.C. and Balch, T., AuRA: principles and practice in review, *J. Exp. Theor. Artif. Intelligence*, 1997.

36. Simmons, R.G., Structured control for autonomous robots, *IEEE Trans. Robotics Automat.*, 10, 34, 1994.

37. Simmons, R.G. et al., XAVIER: experience with a layered robot architecture, *SIGART Bull.*, 22, 1997.

38. Koenig, S. and Simmons, R.G., Xavier: a robot navigation architecture based on partially observable Markov decision process models, in *Artificial Intelligence and Mobile Robots: Case Studies of Successful Robot Systems*, Kortenkamp, D., Bonasso, R.P., and Murphy, R., Eds., AAAI Press/MIT Press, Cambridge, MA, 1998, 91.

Index

A

ADALINE, **VI**-19
    identifier, **VI**-19
    RBF identifier vs., **VI**-23
    structure, **VI**-20
Adaptive control, **VI**-212
    algorithm for optimal, **VI**-13
"Advanced Control Advice for Power Systems with
        Large Scale Integration of Renewable Energy
        Sources," **VI**-58
Alarm(s), **VI**-67 to **VI**-69
    false, **VI**-57
    generalized analysis module, **VI**-116
    messages, **VI**-68
    as a nuisance, **VI**-69
    processing, **VI**-68, **VI**-115
    fault analysis and, **VI**-71
Algorithm(s)
    back propagation, **VI**-25, **VI**-53
    genetic, **VI**-71, **VI**-115, **VI**-138
    for optimal adaptive control, **VI**-13
    self-optimizing pole shifting, **VI**-16
Anomalous load periods, **VI**-160
Application program interfaces, **VI**-75
Architecture(s)
    agent-based, **VI**-356
    blackboard-based, **VI**-356
    comparisons of, **VI**-354
    description language, **VI**-215
    flexible pole-cart balancing system, **VI**-307
    hierarchical, **VI**-355
    hybrid deliberative/reactive, **VI**-356
    inference mechanism, **VI**-231, **VI**-238, **VI**-247
    intelligent coordinated control systems, **VI**-192
    knowledge-based systems, **VI**-93
    neural networks, **VI**-149
    short-term electric load forecasting, **VI**-149
    pole-cart balancing system, **VI**-307
    reactive, **VI**-355
    real-time control, **VI**-215
    SPADE, **VI**-231, **VI**-235, **VI**-244
    equivalence relationship, **VI**-244
    properties, **VI**-229

    subset relationship, **VI**-244
    union of, **VI**-244
    specialist, **VI**-339, **VI**-340, **VI**-357
    STLF, **VI**-149
    3T, **VI**-358
    task control, **VI**-358
    VERITAS, **VI**-97
ARCHON, **VI**-189
ARIMA Box-Jenkins model, **VI**-137
Artificial intelligence, **VI**-65, **VI**-71, *See also*
        Intelligent terms
Artificial neural networks, **VI**-6, *See also*
        Neural networks
    controller, **VI**-24
    day-type characteristics, **VI**-166, **VI**-167
    for fault diagnosis, **VI**-115
    forecast, **VI**-149, **VI**-175, *See also* Short-term elec-
        tric load forecasting (STFL)
    identifier, **VI**-18, **VI**-24
    offline training, **VI**-9
    performance evaluation, **VI**-55
    security assessment based on, **VI**-39, **VI**-54
    unsupervised, **VI**-165
Assumption-based truth maintenance system, **VI**-99
Asynchronous generators, **VI**-44
Auto-regressive model, **VI**-170
Automatic reconfiguration, **VI**-212
Automatic voltage regulator, **VI**-1
Autotuning, **VI**-254
Available transfer capacity, **VI**-46

B

Back propagation algorithm, **VI**-25, **VI**-53
Box-Jenkins statistical analysis, **VI**-137
Breaker failure device, **VI**-118, **VI**-119, **VI**-127
Busbar faults, **VI**-125

C

CARE software, **VI**-58
    installation, **VI**-59

Cartesian product, **VI**-221
Chaos theory, **VI**-138
Component roles, **VI**-219
Compositional operations, **VI**-220
Computers, **VI**-66
Control architecture, comparisons of, **VI**-354
Control centers, **VI**-63
    alarms, **VI**-67 to **VI**-69
    computerization, **VI**-66
    expert systems and, **VI**-75
    functions, **VI**-67
    KBS integration in, **VI**-75
    operators, **VI**-70, **VI**-181
    performance, **VI**-67
    SCADA functions, **VI**-67
    user interface for, **VI**-92
Control system(s), *See also* Power system(s)
    adaptive, **VI**-10, **VI**-212
    automatic, **VI**-212
    intelligent, **VI**-183, **VI**-212
    coordinated, **VI**-190, *See also* Intelligent coordinated
        control systems
    large scale, **VI**-182
    multiagent, **VI**-182
    pole-cart balancing system, **VI**-286
    software engineering, **VI**-183 to **VI**-186
    agent-based design, **VI**-187 to **VI**-188
    agent-oriented methodologies, **VI**-188 to **VI**-191
    approach to, **VI**-183 to **VI**-185
    problems, **VI**-185 to **VI**-186
Controller(s)
    adaptive, **VI**-10
    behavior, desired behavior vs., **VI**-327, **VI**-328,
        VI-329, **VI**-330
    copying existing, **VI**-7
    fixed parameters, **VI**-2
    estimation, **VI**-12
    fuzzy logic, **VI**-4
    with network, **VI**-7
    neural-network based, **VI**-6
    input vector to, **VI**-8
    self-tuning, **VI**-11
Credibility index, of hypothesis, **VI**-122
Critical clearing time, **VI**-46

D

Data preprocessor, **VI**-121
Database, **VI**-256
Decision tree(s), **VI**-46
    application, **VI**-48
    for gas turbine outage, **VI**-49
    for machine-outage disturbance, **VI**-51
    node types, **VI**-47
    performance evaluation, **VI**-47, **VI**-53
    reformulaton into neural network, **VI**-56
    security assessment based on, **VI**-39
    for short-circuit contingency, **VI**-51

Decoupling strategy, **VI**-269
Defuzzier, **VI**-319
Diesel engines, **VI**-43
Differential relays, **VI**-117, **VI**-118
    malfunction analysis, **VI**-123 to **VI**-125
Dispatcher training simulators, **VI**-70
Distance protections, **VI**-116
Distribution networks, **VI**-71
Disturbance, **VI**-313, **VI**-322
    decision tree for machine-outage, **VI**-51
    gas turbines outage, **VI**-49
    response to external, **VI**-313, **VI**-322
Dual features, **VI**-254
Dynamic multilayer network, **VI**-148
Dynamic neurons, with feedback connections, **VI**-146

E

Electric utilities, **VI**-72
Energy management systems, **VI**-69, **VI**-113, **VI**-114
    economic dispatch, **VI**-136
    hydro-thermal coordination, **VI**-136
    load management, **VI**-136
    security functions, **VI**-136
    unit commitment, **VI**-136
Entropy networks, **VI**-56 to **VI**-57
    classification success, **VI**-57
    use, **VI**-57
ESES, **VI**-86 to **VI**-87
Euclidean distance, **VI**-156
EUROSTAG, **VI**-45
Evolution, **VI**-178
Expert systems, **VI**-75, **VI**-115
    functions, **VI**-75
    integration in control centers, **VI**-75
    real-time performance, **VI**-86
    uses, **VI**-115
Explanation module, **VI**-77

F

False alarm, **VI**-57
Fault diagnosis, **VI**-75, **VI**-114, **VI**-115, **VI**-212
Feedforward neural networks, **VI**-55
    layers, **VI**-56
    processing elements of online, **VI**-311
Fluxogram, **VI**-81, **VI**-82
    rule-scheduling, **VI**-83
Fossil fuel power unit, **VI**-191 to **VI**-192
Functional-link networks, **VI**-139
Fuzzy associative memory matrix, **VI**-316 to **VI**-317
Fuzzy-logic based neural network implemented
    controller (FLBNNIC), **VI**-323
    prediction error, **VI**-331
Fuzzy logic system, **VI**-138, **VI**-159, **VI**-175, **VI**-314
    associative memory matrix, **VI**-316, **VI**-317
    chattering, **VI**-275

decoupling strategy, **VI**-269
fine tuning, **VI**-256
multicontroller, **VI**-315
nominal design, **VI**-256, **VI**-258
phase plane techniques, **VI**-257
PID type structure, **VI**-253
derivative effects, **VI**-279, **VI**-280
dual features, **VI**-254
gain tuning, **VI**-254
performance, **VI**-277, **VI**-279
scaling gain design, **VI**-254
Fuzzy rules, **VI**-139

G

Gas turbines, **VI**-43
outage disturbance, **VI**-49
Generalized alarm analysis module (GAAM), **VI**-114
class hierarchies, **VI**-129
computer environment, **VI**-128
data preprocessor, **VI**-121
diagnostic process, **VI**-121
using color technique, **VI**-132
flowchart for functional procedures, **VI**-124
implementation, **VI**-128
logic-based reasoning technique, **VI**-123
man-machine interface of, **VI**-129
salient feature, **VI**-122
software classification, **VI**-128
visual diagnosis using color technique, **VI**-131
Generator, synchronous, **VI**-1, **VI**-3, **VI**-44
Genetic algorithms, **VI**-71, **VI**-115, **VI**-138
Group method of data handling, **VI**-140, **VI**-178
applications, **VI**-143
evolution, **VI**-178

H

Hierarchical distributed control, **VI**-215
Hybrid methodology, **VI**-256
Hybrid solutions, **VI**-139
Hybrid systems, **VI**-138
Hypothesis, credibility index of, **VI**-122

I

Incident analysis, **VI**-77
Inference cells, **VI**-259
regions, **VI**-259
Inference engine, **VI**-77, **VI**-80
fluxogram, **VI**-81, **VI**-82
Intelligent agent paradigm, **VI**-182
Intelligent control, **VI**-212
Intelligent coordinated control systems, **VI**-190
architecture, **VI**-192
context, **VI**-192

function specification of, **VI**-198
fundamental functional structure, **VI**-193
generic intelligence functions, **VI**-196
intelligent functions of, **VI**-195
minimum prototype of, **VI**-200 to **VI**-204
multiagent organization, **VI**-197
organization of, **VI**-195
physical scope of, **VI**-199 to **VI**-200
Intelligent server, **VI**-88
Intelligent tutor, **VI**-100, **VI**-101
interaction between learner and, **VI**-104
oriented sessions, **VI**-103
International Council on Large Electric Systems,
VI-69, **VI**-70
Inverse plant modeling, **VI**-7

K

Kalman filtering, **VI**-138
"Knowledge Based Applications in SCADA/EMS-A
Practical Approach," **VI**-69
Knowledge-based systems (KBS), **VI**-72
advantages, **VI**-76
architecture, **VI**-93
development, **VI**-76
disadvantages, **VI**-74
explanation ability, **VI**-86
forward-chaining, **VI**-88, *See also* SPARSE
rule base, **VI**-72
machine learning techniques for suggesting changes
in, **VI**-73
temporal reasoning, **VI**-83
Kohonen networks, **VI**-138
Kohonen self-organizing map, **VI**-155 to **VI**-156, **VI**-160
as an unsupervised classifier, **VI**-156
associative memory and, **VI**-156
characteristics of implemented, **VI**-167
cluster identification code, **VI**-164
day-type classification activity, **VI**-167
training sets, **VI**-160, **VI**-163

L

Learning theory, **VI**-138
Load(s), **VI**-45
anomalous periods, **VI**-160
demand, **VI**-169
forecasting, **VI**-136, *See also* Short-term electric load
forecasting (STLF)
online, **VI**-145
pole-cart balancing system, **VI**-293
recognition, **VI**-171

M

Machine learning, **VI**-90, **VI**-139
techniques, **VI**-73, **VI**-90

Machine-outage disturbance, **VI**-51, **VI**-54

Maintenance system, **VI**-99

Mean absolute error, **VI**-53, **VI**-158

Mean squared error, **VI**-142

Membership function, **VI**-318

Mobile robot Nomad **VI**-200, **VI**-347

Multi-layer network based systems, **VI**-25

Multi-layer perceptron, **VI**-53, **VI**-138

    interpolation and, **VI**-171

    number of, **VI**-172

    uses, **VI**-139

Multiagent system, **VI**-182

ARCHON development environment for, **VI**-189

    design, **VI**-189

    organization, **VI**-197

Multiple linear regression, **VI**-137

N

Nested hierarchical control, **VI**-215

NEUFOR, **VI**-145

Neural corrector, **VI**-151

Neural networks, **VI**-38, *See also* Artificial
      neural networks

    backpropagation stimulator, **VI**-304

    dynamic neurons, with feedback
        connections, **VI**-146

    feedforward, **VI**-55, **VI**-56

    interpolation problem, **VI**-169

    recognition and, **VI**-171

    load demand, **VI**-169

    perspective, **VI**-303

    physical experiments, **VI**-310

    recurrent, **VI**-139

    reformulaton of decision tree into, **VI**-56

    STLF architecture, **VI**-149

    training and learning in, **VI**-302

Nominal design, **VI**-256

Nominal rule base, **VI**-254

Non-monotonic reasoning, **VI**-80

North American Electric Reliability
      Council, **VI**-37

O

Object-oriented software, **VI**-213

Online corrector, **VI**-152

Online load forecasting, **VI**-145

    correction, **VI**-152

Operating points, **VI**-45, **VI**-46

    acceptable, **VI**-50

    preclassified, **VI**-47

Operator training, **VI**-131

Overcurrent protection, **VI**-117

Overcurrent relays, **VI**-125

P

Pattern recognition, **VI**-171

Phase plane techniques, **VI**-257

PID type, **VI**-253

Pole-cart balancing system, **VI**-286

    cantilever, **VI**-292

    control systems, **VI**-286

    FLBNNIC for, **VI**-323

    flexible, **VI**-288, **VI**-294, **VI**-300

    hardware architecture, **VI**-309

    neural network model, **VI**-308

    physical architecture, **VI**-307

    response to external disturbance, **VI**-313, **VI**-322

    forces acting on cart, **VI**-290

    fuzzy logic controller, **VI**-315

    membership functions for pole deflection
        and velocity, **VI**-296

    online application, **VI**-308

    pole location, **VI**-296

    real-time movement, **VI**-299

    uniform load of pole, **VI**-293

Pole-shift control, **VI**-18

Pole-shifting factor, **VI**-16, **VI**-17

Power plant

    control, **VI**-181, **VI**-191 to **VI**-192, *See also* Control
        system(s)

    coordinated, **VI**-205

    operation, **VI**-191 to **VI**-192

Power security system(s), *See also* Power system(s)

    adaptive-network based fuzzy, **VI**-27, **VI**-28

    assessment, **VI**-75

    ANN and, **VI**-39

    decision trees and, **VI**-39

    identification of problems, **VI**-43

    machine learning techniques for, **VI**-42

    online dynamic, **VI**-48

    control centers, **VI**-63

    defined, **VI**-37

    fixed parameter controllers, **VI**-2

    loads, **VI**-45

    modeling of, **VI**-43

    operating points, **VI**-45, **VI**-46

    operation, **VI**-63

    problems, **VI**-68

    stability, **VI**-, **VI**-38

    frequency, **VI**-38

    rotor angle, **VI**-38

    voltage, **VI**-38

    stabilizers, **VI**-1, *See also* Stabilizer(s)

    adaptive, **VI**-14, **VI**-18, **VI**-24, **VI**-25, **VI**-27

    controllers, **VI**-2, *See also* Controller(s)

    conventional, **VI**-2

    development of, **VI**-7

    pole-shift based adaptive, **VI**-16

    signal, **VI**-2

Power system(s), *See also* Power control system(s)

    computers in, **VI**-66

    efficiency, **VI**-70

fault diagnosis, **VI**-75, **VI**-114, **VI**-115
incident analysis, **VI**-77
operation, **VI**-66, **VI**-75
efficiency of, **VI**-70
problems, **VI**-71
performance, **VI**-67
real-time information, **VI**-66
restoration, **VI**-71, **VI**-74, **VI**-75, **VI**-76
substation monitoring, **VI**-75
Prediction error, **VI**-331
Probabilistic reasoning, **VI**-138
Programmable impulse generator, **VI**-96
Protection device, visual diagnosis of, **VI**-131

R

Radial basis function, **VI**-19
network identifier, **VI**-21
ADALINE identifier vs., **VI**-23
for online application, **VI**-22
Reactivity, **VI**-345
external, **VI**-354
integration of behaviors, **VI**-352
Real-time, **VI**-84
control, **VI**-13, **VI**-64
control architecture, **VI**-215
information, **VI**-66, **VI**-94
performance, of expert systems, **VI**-86
power flow applications, **VI**-67
Recurrent-network based system, **VI**-25
Recursive least squares, **VI**-12
Regression models, **VI**-138, **VI**-178
Restoration, after incident, **VI**-71, **VI**-74, **VI**-75, **VI**-76
Robotics, **VI**-344, **VI**-346, *See also* Specialist(s)
integration of behaviors, **VI**-352, **VI**-353
Robust control, **VI**-212
Root mean square error, **VI**-53
Rule base, **VI**-72, **VI**-73
mathematical model, **VI**-259
inference cell regions, **VI**-259
nominal, **VI**-254
RUNGE, **VI**-297

S

SCADA, *See* Supervisory control and data acquisition (SCADA) systems
Scaling design, **VI**-254
Seasonal classification, **VI**-163
Self-optimizing pole shifting algorithm, **VI**-16
Self-organizing maps, **VI**-139, **VI**-155
to **VI**-156, **VI**-160
Sequence-of-events recorders, **VI**-114, **VI**-121
Short-circuit contingency, **VI**-51, **VI**-54
Short-term electric load forecasting (STLF), **VI**-136, **VI**-137, **VI**-145
criteria for an efficient model, **VI**-136

effect of weather, **VI**-140
error function, **VI**-151
neural network architecture, **VI**-149
dynamic layer, **VI**-148
load shape and, **VI**-148
synapses, **VI**-148
online correction, **VI**-151 to **VI**-157
flow chart of module, **VI**-153
seasonal classification, **VI**-163
Short-term human forecasting, **VI**-137
Single tripping representation, **VI**-107
Sliding control, **VI**-254
features, **VI**-262
Soft computing
constituents, **VI**-138
fuzzy logic, **VI**-138
neural network, **VI**-138
probabilistic reasoning, **VI**-138
conventional computing vs., **VI**-138
Software
CARE, **VI**-58, **VI**-59
classification, **VI**-128
control system, **VI**-183 to **VI**-186
agent-based design, **VI**-187 to **VI**-188
agent-oriented methodologies, **VI**-188 to **VI**-191
approach to, **VI**-183 to **VI**-185
problems, **VI**-185 to **VI**-186
cost, **VI**-184
object-oriented, **VI**-213
SPADE, **VI**-216
advantages, **VI**-249
architectural inference rule, **VI**-231
architectural matching, **VI**-235
architectural patterns, **VI**-244
equivalence relationship, **VI**-244
subset relationship, **VI**-244
union of, **VI**-244
architectural properties, **VI**-229
closure matching, **VI**-237
comprehensive matching, **VI**-235
computational vs. supervisory behaviors, **VI**-217
concepts, **VI**-216
configuration closure, **VI**-233
configuration formula, **VI**-225
components within, **VI**-226
configuration constraints as extensions to, **VI**-227
consistent property, **VI**-229
control system, **VI**-229
design, **VI**-213
development of, **VI**-216
diagrammatic representation of attachments, **VI**-220
function, **VI**-211
hierarchical property, **VI**-230
loop composition, **VI**-222
mathematical formulation in, **VI**-230
monitoring composition, **VI**-224
neat property, **VI**-229
partial matching, **VI**-237
provision for abstractions, **VI**-221

role constraints, **VI**-224
semantics, **VI**-220
signal composition, **VI**-222
supervisory augmentation, **VI**-231
supervisory composition, **VI**-224
supervisory control system, **VI**-229
supervisory framework, **VI**-245
equivalence relationship, **VI**-248
subset relationship, **VI**-248
union, **VI**-248
update composition, **VI**-223
valid binding, **VI**-235
SPARSE, **VI**-72, **VI**-76, **VI**-103, **VI**-104
development, **VI**-78
distinct knowledge bases, **VI**-85
explanation server, **VI**-88
explanations, **VI**-86
inference engine, **VI**-77, **VI**-80
intelligent server, **VI**-88
intelligent tutor, **VI**-101
meta-knowledge embedded in, **VI**-82
modules, **VI**-77
operators, **VI**-106
real-time information, **VI**-94
rule base, meta-knowledge embedded in, **VI**-82
TEMPU's integration with, **VI**-91
time-tagged list, **VI**-83
user interface, **VI**-77, **VI**-93, **VI**-94
validation, **VI**-94
Specialist(s)
architecture, **VI**-339, **VI**-340, **VI**-357
communication between, **VI**-341 to **VI**-342
components, **VI**-340
control agent, **VI**-342
external reactivity, **VI**-354
robotics, **VI**-344
structure, **VI**-343
Specialist-based systems, **VI**-339,
    See also Specialist(s)
Stability, **VI**-1, **VI**-38
Stabilizer(s), **VI**-1
adaptive, **VI**-18, **VI**-24, **VI**-25, **VI**-27
optimal, **VI**-14
controllers, **VI**-2, *See also* Controller(s)
conventional, **VI**-2
development of, **VI**-7
pole-shift based adaptive, **VI**-16
signal, **VI**-2
Steam unit, **VI**-43
Stochastic time series, **VI**-137
Supervisory control and data acquisition
    (SCADA) systems, **VI**-66, **VI**-74, **VI**-113
alarm messages generated by, **VI**-96
control center functions, **VI**-67
for SPARSE systems, **VI**-77, **VI**-83

Synapses, **VI**-148
Synchronous generator, **VI**-1, **VI**-3, **VI**-44

T

Temporal reasoning, **VI**-83
TEMPU, **VI**-91
Training sets, **VI**-147, **VI**-160, **VI**-163
for extrapolation, **VI**-172
management of procedure for, **VI**-173
observations, **VI**-179
TransAlta system, **VI**-15
Transformer protection, **VI**-123
Transmission network, **VI**-76

U

User interface, **VI**-77, **VI**-79, **VI**-91
adaptive, **VI**-93
for control center application, **VI**-92
development, **VI**-92, **VI**-93
SPARSE, **VI**-77, **VI**-93, **VI**-94

V

Variable structural control, **VI**-254
VERITAS, **VI**-96, **VI**-98
architecture, **VI**-97
development of, **VI**-98
Voltage regulator, **VI**-1, **VI**-44

W

Weather variables, **VI**-137, **VI**-140
Widrow-Hoff learning rule, **VI**-20
Wind parks, **VI**-41

Z

Ziegler-Nichols method, **VI**-254, **VI**-268